Ethnobiology

Ethnobiology is the study of the dynamic relationship between plants, animals, people, and the environment. Academic and applied interests include ethnobotany, ethnozoology, linguistics, paleoethnobotany, zooarchaeology, ethnoecology, and many others. The field lies at a dynamic intersection between the social and biological sciences. The major contribution from the biological sciences has come from economic botany, which has a rich historical and scientific tradition. Indeed, the objectives of the colonial enterprise were as much about the quest for "green gold" –herbal medicines, spices, novel cultivars, and others—as it was for precious metals and sources of labor. The view that ethnobiology concerns mostly the discovery of new and useful biota extended into the 20th century. The social sciences have contributed to the field in both descriptive studies but also within quantitative approaches in cognitive anthropology that have led to general principles within ethnobiological classification. Ethnobiological research in recent years has focused increasingly on problem solving and hypothesis testing by means of qualitative and especially quantitative methods. It seeks to understand how culturally relevant biotas are cognitively categorized, ranked, named, and assigned meaning. It investigates the complex strategies employed by traditional societies to manage plant and animal taxa, communities, and landscapes. It explores the degree to which local ecological knowledge promotes or undermines resource conservation, and contributes to the solution of global challenges, such as community health, nutrition, and cultural heritage. It investigates the economic value and environmental sustainability to local communities of non-timber forest products, as well as the strategies through which individual ecological knowledge and practices encourage resilience to change—modernization, climate change, and many others. Most importantly, contemporary ethnobiological research is grounded in respect for all cultures, embracing the principles of prior informed consent, benefit sharing, and general mindfulness.

More information about this series at http://www.springer.com/series/11551

Meredith Welch-Devine • Anne Sourdril
Brian J. Burke
Editors

Changing Climate, Changing Worlds

Local Knowledge and the Challenges of Social and Ecological Change

Springer

Editors
Meredith Welch-Devine
The Graduate School
University of Georgia
Athens, GA, USA

Anne Sourdril
Centre National de la Recherche
Scientifique UMR 7533 Ladyss
Nanterre, France

Brian J. Burke
Department of Sustainable Development
Appalachian State University
Boone, NC, USA

ISSN 2365-7553 ISSN 2365-7561 (electronic)
Ethnobiology
ISBN 978-3-030-37311-5 ISBN 978-3-030-37312-2 (eBook)
https://doi.org/10.1007/978-3-030-37312-2

This Springer imprint is published by the registered company Springer Nature Switzerland AG
The registered company address is: Gewerbestrasse 11, 6330 Cham, Switzerland

Foreword

This book could not be better timed. It hits the printing press as the second decade of this new millennium comes to a close, after a series of dreadful events that dispel any doubt that our planet is in dire peril. Climate change poses a formidable threat to both the Earth and the people who inhabit it, as exemplified by the grim lineup of hurricanes, floods, droughts, and wildfires that we have experienced this year alone. Evidence is piling up in science circles that glaciers are disappearing, oceans are warming and acidifying, and crop yields and food supplies are being threatened at unprecedented rates. The Amazon forest is devastated by fire and, even worse, destroyed for large-scale agriculture, industrial logging, and mining in ways that dangerously approach a tipping point.

Yet, this book is unexpectedly affirming and forward-looking. It encourages us to look beyond the parade of natural disasters and the natural impulse to despair, neither of which are as "natural" as they seem. The authors make this clear by reframing our understanding of global environmental change in ways that place humanity, and the cultures, societies, histories, and narratives it creates, squarely at the center. While this perspective leaves no escape from the recognition of our responsibilities, it equally and positively points to agency and options that remain available to us, as anthropologists, scientists, policy-makers, activists, and citizens, to transform the way we relate to the natural world and our place in it. This framing is supported by a number of critical and timely insights that stem from this volume: two in particular stand out as uniquely generative.

First, the authors made a convincing case for the indivisibility of humans and natural systems. People do not simply observe, understand, manage, or measure environmental phenomena as if they were external to their being. Rather, we exist in and engage with the environment in ways that are mediated by our senses, feelings, thoughts, values, and stories. That is true for momentous as well as mundane manifestations of global change which are felt on the skin, in the soul, and within the mind. Across the wide and rich plurality of such experiences, this reality is shared across humanity, and, as such, it can offer fertile ground for coalition building and resistance movements. It can propel and unite efforts to confront "long histories of inequality" and "uneven geographies of destruction" (as stated in the editors' introduction) and to sustain collective action at multiple scales. The recent school strikes and street protests by young people across the globe to shake world leaders out of their inertia on climate change are an inspirational example. Less widely covered in the popular media but equally energizing is the emergence of worldwide alliances of indigenous people for environmental and climate justice. For example, a Local Communities and Indigenous Peoples Platform was recently convened within the UN Framework Convention Climate Change Conference to promote legitimate representation of indigenous groups in global climate change negotiations. Of course, these actions and voices confront formidable political hurdles and economic interests that stand in the way, but they do renew our faith in the actual possibility of social transformation.

This brings us to a second important insight that infuses most chapters and is articulated in the conclusion to the book. Changing the way we engage with nature and with other human and nonhuman beings in nature requires more than improving policies and practices (though doing so would surely be a welcome head start). It calls for a more radical shift in how we gain and use environmental knowledge, turning away from colonizing epistemologies toward politically engaged, diversity-embracing ways of knowing. Anthropology's genesis in colonial times and places, and its historical record of manufacturing knowledge that reifies research subjects, erases difference, and upholds the status quo calls for intentional reflexivity by those of us who have been trained in this discipline. Auspiciously, as outlined in the concluding chapter, growing numbers of anthropologists are breaking out of the mold and leading the way in co-producing streams of research which document, contextualize, include, and mobilize other ways of knowing our changing climates. Even then, in addition to entrenched interests and ideologies, engaged anthropologists and academics face powerful challenges that stem from the established systems and criteria that govern performance evaluation and career advancement in US universities. These structures – which continue to promote disenfranchising epistemologies – must be decolonized at last.

Published a decade after the landmark volume *Anthropology and Climate Change: From Actions to Transformations* (Crate and Nuttall eds. 2009) – which foregrounded the vast potential offered by the discipline to enrich analyses, policies, and practices centered on global change – the current collection strides even further. It challenges us to critically examine the knowledge production processes and contexts we engage in. It equally urges us to balance the pursuit of academic excellence and theoretical rigor with a commitment to public scholarships that nurtures a vibrant and visionary environmental citizenship.

Atlanta, Geogia, USA Carla Roncoli

Contents

About the Editors

Meredith Welch-Devine is Director of Interdisciplinary Graduate Programs at the UGA Graduate School and an Adjunct Faculty Member in the Department of Anthropology. Her primary research interests include climate change perceptions and adaptation, impacts of sea level rise and extreme weather events on coastal populations, collective management of common-pool resources, and policy and practice related to conservation and sustainability. Her dissertation research – set in the Basque region of France – centered on the implementation of the European Union's Habitats Directive in an agricultural area with a strong common property regime. This directive created a pan-European network of conservation sites, called Natura 2000, and was the subject of intense resistance by the local population. Since that time, she has continued to collaborate with Basque farmers on research related to sustainable agriculture. Closer to home, she has been associated with the Coweeta Long-Term Ecological Research program since 2003, where she has sought to understand how people observe and make sense of changes in their environment, including climate change. Most recently, she has worked on the Georgia coast to understand how people view their adaptation options, including migration away from the coast, in the face of increasing storm severity. This work also provides a starting point for a collaboration with faculty in history and in design that will use augmented and virtual realities to explore environmental pasts, presents, and futures with research participants.

Anne Sourdril is a Social Anthropologist, Research Fellow at the CNRS, and Member of the Laboratoire Dynamiques Sociales et Recomposition des Espaces (UMR 7533 Ladyss). Her research focuses on the dynamics of socio-ecological systems in the context of long-term environmental change. She is working on the processes of spatial greening, the multiplication of measures to protect and conserve nature or biodiversity, and the reconfiguration of landscapes. She is interested in the dynamics of everyday biodiversity and its associated knowledge and representations. Her latest work focuses on the sounds of nature and soundscapes seen as indicators of environmental change and notably climate changes. She addresses these issues in interdisciplinary and comparative research programs. Her research sites are located in southwestern France, western North Carolina, and more recently southern Arizona.

Brian J. Burke is an Associate Professor in the Goodnight Family Sustainable Development Department at Appalachian State University. His research examines popular understandings of environmental and economic issues, as well as grassroots mobilization to address these issues. Since 2012, he has focused largely on local environmental knowledge and strategies for scientist/nonscientist collaboration in Southern Appalachia, including via work with the Coweeta Long-Term Ecological Research Program, the Coweeta Listening Project, the PIAF Project, and collaborative research with Meredith Welch-Devine. His research has also included a variety of short- and long-term ethnographic engagements examining sustainable development, environmental health activism, ecovillages, and noncapitalist/solidarity economies (alternative currencies, barter systems, fair trade, and cooperatives) in Latin America.

Chapter 1
Understanding Microexperiences of Climate Change: How Climate Ethnography Informs Collaboration, Adaptation, and Effective Responses

Brian J. Burke, Meredith Welch-Devine, and Anne Sourdril

Abstract Projections of climate change, biodiversity loss, and associated socioeconomic impacts are increasingly dire. In this volume, we turn our attention from the spectacular scenes of climate disruption to the slow and subtle, the small but consequential shifts in the species and landscapes that we humans interact with on a constant basis. This introductory chapter offers an analytical framework for the chapters that follow. Synthesizing lessons from environmental anthropology, we argue that *microexperiences of change* offer a critical but neglected lens for understanding the Anthropocene as a new geological, cultural, and political era. Focusing on microexperiences allows us to examine how individuals and communities are experiencing climate change in intimately meaningful ways, how they are constructing knowledge based on these experiences, and how that knowledge shapes their responses. This in turn provides unique insights into the diverse ways that people are embedded in their environments; the dynamics of differentiation, inequality, and violence that result from that; and how these affect knowledge, denialism, and climate responses. Perhaps most importantly, examining climate change at the resolution of microexperiences has the advantage of showing us change where many people—perhaps especially those whose livelihoods, social relations, and cultures are most intimately linked to the environment—see it, feel it, and make sense of it. Careful analysis and appreciation of these microexperiences and the resulting knowledge systems may therefore broaden the foundation for shared understanding and collaborative action to address climate change in an inclusive and effective manner.

B. J. Burke (✉)
Department of Sustainable Development, Appalachian State University, Boone, NC, USA
e-mail: burkebj@appstate.edu

M. Welch-Devine
The Graduate School, University of Georgia, Athens, GA, USA

A. Sourdril
Centre National de la Recherche Scientifique, UMR 7533 Ladyss, Nanterre, France

M. Welch-Devine et al. (eds.), *Changing Climate, Changing Worlds*, Ethnobiology, https://doi.org/10.1007/978-3-030-37312-2_1

Keywords Climate change · Local ecological knowledge · Adaptation · Environmental anthropology · Slow violence · Biodiversity · Climate ethnography · Inequality · Ethnoecology · Political ecology · Democratization of science · Human dimensions of climate change

Global climate change activism took on new urgency in October 2018, when the Intergovernmental Panel on Climate Change (IPCC) released a special report warning that the Earth will suffer severe consequences if carbon emissions are not cut more deeply and on a much faster timeline than previously thought necessary. Commissioned during the 2016 Paris Climate Conference at the urging of small island states and other highly vulnerable countries, the report detailed the impacts that would result from only 1.5 °C of warming above preindustrial levels rather than the 2 °C that has been the focus of international policy since Kyoto. The results were sobering. Even at this lower threshold, we will face reduced food security, increased exposure to extreme weather events, and dramatic loss of biodiversity as soon as 2040. Maize, rice, and wheat yields will decline globally, storms will continue to become more violent and damaging, and 70–90% of tropical coral reefs are expected to die (Hoegh-Guldberg et al. 2018).

The links between climate, biodiversity, and human well-being are clear. In May 2019, the global assessment for biodiversity and ecosystem services[1] warned that more than one million species are at risk of extinction due to five key drivers—land and sea use change, direct exploitation of species, climate change, pollution, and invasion of non-native species—and that climate change is increasingly exacerbating the impact of these other drivers (IPBES 2019: 3). The IPBES report declares current biodiversity conservation measures inadequate and illustrates that extirpation and extinction have highly significant social, economic, political, and moral consequences. Among other impacts, the continued degradation of nature "will undermine progress toward 80 percent (35 out of 44) of the [Sustainable Development Goals] related to poverty, hunger, heath, water, cities, climate, oceans and land" (IPBES 2019: 6). The global assessment points to dramatic impacts that have already happened under only 1 °C (+/− 0.2) of warming, including heat waves, storms, and droughts, as well as more subtle shifts such as earlier springs and changes in phenology and primary production. Climate change is likely to interact with other drivers in unexpected ways, potentially exacerbating species loss and leading to irreparable damage to the Earth and those who dwell here.

To avert catastrophe, the authors of both reports call for nothing short of transforming the world's economy. According to the IPCC, "pathways limiting global warming to 1.5 °C… require rapid and far-reaching transitions in energy, land, urban and infrastructure…, and industrial systems" (IPCC 2018: 21) so that greenhouse gas emissions can be reduced by 45% from 2010 levels by 2030 and to net zero by around

[1] As of this writing, the global assessment was available in draft form, with the final chapters to be released in late 2019. We draw here on an advance unedited copy of the Summary for Policymakers.

2050. IPBES additionally calls for a significant democratization of environmental governance, the establishment of socio-ecological well-being as a central criterion for resource use planning, and the widespread implementation of policies to eliminate negative environmental externalities. These tasks may seem daunting, but it is increasingly clear that we must dedicate ourselves fully to "transformative change" to protect human and ecological well-being through nature conservation, sustainable resource use, climate adaptation, and greenhouse gas mitigation. Fortunately, these transformative visions are not shots in the dark: they have been thoroughly evaluated in terms of both their ability to improve well-being and their cost-effectiveness (see, e.g., Brown et al. 2018, Hawken 2017, Jacobson et al. 2018, and UNEP 2012–2018).

In the United States, the IPCC report was released during one of the worst years on record for "natural" disasters. The year began with mudslides killing 21 people in Montecito, California; these slides were triggered by rainfall that exceeded half an inch in only 5 minutes. Later that spring, Maryland experienced flash flooding when more than 8 inches of rain fell in a few hours in Howard County. In the fall, California was hit by the deadliest wildfire in its history—the Camp Fire—and the East Coast saw Hurricanes Florence and Michael, which caused record-breaking flooding and damage. The United States was no exception. In the same time period, South Africa faced a critical drought that left Cape Town's four million residents just months away from running out of water. Early in 2019, Tropical Cyclone Idai pummeled Mozambique, Zimbabwe, Madagascar, and Malawi, killing more than 1000 people and causing more than $2 billion in damages, only to be followed by Tropical Cyclone Kenneth just 5 weeks later. Perhaps influenced by watching those epic storms unfold, in early May of 2019, India managed the gargantuan task of evacuating more than 1 million people from the path of Cyclone Fani. In September, northern islands of the Bahamas were devastated by Hurricane Dorian. In each of these disasters, some escaped the worst harm, while others had their lives completely and permanently upended. The uneven geography of destruction underscores yet again the role that political-economic factors play in delivering ruin to some while allowing others simply a "close call" (Taylor 2014). Disasters do not "fall from the sky" (Ribot 2010; Smith 2006); they arise from long histories of inequality that have created institutional and social structures that ensure that some among us will always have a harder time preparing for, weathering, and recovering from extreme events.

The horrific scale of today's climate disasters commands our attention, prompts philanthropy and policy discussions, and has even given rise to a new genre of literature and film: climate fiction, or "Cli-Fi." But in many ways, the focus on the dramatic obscures other, more subtle, ways that people are experiencing climate change in their everyday lives. Though these effects may not grab headlines, unpredictable rains, less reliable harvests, and disappearing animals can be devastating to those whose livelihoods depend most directly on the environment. In southern Europe, for example, unusually heavy rains in 2018 triggered erosion and mudslides that ruined newly sown fields, while strong storms destroyed fruit and vegetable harvests. In each case, community solidarity was key to recovery, highlighting the crucial role of social institutions and cultural practice in resilience, adaptation, and transformation.

It is also evident that watching the loss and permanent change of our landscapes and homes causes real grief (Cunsolo and Ellis 2018). The salamanders we used to catch as kids, the cornflowers, poppies, or orchids that sprinkled hillsides with color, or the songbird we first learned to imitate, when they disappear, it hurts. Beyond causing distress, the loss of species and associated memories also erodes the local representations of nature that contribute to cultural diversity. Local biodiversity losses coupled with an increase in invasive species can disturb processes of building, maintaining, and transmitting the high specialized vernacular knowledges essential to understanding and managing environments in a context of global change (Cruikshank 2005; Nadasdy 2003).

With this volume, we turn our attention from the spectacular scenes of climate disruption to the slow and subtle. We shift focus from the extreme events, catastrophes, and chaos that dominate public representations of climate change— even in the usually sober realms of international policy and scientific synthesis—to small but consequential shifts in the species and landscapes that we humans interact with on a constant basis. Here, we foreground the *microexperiences* of change, suffering, and risk, illustrating how individuals and communities are experiencing climate change in intimately meaningful ways, how they are constructing knowledge based on these experiences, and how that knowledge shapes their responses.

These microexperiences of climate change are, we argue, a critical but neglected component of our understanding of the Anthropocene as a new geological, cultural, and political era. Of course, macroexperiences are important, too. Without the careful construction of global models, we would not understand the multiple drivers, trajectories, and effects of climate change, and many people have come to understand and respond to climate change through the globally scaled imagery and narratives that emanate from scientific synthesis, international policy, and activism. For many others, however, stories of global dynamics and distant impacts seem like mere abstractions. Examining climate change at the resolution of microexperiences has the advantage of showing us change where many people—perhaps especially those whose livelihoods, social relations, and cultures are most intimately linked to the environment—see it, feel it, and make sense of it. Careful analysis and appreciation of these microexperiences and the resulting knowledge systems may therefore broaden the foundation for shared understanding and collaborative action to address climate change in an inclusive manner.

1.1 Key Insights for Climate Ethnography

This volume's focus on microexperiences contributes to what Susan Crate calls "climate ethnography." Crate argues that ethnography offers a particularly valuable complement to other ways of knowing climate change because ethnography maintains an analytical focus on both "the specifics and multilayered complexities of local human experience" and "the generalities and abstractions of… the global" (2011: 176). Because ethnography "is a comprehensive method that capitalizes on anthropologists' skill in 'being there,' [it] has the methodological power to bridge local understandings beyond the local" so that a "multitude of stakeholders… on a multitude of scales" can develop effective, coordinated policies and responses (177). However, she also underscores an important point: not all ethnography related to climate change counts equally as climate ethnography. True climate ethnography, she argues, should be critical, collaborative, and multi-sited. We therefore use this introduction to frame our diverse case studies of microexperiences of climate change within the scope of critical theory and to begin drawing lessons across our international cases. We believe that a deeper understanding of potential adaptations to climate change will arise from thinking across multiple local ethnographies, and we hope readers will contribute to this cross-pollination.

While contributors to this volume were trained in a range of disciplines, our approach is primarily anthropological. Four key themes from the field of environmental anthropology are present—sometimes explicitly but often implicitly—in the chapters included here. The first theme is a central tenet of environmental anthropology: that the environment is not just something "out there" that we draw on for resources, or that we pollute and disturb, or that threatens us from time to time. Rather, the chapters within this book underline that we are part of the environment, and we must understand it as the matrix of our everyday existence in a holistic and integrative way. Indeed, even this language of "we" and "it" does not do justice to the indivisibility of nature-culture and the diverse ways that people understand this connection (Descola 2013). Three broad areas of research have flowed from this. The first (ethnoecology, cultural ecology, ecological knowledge) examines people's use of the environment, cultural adaptations to different environments, and the knowledge that people develop as they devise livelihood and social systems through unique assemblages of human and nonhuman elements (Balée 2002, 2013; Berkes 1999; Menzies 2006; Nadasdy 1999; Rappaport 1979; Steward 1972). The second (political ecology) examines how power and exploitation are enacted through the environment and reflected in the environment, for example, via rules of access, the distribution of environmental benefits and harms, control over environmental governance, or the relegation of certain groups to "sacrifice zones" (Wolf 1972; Bryant and Bailey 1997; Greenberg and Park 1994; Robbins 2012; Rocheleau et al. 1996). The third (anthropology of nature) explores diverse ontologies, arguing that every cultural group classifies and understands the world in fundamentally different ways and that we therefore live in actually different worlds, the "truth" of which we cannot judge because there is no unfiltered perspective (de la Cadena 2015; Descola 1994, 2013; Escobar 2006; Kohn 2013; Latour 2004; Viveiros de Castro 1998). All three of these areas of study also speak to the construction of identities, values, and visions for the world in and through the socio-environment.

The second theme we want to underscore is that each of the areas of study mentioned above includes a focus on differentiation and inequality. Together, they consider the different and unequal ways in which people use the environment, develop understandings of it, are impacted by changes in it, and have influence over it. In the case of the ontological turn in anthropology, they also interrogate the very different environments that people experience. This focus on difference is an important complement to the generalizations created by global perspectives on climate change—understanding the full range of human responses to climate change and the implications of those responses requires that scholars think beyond generalizations and population trends to also discern meaningful variation (Adger et al. 2013; Barnes et al. 2013; Roncoli et al. 2009). Perhaps most importantly, understanding this diversity allows us to examine relations of complementarity (of groups of people, knowledge systems, and responses) and competition, coercion, and oppression (due to the prioritization of certain knowledges over others, the normalization of certain cultural values or ontologies over others, etc.) (Berkes et al. 2000; Goulden et al. 2009; Lazrus 2016; Marino 2018; Nadasdy 1999; Nelson and Finan 2009; Rice et al. 2015). Thus, we would argue, a micro-experiential perspective is necessary (though not sufficient) for thinking about and planning for a just response to climate change.

Third, one consequence of diversity and inequality is violence, and we would especially like to highlight the multiplicative violences that arise as climate change intersects with other environmental, economic, and sociopolitical burdens (Bunce et al. 2009; Djoudi et al. 2016; O'Brien et al. 2004; Watts 1983). A changing climate brings with it both the "slow violence" (Nixon 2011) of rising seas and increasing weather variability and abrupt catastrophes like monster wildfires and powerful hurricanes (Rahmstorf and Coumou 2011). The unnaturalness of both types of disaster is one of the reasons that ethnographic approaches are valuable for discerning how and why violence manifests in particular ways for particular populations (Oliver-Smith 2013). At face value, slow violence may seem analogous to microexperiences, while rapid disasters parallel macrophenomena. A strength of ethnography, however, is revealing how processes interact across different temporal and spatial scales (Crate 2011). Marino's (2015) analysis of climate change, vulnerability, and environmental displacement in Alaska offers a strong example in this regard. She illustrates how slow changes like global warming and erosion, punctuated impacts like floods, cyclical dynamics like seasonal subsistence calendars, and historical dynamics like settler colonialism and the construction of the "fourth world" in the Americas collectively shape one another and shape the local and global experiences of climate refugees. Anthropological analysis also offers the critical interrogation of "solutions," including attention to the possible violences of our climate change responses (Marino and Ribot 2012; Nightingale et al. 2019).

Fourth, environmental anthropology helps us understand the social organization of knowledge and denial. Ethnoecologists have long documented the division of labor in the environment and the resulting distribution of environmental knowledge. A number of scholars using ethnographic approaches have also offered extraordinarily insightful analyses of how denialism, skepticism, and confusion are constructed and maintained in the realms of climate change (Norgaard 2011), environmental injustice (Auyero and Swistun 2009), and other environmental public health problems (Kleinman 1998; Kleinman and Suryanarayanan 2012). Two key insights emerge from this literature. First, if environmental knowledge is distributed across society, then successfully addressing problems of human survival and adaptation often requires effective collaboration across broad networks. Political ecology teaches us that such collaboration is no small feat given that power is often exercised via the monopolization of knowledge about the environment and control over the environment (Burke and Heynen 2014; Forsyth 2003; Goldman et al. 2010; Nadasdy 1999). Detailing the co-organization of knowledge and power may aid in developing innovative strategies for establishing solidarity across radical difference. Second, this literature reveals that knowledge does not simply exist and spread, and knowledge does not lead in any straightforward way to action. Rather, knowledge, non-knowledge, and action are all products of particular forms of social organization and are shaped by cultural and communicative norms (see especially Norgaard). As we attempt to understand and confront what is likely the greatest challenge humanity has ever faced, it is therefore important to interrogate not only what is (is climate change happening, who will be affected, how vulnerable are they, etc.) but also how we come to know and think about what is (i.e., what is the social and cultural context that shapes our understanding and response). This is perhaps one of the most valuable contributions of these chapters, and it is one of the places where attention to the diverse microexperiences of climate change is most critical.

Collectively, these four themes in environmental anthropology remind us of a critically important lesson laid out so clearly by Marcus Taylor (2014): climate change adaptation is not a "self-evident analytical framework and normative goal" (xi). Adaptation, like climate change itself, is perceived and theorized differently by different people, provoking different types of responses with different levels of effectiveness and different distributions of socio-ecological benefits and harms. Adger et al. (2011) and McDowell and Hess (2012) point out that adaptation to one stressor can reduce capacity to respond to other stressors, and Atteridge and Remling (2018) add that adaptation, rather than reducing vulnerability, can simply serve to redistribute it to others. Taylor goes farther, though, arguing that a focus on vulnerability and adaptation can itself obscure questions of power and sustainability and impede critical thinking about our changing climate. We would argue that an ethnographic lens and a focus on microexperiences can help speak to those concerns by highlighting particular people's knowledges and worldviews and demonstrating how specific responses have concrete impacts on social organization, equality, suffering, and violence. Importantly, such an approach tells a subtle and nuanced story, highlighting triumphs, failures, and the much more common "mixed results" in between.

1.2 This Volume

The chapters in this volume take us to diverse settings in South America, Africa, Europe, Asia, and North America—from cities to farms, alpine meadows to lowland jungles. In each, we see the commonalities of human experience, the connections we make with our environments, the ability of local societies to perceive and face change, and the emotions that emerge from change. In them we also see difference. What matters to whom, when, and why is inflected with history and shaped by culture, place, and circumstance. We see disconnection, misunderstanding, and fear, along with deep ties to place, hope, and collective action. Perhaps most importantly, these chapters show that our diverse histories and experiences give rise to different ways of knowing and being in our environment and highlight the value that comes from bringing those different perspectives into conversation about our shared future.

This book opens with a contribution from Puerto Rico, where Seara, Pollnac, and Jakubowski examine the impact of environmental degradation and climate change on a small-scale fishery. Focusing on perceptions as motivators of behavior, the authors investigate how fishers perceive their changing environment and seek to understand what influences those perceptions. They find that fishers have diverse ways of experiencing their environment and that education, age, and time in the profession all influence perceptions of degradation. Fishers have responded to change in many ways, and the authors underscore the many factors that influence willingness and ability to adapt and what responses fishers choose. Some of the changes fishers have made (e.g., fishing farther from shore and deeper) have serious potential consequences for health and safety. Dervieux and Belgherbi's study focuses on residents of communal lands near Hwange National Park in Zimbabwe. They find that residents have identified a very broad range of changes in their environment—in weather, species, and landscape. Dervieux and Belgherbi link these observations and interpretations to a violent history of oppression, dispossession, and relocation. Importantly, the discourses that villagers use to discuss environmental changes are also "an expression of their own deprivation." Having been "dispossessed of their rights over the land and the natural resources of their environment for more than a century," the discourses of science and law are of little use to them. Cultural explanations focused on the anger of ancestors help fill the gap in expressing moral outrage, fear, and concern. Together, these two chapters underscore the point that violence arises not only from the effects of a changing climate but also from our responses to it.

The third chapter takes us to the Eastern Himalaya. Salick, Staver, and Hart interrogate the interrelationship between climate change, vegetation change, and human adaptation. This chapter in particular shows how tightly linked human and natural systems are, laying out the interplay and feedback loops between changes in plant communities and human responses to that change. Sourdril and colleagues also focus on plants, this time in southern France, where their investigation of "weeds" and how to address them opens questions of social conflict, territorial reform, and rural migration. This is followed by a contribution from Raimond and colleagues, in which they address experimentation by farmers in the Sudano-Sahelian area of Cameroon. The authors show that some changes characterized by scientists as climate impacts—such as variations in temperature, rains, or wildlife populations—are explained by local communities as resulting from deforestation, changing agricultural practices, or population increases. These three chapters illustrate how climate change is intertwined with demographic shifts, changes in traditional practices, and other forms of environmental and social change. Much as humans cannot be analytically "isolated" from their environment, climate cannot be completely disentangled from other drivers of change. Indeed, one benefit of examining microexperiences of climate change is that people often interpret these experiences in an already-integrated manner. Popular knowledge may thus provide important clues for scientists pursuing socio-ecological synthesis or research on coupled natural-human systems.

In the sixth chapter, Katz, Lammel, and Bonnet write of seasonal water levels and flooding in the Brazilian Amazon, comparing the perceptions of local residents with those of natural scientists. Their work illustrates both the conflict and complementarity of these different bodies of knowledge, highlighting the role of scale and occupation in determining knowledge and perception. The two following chapters, from Roque de Pinho and Reyes-Garcia and colleagues, point to ways to bring different types of knowledge together. Roque de Pinho uses PhotoVoice with Maasai pastoralists to understand climatic variability, extreme drought, and animal response, and Reyes-Garcia and colleagues detail a citizen science initiative to gather and share information related to climate change impacts. Drawing on a similar project dedicated to traditional knowledge in Spain, the authors are developing a global platform to collect indigenous and local knowledge on indicators of climate change, complementing instrumental weather and climate observations. These three cases thus highlight multiple ways to integrate the knowledge of scientists and nonscientists, showing the clear benefits that can be derived from these collaborations but also detailing the challenges to achieving authentic participation. As Pennesi argues in the conclusion, these are valuable contributions to a long-term project to decolonize and democratize knowledge and environmental governance.

The final two chapters also address collaboration across social difference, focusing not only on the complementarity of different knowledge systems but also prospects for more inclusive and democratic decision-making about how to craft the socio-ecological futures we want. In the first, Burke and colleagues write of the potential for finding common ground in climate discussions in southern Appalachia (southeastern United States) by focusing on local observations of change rather than far-off and abstract images. By coming together around what different people have seen and experienced first-hand, there is potential to leave behind political polarization and group stereotypes, instead viewing differences in knowledge and experience as strengths to be drawn upon. In the second, we return to east Africa, where Galvin and colleagues participated in collaborative workshops with other scientists and with pastoralists focused on responding to challenges and improving well-being. The workshops were pastoralist-led and paid careful attention to differences in knowledge generated by social position or occupation. These workshops illustrate how diverse participants can collectively identify pathways toward a healthier and more secure future.

In her conclusion, Karen Pennesi considers how these chapters contribute to four goals of anthropological research on climate change: documentation, connection, collaboration, and social transformation. She summarizes the numerous impacts and adaptations that are documented throughout this book and notes, importantly, that people's "observations are made with both bodies and minds, and the effects are felt in profound emotional and psychological—not just physical—ways." Appreciating the multifaceted experience of climate change is one of the central goals of climate ethnography. Pennesi teases out two ways that these chapters contribute to building fruitful connections—by illustrating how scientists and nonscientists can connect in mutually enriching collaboration and by illustrating how we all may think more carefully about the connections between climate and other social and political-economic changes. Finally, she discusses how the chapters here contribute to activism for social transformation. "This book," she writes, "helps us see the importance of developing adaptive strategies for climate change that take into account sociocultural factors influencing relationships between people and their environment, as well as relationships among people that are mediated by the environment." But we must do significantly more if we are to achieve the transformation necessary to avert the multiple environmental and human crises that we currently face. Pennesi therefore concludes by suggesting how these authors might go further and by offering a vision for a radically democratized and anti-colonial/de-colonial system of environmental knowledge and governance.

Like Pennesi, we believe that effective and just responses to the intersecting crises and violences of the present—which include the local and the global, the dramatic and the mundane, and the ecological and the sociopolitical—demand decolonization and democratization. Indeed, this is often neglected in the recommendations of international science and policy communities. In this book, we contribute to this democratization by highlighting people's diverse experiences of climate change, the multiple knowledge systems they employ in responding to local and global changes, and strategies for honoring this plurality in our collective responses. We hope that these chapters will be useful to those who aim to carry this project further, toward a more complete democratization and decolonization of knowledge, politics, and economics, and toward the construction of more just and sustainable societies.

References

Adger WN, Brown K, Nelson DR, Berkes F, Eakin H, Folke C, Galvin K, Gunderson L, Goulden M, O'Brien K, Ruitenbeek J, Tompkins EL (2011) Resilience implications of policy responses to climate change. Wiley Interdiscip Rev Clim Change 2(5):757–766

Adger WN, Barnett J, Brown K, Marshall N, O'Brien K (2013) Cultural dimensions of climate change impacts and adaptation. Nat Clim Chang 3(2):112–117

Atteridge A, Remling E (2018) Is adaptation reducing vulnerability or redistributing it? Wiley Interdiscip Rev Clim Change 9(1):e500

Auyero J, Swistun DA (2009) Flammable: environmental suffering in an argentine shantytown. Oxford University Press, Oxford

Balée WL (2002) Advances in historical ecology. Columbia University Press, New York

Balée WL (2013) Cultural forests of the Amazon: a historical ecology of people and their landscapes. University of Alabama Press, Tuscaloosa

Barnes J, Dove M, Lahsen M, Mathews A, McElwee P, McIntosh R, Moore F, O'Reilly J, Orlove B, Puri R, Weiss H, Yager K (2013) Contribution of anthropology to the study of climate change. Nat Clim Chang 3:541–544

Berkes F (1999) Sacred ecology: traditional ecological knowledge and resource management. Taylor and Francis, New York

Berkes F, Colding J, Folke C (2000) Rediscovery of traditional ecological knowledge as adaptive management. Ecol Appl 10:1251–1262

Brown TW, Bischof-Niemz T, Blok K, Breyer C, Lund H, Mathiesen BV (2018) Response to 'burden of proof: a comprehensive review of the feasibility of 100% renewable-electricity systems'. Renew Sust Energ Rev 92:834–847

Bryant RL, Bailey S (1997) Third world political ecology. Routledge, New York

Bunce M, Rosendo S, Brown K (2009) Perceptions of climate change, multiple stressors and livelihoods on marginal African coasts. Environ Dev Sustain 12(3):407–440. https://doi.org/10.1007/s10668-009-9203-6

Burke BJ, Heynen N (2014) Transforming participatory science into socio-ecological praxis: valuing marginalized environmental knowledges in the face of the neoliberalization of nature and science. Environ Soc 5:7–27

Crate SA (2011) Climate and culture: anthropology in the era of contemporary climate change. Annu Rev Anthropol 40:175–194

Cruikshank J (2005) Do glaciers listen? Local knowledge, colonial encounters, and social imagination. UBC Press, Vancouver

Cunsolo A, Ellis NR (2018) Ecological grief as a mental health response to climate change-related loss. Nat Clim Chang 8:275–281

De la Cadena M (2015) Earth beings: ecologies of practice across Andean worlds. Duke University Press, Durham

Descola P (1994) In the society of nature: a native ecology in Amazonia, N. Scott, trans. Cambridge University Press, Cambridge

Descola P (2013) Beyond nature and culture. University of Chicago Press, Chicago

Djoudi H, Locatelli B, Vaast C, Asher K, Brockhaus M, Sijapati BB (2016) Beyond dichotomies: gender and intersecting inequalities in climate change studies. Ambio 45(Supplement 3):248–262

Escobar A (2006) Discourse and conflict in the struggle over natural resources: a political ecology framework. Development 49(3):6–13

Forsyth T (2003) Critical political ecology: the politics of environmental science. Routledge, London

Goldman MJ, Nadasdy P, Turner MD (eds) (2010) Knowing nature: conversations at the intersection of political ecology and science studies. University of Chicago Press, Chicago

Goulden M, Noess LO, Vincent K, Adger WN (2009) Accessing diversification, networks and traditional resource management as adaptations to climate extremes. In: Adger WN, Lorenzoni I, O'Brien K (eds) Adapting to climate change: thresholds, values, governance. Cambridge University Press, Cambridge, pp 448–464

Greenberg JB, Park TK (1994) Political Ecology. Journal of Political Ecology 1(1):12

Hawken P (ed) (2017) Drawdown: the most comprehensive plan ever proposed to reverse global warming. Penguin Books, New York

Hoegh-Guldberg O, Jacob D, Taylor M, Bindi M, Brown S, Camilloni I, Diedhiou A, Djalante R, Ebi KL, Engelbrecht F, Guiot J, Hijioka Y, Mehrotra S, Payne A, Seneviratne SI, Thomas A, Warren R, Zhou G (2018) Impacts of 1.5 °C global warming on natural and human systems. In: Masson-Delmotte V, Zhai P, Pörtner H-O, Roberts D, Skea J, Shukla PR, Pirani A, Moufouma-Okia W, Péan C, Pidcock R, Connors S, Matthews JBR, Chen Y, Zhou X, Gomis MI, Lonnoy E, Maycock T, Tignor M, Waterfield T (eds) Global warming of 1.5 °C. An IPCC special report on the impacts of global warming of 1.5°C above pre-industrial levels and related global greenhouse gas emission pathways, in the context of strengthening the global response to the threat of climate change, sustainable development, and efforts to eradicate poverty. https://www.ipcc.ch/sr15/chapter/chapter-3/

IPBES (2019) Summary for policymakers of the global assessment report on biodiversity and ecosystem services of the Intergovernmental Science-Policy Platform on Biodiversity and Ecosystem Services. Advance Unedited Version. May 6, 2019. Díaz S, Settele J, Brondízio E (Co-Chairs). IPBES Secretariat, Bonn, Germany

IPCC (2018) Summary for policymakers. In: Masson-Delmotte V, Zhai P, Pörtner HO, Roberts D, Skea J, Shukla PR, Pirani A, Moufouma-Okia W, Péan C, Pidcock R, Connors S, Matthews JBR, Chen Y, Zhou X, Gomis MI, Lonnoy E, Maycock T, Tignor M, Waterfield T (eds) Global warming of 1.5 °C. An IPCC special report on the impacts of global warming of 1.5 °C above pre-industrial levels and related global greenhouse gas emission pathways, in the context of strengthening the global response to the threat of climate change, sustainable development, and efforts to eradicate poverty. World Meteorological Organization, Switzerland, p 32

Jacobson MZ, Delucchi MA, Cameron MA, Mathiesen BV (2018) Matching demand with supply at low cost in 139 countries among 20 world regions with 100% intermittent wind, water, and sunlight (WWS) for all purposes. Renew Energy 123:236–248
Kleinman DL (1998) Beyond the science wars: contemplating the democratization of science. Politics Life Sci 17(2):133–145
Kleinman DL, Suryanarayanan S (2012) Dying bees and the social production of ignorance. Sci Technol Human Values 38(4):492–517
Kohn E (2013) How forests think: toward an anthropology beyond the human. University of California Press, Berkeley
Latour B (2004) Politics of nature. Harvard University Press, Cambridge, MA
Lazrus H (2016) The governance of vulnerability: climate change and agency in Tuvalu, South Pacific. In: Crate SA, Nuttall M (eds) Anthropology and climate change: From encounters to actions. Routledge, London
Marino E (2015) Fierce climate, sacred ground: an ethnography of climate change in Shishmaref. University of Alaska Press, Alaska
Marino E (2018) Adaptation privilege and voluntary buyouts: perspectives on ethnocentrism in sea level rise relocation and retreat policies in the US. Glob Environ Chang 49:10–13
Marino E, Ribot J (2012) Special issue introduction: adding insult to injury: climate change and the inequities of climate intervention. Glob Environ Chang 22(2):323–328
McDowell J, Hess J (2012) Accessing adaptation: multiple stressors on livelihoods in the Bolivian highlands under a changing climate. Glob Environ Chang 22(2):342–353
Menzies CR (ed) (2006) Traditional ecological knowledge and natural resource management. University of Nebraska Press, Lincoln
Nadasdy P (1999) The politics of TEK: power and the" integration" of knowledge. Arct Anthropol 36:1–18
Nadasdy P (2003) Hunters and bureaucrats: power, knowledge, and aboriginal-state relations in the Southwest Yukon. University of British Columbia Press, Vancouver
Nelson DR, Finan TJ (2009) Praying for drought: persistent vulnerability and the politics of patronage in Ceara, Northeast Brazil. Am Anthropol 111(3):302–316
Nightingale AJ, Eriksen S, Taylor M, Forsyth T, Pelling M, Newsham A, Boyd E, Brown K, Harvey B, Jones L, Kerr RB, Mehta L, Naess LO, Ockwell D, Scoones I, Tanner T, Whitfield S (2019) Beyond technical fixes: climate solutions and the great derangement. Clim Dev. https://doi.org/10.1080/17565529.2019.1624495
Nixon R (2011) Slow violence and the environmentalism of the poor. Harvard University Press, Cambridge
Norgaard KM (2011) Living in denial: climate change, emotions, and everyday life. MIT Press, Cambridge
O'Brien K, Leichencko R, Kelkar U, Venema H, Aandah G, Tompkins H, Javed A, Bhadwal S, Barg S, Nygaard L, West J (2004) Mapping vulnerability to multiple stressors: climate change and globalization in India. Glob Environ Chang 14(4):303–313
Oliver-Smith A (2013) 2013 Malinoskwi Award Lecture. Disaster risk reduction and climate change adaptation: the view from applied anthropology. Hum Organ 72(4):275–282
Rahmstorf S, Coumou D (2011) Increase of extreme events in a warming world. Proc Natl Acad Sci 108(44):17905–17909
Rappaport RA (1979) Ecology, meaning, and religion. North Atlantic Books, Berkeley
Ribot J (2010) Vulnerability does not fall from the sky: toward multiscale, pro-poor climate policy. In: Mearns R, Norton A (eds) Social dimensions of climate change: equity and vulnerability in a warming world. The World Bank, Washington, DC, pp 47–74
Rice JL, Burke BJ, Heynen N (2015) Knowing climate change, embodying climate praxis: experiential knowledge in Southern Appalachia. Ann Assoc Am Geogr 105(2):253–262
Robbins P (2012) Political ecology: a critical introduction. Blackwell, New York
Rocheleau D, Thomas-Slayter B, Wangari E (1996) Feminist political ecology: global issues and local experiences. Routledge, London

Roncoli C, Crane T, Orlove B (2009) Fielding climate change in cultural anthropology. In: Crate SA, Nuttall M (eds) Anthropology and climate change: from encounters to actions. Routledge, London, pp 87–115

Smith N (2006) There's no such thing as a natural disaster. Essay contributed to the Learning from Katrina Program of the Social Science Research Council. http://understandingkatrina.ssrc.org/Smith

Steward JH (1972) Theory of culture change: the methodology of multilinear evolution. University of Illinois Press, Urbana

Taylor M (2014) The political ecology of climate change adaptation. Livelihoods, agrarian change, and the conflicts of development. Routledge, London

United Nations Environment Programme (UNEP) (2012–2018) Emissions gap reports. United Nations Environment Programme, Nairobi

Viveiros de Castro E (1998) Cosmological deixis and Amerindian perspectivism. J R Anthropol Inst 4(3):469–488

Watts M (1983) Silent violence: food, famine and peasantry in Northern Nigeria. University of California Press, Berkeley

Wolf E (1972) Ownership and political ecology. Anthropol Q 45(3):201–205

Chapter 2
Fishers' Perceptions of Environmental and Climate Change in Puerto Rico: Implications for Adaptation and Sustainability

Tarsila Seara, Richard Pollnac, and Karin Jakubowski

Abstract Marine fisheries in the Caribbean are vulnerable to a wide range of environmental and climatic change impacts. Direct and indirect effects of these impacts on fish species affect the ability of fishers to harvest them resulting in reductions in revenue and food security. Understanding factors impacting and transforming fisheries from the viewpoint of the fishers is crucial for developing adequate strategies to maximize coastal communities' resilience and adaptation to change, particularly under future climate change scenarios. This study uses qualitative and quantitative data collected from 212 surveys with Puerto Rican fishers to explore aspects of fishers' subjective perceptions of environmental and climate change and investigate factors influencing these perceptions. Our findings show that fishers perceive the local environment and climate to have undergone significant changes in the past couple of decades and they believe these changes have been affecting the fishery and consequentially leading them to adapt. Adaptations to these impacts, which consist mostly of seeking new fishing grounds, have led them to increase their exposure to risks, particularly among SCUBA divers fishing in deeper waters and farther away from the coast. Results also show important relationships between fishers' perceptions of the status of fishery resources, demographics, levels of environmental awareness, and concern about climate change. These findings have significant implications for the development of policy and educational strategies aimed at increasing sustainability and well-being in fishing communities.

T. Seara (✉) · K. Jakubowski
Department of Biology & Environmental Science, University of New Haven, West Haven, CT, USA
e-mail: tseara@newhaven.edu

R. Pollnac
Department of Marine Affairs, University of Rhode Island, Kingston, RI, USA

M. Welch-Devine et al. (eds.), *Changing Climate, Changing Worlds*, Ethnobiology, https://doi.org/10.1007/978-3-030-37312-2_2

Keywords Fisheries · Puerto Rico · Climate change · Adaptive capacity · Caribbean

2.1 Introduction

Marine ecosystems and resources worldwide are being transformed and threatened by human activity at an increasing pace. In the Caribbean region, economic development and population growth have generally occurred without effective policies to safeguard the sustainability of natural resources, resulting in rapid degradation of coastal waters and critical ecosystems (Valdés-Pizzini et al. 2012). Ecological deterioration and overexploitation and overfishing have contributed to the collapse of many important fishery resources in the region, reducing biodiversity, revenues, and the resilience of fishers, their families, and communities. Marine fisheries in the Caribbean are also vulnerable to a wide range of climate change impacts. Among the most significant are loss of critical habitat such as mangrove forests and seagrass beds (Short et al. 2016; Alongi 2015), coral bleaching and disease (Baker 2014; Randall and Woesik 2015), changes in patterns of freshwater flows (Holding and Allen 2015), and ocean acidification impacting shell formation for corals, plankton, and shellfish (Rhein et al. 2013). Climate change impacts also affect the life cycles, abundances, and distributions of fish species (Perry et al. 2005). Fishing is an important cultural and socioeconomic component in the Caribbean context. Understanding factors impacting and transforming this activity from the viewpoint of those directly involved, that is, the fishers, is crucial for developing strategies that will maximize coastal communities' resilience and adaptation to change in the region, particularly under future climate change scenarios. This study uses data collected from surveys with Puerto Rican fishers to explore aspects of fishers' subjective perceptions of environmental and climate change and investigate factors influencing these perceptions.

2.1.1 Climate Change and Fisheries

In fishing communities, direct and indirect impacts of environmental degradation and climate change on species that are important for income and subsistence affect the ability of fishers to harvest them (Sumaila et al. 2011; Pinsky and Mantua 2014; McCay et al. 2011; Allison et al. 2009; Weatherdon et al. 2016) resulting in reductions in revenue and food security. Declines in revenue impact fishers and their families as well as local economies by negatively affecting supporting businesses such as seafood dealers and distributors, fish markets, and restaurants and associated sectors such as tourism. Impacts on the ability of fishers to catch fish for their subsistence result in reduced food security. In small-scale fishing communities, and as evidenced in Puerto Rico (Garcia-Quijano et al. 2015; Griffth and Valdés-Pizzini

2002), it is common practice for fishers to give a portion of their catch to family and other community members. Thus, reductions in catch also affect practices that foster the development and maintenance of social ties that are important for generating social capital. Natural resource decline also influences changes in fishery management, for example, prompting reduction of allowable catches and extended closures, contributing to pressures associated with resource scarcity and further affecting fishing communities' socioeconomic well-being. In addition, from a human dimensions perspective, fishing communities are typically located in low-lying zones which are at risk from sea level rise and present high exposure to extreme weather events (Nicholls and Cazenave 2010), potentially compounding their socioeconomic vulnerability under climate change.

Transformations associated with environmental and climatic change and the potential impacts of indirect effects on range and productivity of commercially important species require that fishers adopt strategies to respond adaptively. Adaptive responses may include *within-fishing* adaptations, for example, finding new fishing grounds and exploiting different species, or *outside-fishing* adaptations, that is, finding alternative income or employment (Pinsky and Mantua 2014; Cinner et al. 2011; McCay et al. 2011). Poverty and other types of marginalization in fishing communities may reduce the ability of fishers to adapt to changes (Daw et al. 2009), particularly through *outside-fishing* adaptations. In addition, a great deal of research suggests that fishers are reluctant to leave the occupation of fishing even under economic hardship due to a combination of sociocultural and psychological factors (Smith and Clay 2010; Pollnac et al. 2015). Job satisfaction in fisheries has long been recognized as an important aspect related to fishers' adaptation to change (Pollnac and Poggie 1988). Although satisfaction with aspects of one's job is important in any occupation, it is especially significant in fishing jobs. Among fishers, the structure of job satisfaction includes attributes of "adventure," "challenge," and "being outdoors" that are infrequently found in other occupations (Apostle et al. 1985; Pollnac and Poggie 1988, 2008; Binkley 1995; Pollnac et al. 2008; Seara et al. 2017a, b). Understanding these satisfactions derived from fishing is important because the more attached people are to their jobs, the more difficult it is to either leave or deal with significant changes to their occupation. For people presenting strong occupational attachment, the prospect of losing their jobs may represent not only the loss of income but of part of their self-identity (Marshall et al. 2007). Therefore, leaving the occupation of fishing altogether as an adaptation strategy is unlikely and of particular concern for individual, familial, and community well-being (Pollnac and Poggie 2008; Pollnac et al. 2015). In a detailed ethnography of Puerto Rican fishers, Griffith and Valdés-Pizzini (2002) describe that a great deal of them consider fishing as "therapy." Many Puerto Rican fishers who work on land jobs (e.g., farming) during the fishery off seasons return to the sea, to what they describe as a healthy activity that keeps their minds occupied on useful things and that provides relief from stress (Griffith and Valdés-Pizzini 2002). In a study comparing Southeast Puerto Rico and other fishing communities in the USA and the wider Caribbean region, Seara et al. (2017b) found that Puerto Rican fishers presented the highest levels of job satisfaction among compared samples. The authors argue that high job satisfaction among Puerto Rican fishers is associated with freedom to pursue their own inclinations in fishing and other livelihood activities and spend time with friends and family, coupled with a sufficient material well-being and strong social ties within their communities (Seara et al. 2017b). Therefore, changes resulting from climate change and other anthropogenic impacts affecting the fisheries in Puerto Rico will have significant impacts on the psychological and social well-being of fishers and their families. According to Daw et al. (2009), "climate change impacts on fisheries will occur in the context of, and interact with existing drivers, trends and status of fisheries." Specifically in Puerto Rico, evidence of the significance of fisheries to economic, cultural, psychological, and basic subsistence aspects further emphasizes the challenges associated with fishers' adaptation to environmental and climate change.

2.1.2 *Puerto Rico Fisheries*

Fishing activities in Puerto Rico are predominantly dependent on nearshore coral reef systems (Appeldoorn 2008) and adjacent ecosystems such as seagrass beds and mangrove forests. Coral reef ecosystems in the Caribbean have been declining for at least the last 40 years, although pinpointing the beginning of the decline has been difficult (Appeldoorn et al. 2009). Overfishing and climate change are considered two of the most significant threats for the great majority of Puerto Rico's reefs and marine ecosystems (García-Sais et al. 2008; Rogers 2009; Ramos-Scharrón et al. 2015; Loh et al. 2015; Hernández-Delgado et al. 2014). In 2005, a widespread coral bleaching event associated with record high seawater temperatures in the Caribbean region heavily impacted Puerto Rico's coral reefs (Wilkinson and Souter 2008). Following the bleaching event, researchers in the region reported an average of 50% decline in live coral cover and up to 90% mortality of coral colonies at specific monitoring sites (Miller et al. 2006; García-Sais et al. 2006; Woody et al. 2008). Donner et al. (2007) attributed this particular event to anthropogenic warming. The study suggests that greenhouse gas emissions increase the probability of events of extreme thermal stress in the region by an order of magnitude, which could result in events, such as the one that occurred in 2005, becoming biennial occurrences within the next 30 years (Donner et al. 2007). The authors further stated that expected increase in hurricane activity in the region, also as a result of human-induced climate change, would critically damage and endanger corals already weakened by bleaching events (Donner et al. 2007). The 2017 Atlantic Hurricane Season (AHS) was one of the most active in history producing six hurricanes above category 3 and breaking the record for most consecutive storms in the satellite era (NOAA 2018a). The most intense hurricane of the 2017 season, category 5 Maria, made landfall in Puerto Rico in September and was the most intense storm to hit US territory in recorded history (NOAA 2018a). The 2017 AHS, particularly the two most intense storms Irma and Maria, caused catastrophic damage to Puerto Rico communities and resulted in substantial damage to the island's shallow water coral reefs (NOAA

2018b). Storm damage to Puerto Rico coastal communities' infrastructure and to coral reefs and adjacent ecosystems can significantly impact productivity in the island's marine fisheries.

The Puerto Rican fishery sector is comprised mainly of small-scale operations that often consist of an owner and a crew of typically no more than three members that fish from small vessels with limited horsepower. According to Garcia-Quijano (2009), the Puerto Rican fisheries' socioeconomic context can be typified by "heterogeneity and unpredictability of opportunities for employment and for covering the basic needs of subsistence." Research suggests that, even though the fishery sector in Puerto Rico is generally characterized by occupational multiplicity, fisheries and coastal-based resource extraction constitute important components of the cultural and socioeconomic context of most coastal communities (see Garcia-Quijano et al. 2015). According to the latest census of Puerto Rican fishers conducted in 2008, approximately 900 individuals engage in the activity throughout the island (Matos-Caraballo and Agar 2011). There is indication, however, that the actual number is higher, ranging between 1500 and 2000 as suggested by qualitative data, since many fishers in the island are not fully licensed (Tonioli and Agar 2011). The licensing system in Puerto Rico requires that fishers provide proof of income deriving from fishing in the form of tax documentation in order to be considered eligible for either part- or full-time fishing licenses (Matos-Caraballo 2009). As mentioned above, the fishery sector in Puerto Rico has been historically characterized by occupational multiplicity (Garcia-Quijano et al. 2015; Griffith and Valdés-Pizzini 2002) which may be in conflict with the current licensing system, therefore unintentionally creating an incentive for unlicensed and, thus, under- or unreported fishing activity.

Fishers in Puerto Rico are typically organized around fishery centers/associations (*villas pesqueras*) or along coastal townships. Catches consist of relatively low quantities of diverse species (Griffith and Valdés-Pizzini 2002) including snappers (Lutjanidae), groupers (Serranidae), and a variety of crustaceans and mollusks that are sold to the *villas pesqueras* and private fish markets (*pescaderias*), or more informally from fishers' houses or directly to buyers or restaurants. Puerto Rican fishers typically engage in multiple gear activities including nets, traps, hook and line, and harpoon. In recent years, the percentage of fishers practicing SCUBA diving has increased considerably: from approximately 35% in the mid-1990s to between 40 and 50% in 2008 (Matos-Caraballo and Agar 2011). According to Matos-Caraballo and Agar (2011), this reflects a necessity for fishers to become more specialized as a result of declines in coastal catches. Down trends in commercial fish landings in Puerto Rico have been reported since the 1970s (Matos-Caraballo 2009; Bryan et al. 2016) with an alarming decline of approximately 69% occurring between 1970 and 1990 (Appeldoorn et al. 1992). The species predominantly targeted by SCUBA divers are spiny lobster (*Panulirus argus*) and queen conch (*Strombus gigas*).

Awareness of the importance of understanding human dimensions of fisheries has been increasing considerably in the past few decades (see Weeratunge et al. 2014), and, with that, a new focus is given on fisheries *governance* as opposed to merely management (Daw et al. 2009; Pittman et al. 2015). A focus on governance provides a more holistic approach in which both formal and informal arrangements and institutions are involved in decisions affecting the use of natural resources (Juda 1999), thus allowing for more opportunities to incorporate human dimensions and resource users' involvement in decision making. In the face of environmental degradation, resource decline, and climate change, successful adaptation in fishing communities will require the implementation of policy strategies that are informed by human dimensions in order to properly consider crucial aspects such as job satisfaction, food security, social networks, and overall well-being, while safeguarding natural resource sustainability and ecological resilience. Government intervention and policies must facilitate adaptive capacity, particularly within vulnerable communities (Adger 2003; Daw et al. 2009). In general, the *objective* capacity of individuals or societies to adapt to change is determined by availability of resources and an individual's or group's access to those resources. On the other hand, *subjective* aspects of adaptation are associated with peoples' perceptions of the adequacy of available resources in aiding them to cope and adapt, as well as the extent to which they feel prepared to endure such changes or impacts and actively cope with them, that is, their perceived vulnerability. The focus of this study is on the latter.

Climate change vulnerability and adaptive capacity frameworks often either fail to or poorly incorporate aspects of subjective perception (Adger et al. 2013; Limuwa et al. 2018). Perceptions are largely shaped by one's experiences and, along with values, beliefs, knowledge, and culture, are important motivators of people's behaviors (Taylor et al. 1988; Grothmann and Patt 2005). Understanding people's perceptions of and vulnerability to environmental and climate change is crucial since in the face of stressors, people often act upon their subjective perceptions rather than objective measures (Grothmann and Patt 2005; Smith and Clay 2010; Seara et al. 2016). Programs directed at increasing human resilience and adaptive capacity need to seriously consider these perceptions (Aswani et al. 2015; Cinner et al. 2018). In the case of fishers and fishing communities, adaptive capacity is also closely linked to observed and future changes to marine ecosystems (Daw et al. 2009). This study analyzes the perceptions of fishers in Puerto Rico communities toward environmental and climate change as well as factors potentially influencing these perceptions and their ability to adapt. It is proposed that literature reviewed and the relationships discovered in the data can be used to generate a heuristic model to function as a general hypothesis which will be tested here using path analysis.

2.2 Methods of Data Collection

This study used surveys to collect quantitative and qualitative information from fishers in ten different municipalities in Puerto Rico (Fig. 2.1). Between August 2016 and January 2017, a total of 212 fishers were surveyed using an intercept sampling method that consisted of approaching fishers at different fishing associations (*villas pesqueras*) and other locations where they were known to land their catches

Fig. 2.1 Map of Puerto Rico showing municipalities included in data collection. (Adapted from USDA 2012)

or congregate. This sampling technique was considered the most effective to maximize sampling of the studied universe since no comprehensive list or directory of Puerto Rico fishers is readily available from which to draw a random sample (e.g., Seara et al. 2017a, Pollnac et al. 2015). The questionnaire included demographic and fishery attributes, job satisfaction, perceptions of the status of fishery resources and the environment, and climate change. Questions to collect information about fishers' perceptions of changes in the marine environment and climate used a combination of dichotomous, Likert scale, and open-ended methods to collect data that could be analyzed quantitatively while allowing respondents to elaborate on their views and experiences to aid in interpretation of results.

2.3 Measurements and Analyses

2.3.1 Fishers' Characteristics

For the majority of fishers interviewed (69.8%), fishing was their only source of income. For those who had additional occupations (24.1%) or who did not consider fishing an occupation (subsistence and/or recreational) (6.1%), the most common types of employment were construction (10.9%), owning a business (9.1%), and farming (9.1%). Fishers were on average 50.5 years of age (sd = 13.96) with 35.9 years of fishing experience (sd = 16.07) and 9.1 (sd = 3.72) years of formal education. The majority of the fishers interviewed were married (71.7%) and the average household size was 3.0 people (sd = 1.37).

The sample is characterized by a majority of SCUBA (32.5%[R1]) and trap (*nasas*) (14%[R]) fishers. The species most important for fishers' incomes were spiny lobster (*Panulirus argus*) (36.7% [R]), silk snapper (*Lutjanus vivanus*) (14.3%[R]), queen conch (*Strombus gigas*) (9%[R]), and yellowtail snapper (*Ocyurus chrysurus*) (8.1% [R]). The

[1] Refers to percent of responses as each fisher could indicate more than one option per response.

majority of fishers (72.3%) owned the boats they used to fish which were on average 20.4 feet long (sd = 3.43). Fishers surveyed spent on average 10.9 months out of the year (sd = 2.17), 14.9 days out of the month (sd = 7.41), and 6.6 hours a day (sd = 2.13) fishing.

2.3.1.1 Job Satisfaction

Job satisfaction was measured using a 9-item scale (Table 2.1) derived from a 22-item scale developed by Pollnac and Poggie (1988). The 9-item scale was developed by using a principal component analysis (PCA) from a geographically diverse data set and selecting the three items with the highest loadings on each of the three components commonly derived from the PCA. In the present sample, mean values for all job satisfaction variables, measured on a Likert scale of 1 (*very dissatisfied*) to 5 (*very satisfied*), are above 3.7 suggesting that, in general, fishers present high levels of job satisfaction. Aspects of the job with which fishers are most satisfied relate to independence and adventure.

2.3.1.2 Environmental Ethic

A list of 10 items (Table 2.2) were used to assess aspects of environmental beliefs among fishers by asking their level of agreement with each item on a 5-point Likert scale ranging from 1 (*strongly disagree*) to 5 (*strongly agree*), reversed, reversed for negatively worded items. Mean scores were above three for all items with the exception of the item concerning whether or not fishers agreed that the presence of houses near the coast had an effect on the fishery. Items with the highest scores concerned the need to take care of the land and sea to safeguard sustainability and the impacts of coral death on fishing (see Table 2.2 for all mean scores). A *total environmental ethic* scale ranging from 10 to 50 was created by summing up the responses to all ten items.

Table 2.1 Job satisfaction variables with mean values and standard deviation measured on a 5-point Likert scale ranging from very dissatisfied to very satisfied

Item	Mean	Std. deviation
Opportunity to be own boss	4.64	0.624
Adventure	4.45	0.528
Time spent fishing	4.34	0.676
Challenge	4.25	0.606
Earnings	4.08	0.843
Safety	3.97	0.737
Healthfulness	3.92	0.735
Predictability of earnings	3.83	0.759
Fatigue	3.70	0.752

Table 2.2 Environmental ethic variables with mean values and standard deviation measured on a 5-point Likert scale ranging from strongly disagree to strongly agree except for items marked with an asterisk for which scale was reversed

Item	Mean	Std. deviation
We have to take care of the land and the sea or they will not provide for us in the future	4.43	0.496
If the corals die, it will make a difference for fishing	4.41	0.628
If our community works together, we will be able to protect our resources	4.23	0.704
Tourism around/near the coast can have an effect on the fish	3.38	1.105
Agriculture around/near the coast can have an effect on the fish	3.11	1.036
Industry around/near the coast can have an effect on the fish	4.04	0.897
Houses around/near the coast can have an effect on the fish	2.91	1.142
*If we throw our garbage on the beach, the ocean takes it away and it causes no harm	4.34	0.848
Unless mangroves are protected, we will not have any fish to catch	4.17	0.766
*There are so many fish in the ocean that no matter how many we catch, there will always be enough for our needs	3.44	1.255

2.3.2 Perceptions on Status of Fishery Resources

Results indicate that in general fishers perceive the fishery resources in the region to be in good shape. On a *current status of fishery resources* scale of 1 (*very bad*) to 5 (*very good*) responses averaged 3.51 (sd = 0.84), with 50.2% of the fishers indicating resources to be in *good* shape. When asked if resources are currently in worse, equal, or better shape (a scale of 1 to 3) when compared to 10 years ago, the average was 1.70 (sd = 0.63), with 51.7% of the fishers stating resources are *equal* and 39.3% stating they are *worse*. The most frequent reasons stated by fishers for resources to have worsened during the past decade were pollution (40.4%^{R}), over-exploitation (19.2%^{R}), changes in climate (15.4%^{R}), government regulations (6.7%^{R}), and environmental degradation (5.9%^{R}).

2.3.3 Perceptions of Climate Change and Other Anthropogenic Impacts

The majority of fishers surveyed (72.6%) believe the local climate to be changing, and in general they perceived these changes to be negative, averaging 2.35 (sd = 0.76) on a scale ranging from 1 (*very bad*) to 5 (*very good*) (*climate change status*). Overall, fishers' level of concern with climate change can be categorized as moderate to high, averaging 6.67 (sd = 2.50) on a 10-point *concern over climate change* scale (1 = not worried and 10 = very worried). Just under half (46.0%) of fishers said they have observed changes to the fish, shellfish, and/or local marine

Table 2.3 Factors potentially affecting fishery and percent of fishers who responded affirmatively

Factors	%
Pollution	92.9
Coral bleaching	84.4
Increase in water temperature	59.7
Overfishing	36.8
Increase in frequency and intensity of storms	34.4
Increase in seaweed/algal blooms	32.9
Sea level rise	31.9
Change in behavior of marine animals	30.0
Increase in air temperature	21.8
Increase in droughts	17.5

environment that they believe to be related to climate change. The most common changes observed were resource stock decline (34.1%^{R}), habitat shifts (14.6%^{R}) which include perceptions of species moving into deeper waters or to areas farther away from the coast, change in composition of fish species (11.4%^{R}), and changes in water temperature (6.5%^{R}). Fishers were also asked about whether or not they believe a series of different factors associated with anthropogenic impacts and climate change to be a threat to fisheries. The factors with the highest frequency of affirmative responses were pollution (92.9%), coral bleaching (84.4%), and increase in sea temperature (59.7%) (Table 2.3). The sum of dichotomous responses, no = 0 and yes = 1, to the 10 factors included in Table 2.3 was used to create a scale of perceptions on *anthropogenic impacts* ranging from zero to 10.

2.3.4 Factors Influencing Perceptions of Climate Change

The variables and analyses used in this study deal with a number of important factors that can influence perceptions of the potential negative impacts of climate change on the marine environment and the fishers. We would like to suggest a simple model that is proposed to reflect the interrelationships between these factors to provide an understanding that will facilitate developing targeted policies and educational programs to aid in helping fishers and other coastal dwellers cope with and become more resilient to the impacts of climate change. In general it is proposed that age, education, and exposure (here exposure is assumed to increase through the act of fishing) have an influence on perceptions of the environment, including *current status of fishery resources* and the different ways human behaviors can influence the resource (*anthropogenic impacts*) and the status of the fishery. All this is expected to influence beliefs about the impacts of human behavior on the coastal environment (*total environmental ethic*). Years of formal *education* will also have an independent effect on items forming the *total environmental ethic* scale. It is proposed that these resource beliefs (*anthropogenic impacts* and *total environmental ethic*), along with *education,* will influence perceptions of ongoing climate change and its impact, positive or negative (*climate change status*), and this will be related to concern about climate change (*concern over climate change*), a concern that could lead to human action to reduce its impacts and increase the resilience of the coastal communities.

As a means of more thoroughly testing the causal relationships suggested by the proposed model (Fig. 2.2), we use path analysis; more specifically we use RAMONA, which is based on a modification of the McArdle and McDonald (1984) reticular action model which can fit a path model to a correlation matrix allowing both endogenous and exogenous latent variables to have unit variance producing estimates of standardized path coefficients (Browne 2009). Dependence paths include path coefficients (nonzero) reflecting degree of influence of the variable emitting the path on the variable at the arrow end of the path; for example, the path coefficient associated with the arrow between *climate change status* and *concern over climate change* is −0.404, and like a standardized regression coefficient, it indicates that a change of 1.0 in the *climate change status* score would result in a decrease of 0.404 in the standardized *concern over climate change* score. The model was tested using the maximum Wishart likelihood method. Goodness of fit was evaluated using the root mean square error of approximation (RMSEA). For our model, RMSEA is equal to 0.047 (90% confidence interval 0.000 to 0.100). According to Browne (2009) an RMSEA less than or equal to 0.05 is a "close fit." All path coefficients are strongly statistically significant (most $p < 0.001$) except for the paths between *current status of fishery resources* and the two variables *fishing experience* ($p = 0.02$) and *total environmental ethic* ($p > 0.05$). The number in parentheses next to the path between *fishing experience* and *current status of the resource* is the zero-order correlation between the two variables ($p < 0.05$). This relationship is important in the analysis, and the path coefficient could be misleading. It is both

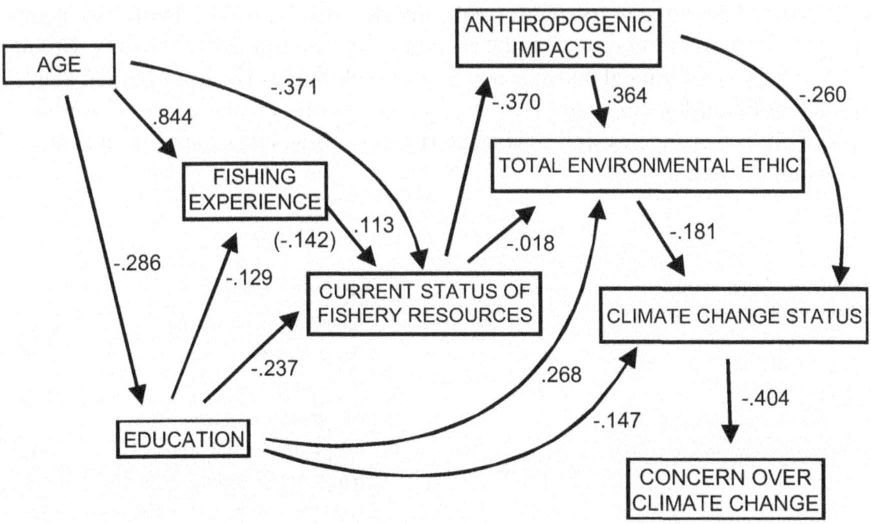

Fig. 2.2 Model of interrelationships between variables tested using the RAMONA path analysis method

weak (90% confidence interval between −0.177 and 0.403) and is the value with all other variables in the model impacting *current state of the resource* controlled.

2.3.5 Fishers' Adaptations to Change

Turning to currently reported adaptation to change, fishers who have observed changes to the local marine species and/or environment that they believe to be related to climate change were asked whether or not they have changed any aspect of their fishing activity to adapt to the changes observed. These responses may provide some indication of adaptation to future changes. The majority (62.5%) have changed some aspect of their fishing activity including changing fishing grounds (42.5%^{R}), gear changes (13.8%^{R}), and fishing in deeper waters (11.3%^{R}) and further away from shore (8.8%^{R}) (Table 2.4).

2.4 Discussion

This study examined the perceptions of fishers in Puerto Rico toward changes in the coastal and marine environment and resources, as well as changes specifically related to climate change. Puerto Rican fishers are highly dependent on marine resources for income and subsistence while also presenting strong attachment to the occupation. Therefore, they not only hold important knowledge about environmental change and its impacts, but it becomes crucial to understand how these changes affect natural resource users to inform the development of more effective strategies and policies to address overexploitation and other anthropogenic impacts without disregard for environmental justice and human well-being. The model tested identifies interrelationships between factors associated with perceptions of the seriousness of climate change as well as characteristics of fishers that can be used to target extension activities.

Table 2.4 Adaptation strategies adopted by fishers in response to changes observed in marine resources and environment

Adaptation strategy	%^{R}
Change fishing grounds	42.5
Change/diversify gear	15.0
Fish in deeper water	11.3
Fish farther away from shore	8.8
Change time of day to fish	5.0
Change target species	3.8
Use better equipment/technology	2.5
Increase effort	2.5
Others ($N < 1$)	8.8

2.4.1 Perceptions of Change and Adaptations

Our findings suggest that Puerto Rican fishers present high levels of environmental awareness and, in general, perceive the local marine environment and climate to have undergone significant transformations which they believe have been affecting fisheries in different ways. The climate change impacts fishers have observed and experienced relate mostly to resource population declines in general, as evidenced by mentions of stock decline, habitat shifts, and changes in species composition. Changes in species composition were sometimes mentioned as the result of the introduction of non-native species. For instance, some fishers mentioned the increasingly large population of lionfish (*Pterois*) in Puerto Rico waters: "Now we see different types of fish not common here like the lionfish." Lionfish species, native to the Indo-Pacific region, have in the past couple of decades been introduced by human activity (e.g., intentional or inadvertent release of aquarium species) and spread into different shallow and deep habitats in the Western North Atlantic and the Caribbean (Whitfield et al. 2007; Betancur-R et al. 2011). Although their invasion is not linked directly to changes in climate, lionfish are voracious predators and there is evidence that this behavior is contributing to the rapid degradation of already stressed (e.g., bleached and overfished) coral reef habitats (Albins and Hixon 2013), thus further affecting species composition and consequently fisheries in the Caribbean region.

Besides fishery resource decline and changes in species composition, fishers frequently mentioned increase in seawater temperature as a significant effect of climate change. One fisher explained: "Usually from October to December, the water gets colder and there are more lobsters. Now, the water is warmer and it is affecting the lobsters." Increased water temperature was also one of the major factors in the anthropogenic impact scale fishers believed to be affecting the fishery. They were also highly aware about the impacts of coral bleaching on the fishery, a phenomenon that is related to warmer sea temperatures. It is important to note that our findings show that fishers have a clear understanding of the importance of healthy coral reefs for the local fishery. One of the items with the highest scores in the environmental ethic measure was the idea that if corals die, it will affect the fishery negatively. This awareness of the ecological importance of coral reefs can be included in conservation and restoration strategies in the region in addition to minimizing potentially impactful and destructive fishing practices.

Overall, fishers expressed overwhelming concern over increased pollution affecting the marine environment and the fishery. Responses to the items included in the *environmental ethic* scale demonstrate that fishers are particularly aware of the impacts of solid waste and industrial activity near the coast. Even though fishers overall perceive the local marine resources to be currently in reasonably good shape, many have expressed concern over declines in the population size and health of these resources over the past decade. Fishers most frequently linked the observed declines to pollution, but many attributed declines to overfishing, climate change, and, at a lower degree, regulations and environmental degradation in general. Our results are in line with previous research conducted in Puerto Rico (Griffth et al. 2007; Matos-Caraballo and Agar 2011; Valdés-Pizzini and Umpierre 2014). Griffth et al. (2007) found that the majority of fishers they interviewed shared the opinion that fishery resources declined through time and the most commonly attributed cause was pollution and to a lesser degree overfishing and fishery regulations. Similarly, in interviews conducted in 2008, Matos-Caraballo and Agar (2011) found that the majority of Puerto Rican fishers believed the island's fishery resources to be worse off at the time of the study, when compared to "other years." Among the most important reasons fishers in that study mentioned as causes for declining fish stocks were overfishing, pollution, habitat degradation, regulations, weather and ocean conditions, and climate change (Matos-Caraballo and Agar 2011). These studies support our findings indicating that fishers in the region have been observing these changes and dealing with their impacts for the past couple of decades.

Changes experienced by fishers have triggered adaptations transforming aspects of their fishing activities. The most frequent adaptation mentioned by fishers relates to changes in fishing grounds, with fishers having to seek other more productive areas. The number of fishers who mentioned specifically having to fish in deeper water and farther away from the shore also indicates that changes in location have been a significant way fishers have adapted to changes. Fishers also mentioned the need to change or diversify gear to increase productivity and target different species. The significant increase in the number of fishers using SCUBA in recent decades in Puerto Rico has been documented by Matos-Caraballo and Agar (2011) as a response to declines in resource abundance. Adaptations that involve traveling farther from the coast and fishing in deeper waters present serious implications for the safety and health of SCUBA diving fishers, particularly the risk of decompression sickness. One study suggests that decompression sickness among artisanal fishers is much higher worldwide than that of recreational or military divers (Huchim-Lara et al. 2015). Research suggests that fishers typically manifest high risk-taking behavior (Poggie et al. 1995, 1996; Pollnac et al. 1998; Davis 2012; Pfeiffer and Gratz 2016; Huchim-Lara et al. 2016). Pollnac and Poggie (2008) attribute reduced perceptions of risk among fishers to a personality type that can be characterized as being active, adventurous, aggressive, and courageous. Our findings show that Puerto Rican fishers present high levels of satisfaction with aspects of independence, adventure, and challenges associated with their job. These factors are likely to result in increased exposure of fishers to potentially dangerous situations, which combined with the need to travel farther and dive into deeper waters, increase the likelihood of decompression sickness. Moreover, SCUBA diving fishers do not receive formal training and Puerto Rico coastal communities have limited or no availability of resources and infrastructure for proper treatment. It is also important to note that in this study, aspects of the job relating to safety, healthfulness, and fatigue were among the ones with which fishers expressed the lowest satisfaction levels, possibly reflecting concern over these increased risks associated with newly adopted adaptation strategies.

Many of the questions about impacts of climate change on the marine environment and resources were met with answers that referred to declines in resource abundance. Similarly, responses referring to adaptations (seeking new fishing grounds, traveling farther from shore, and fishing in deeper waters) allude to resource decline and scarcity, particularly in nearshore areas. Marine resources are heavily impacted by changes in climate, as well as pollution, and the changes fishers are experiencing are likely in part due to these factors. Research conducted on the causes of reef-estuarine ecosystems degradation indicates that other factors along with overfishing can significantly contribute to fish mortality (Mora 2008; Mumby et al. 2004; Nellemann et al. 2008). However, these observations also evidence the occurrence of overfishing in Puerto Rico's marine ecosystems. The relatively high percentage of fishers who stated overfishing to be affecting the fisheries when asked directly about different anthropogenic impacts supports the idea that many fishers are particularly concerned about the effects of overexploitation. However, it is difficult to differentiate fishers' perceptions of the different causes of the changes observed, that is, climate change impacts versus overfishing. This may be due to limitations of the data collection methods or because fishers themselves are not able to clearly discern between these different experiences and perceptions.

2.4.2 Factors Influencing Fishers' Perceptions

The model tested in this study (Fig. 2.2) shows that higher levels of education, older age, and more fishing experience among fishers lead to more negative perceptions of the current state of the fishery resources. Those with more negative perceptions of the state of the resources scored higher on the anthropogenic impact scale, which means they tended to perceive more items in the scale as affecting the fisheries (see Table 2.3 for the complete list of items). Higher scores on the anthropogenic impact scale are associated with higher levels of environmental ethic, which, in turn, leads to more negative perceptions of climate change status, that is, those with higher levels of environmental ethic tend to perceive climatic changes to be more negative than positive. Negative perception of climate change is strongly correlated with expressing more concern about climate change in general. The model also indicates that higher levels of education lead to higher scores on the environmental ethic scale, and both higher education and environmental ethic levels result in more negative perceptions of climate change. The relationships tested in the model emphasize the importance of education as a factor leading to higher environmental and climate change awareness. It is also important to note that, although not strong, a statistically significant zero-order correlation exists between fishing experience and more negative perceptions of the current state of fishery resources. In this sample, more experienced fishers present lower levels of education; thus, the relationship between fishing experience and perceptions of status of fishing resources indicates that perceptions are also shaped by fishers' gained knowledge through experience and involvement, that is, local environmental knowledge (LEK). This is particularly relevant given the fact that the model indicates that fishers' awareness of degradation of local marine resources is an important factor leading to awareness of anthropogenic impacts and ultimately concern about climate change.

Negative perceptions of the status of fishery resources and climate change as well as concern about climate change can be important drivers of changes in behavior that could lead to increased support for conservation strategies as well as climate change mitigation and adaptation plans. Understanding factors influencing fishers' perceptions and experiences will aid in the development of policy strategies and educational programs that focus on factors that are appropriate within the context of fishing communities. For instance, in Puerto Rico, programs to address fishery sustainability may benefit from incorporating aspects of pollution and climate change impacts, such as coral bleaching, in addition to overfishing, to increase the likelihood for acceptance of and compliance with fishery management strategies. Our findings also further highlight the importance of education and the value of LEK in shaping fishers' perceptions of environmental and climate change. This supports the idea that policy strategies that emphasize education and adequately incorporate LEK may lead to more successful long-term outcomes of sustainability and climate change adaptation in fishing communities.

2.5 Conclusion

Frameworks to conceptualize vulnerability and adaptive capacity to environmental and climate change often fail to incorporate aspects of subjective perceptions of those impacted and undergoing adaptation processes. Subjective perceptions are important drivers of behavior and are crucial in determining success of adaptation strategies as people are likely to act on the basis of their perceptions, beliefs, and cultural context, particularly under stressful situations. Understanding factors affecting perceptions of environmental and climate change in fishing communities will aid in the development of strategies for sustainability and adaptive capacity of vulnerable fishing communities. Fishers are highly dependent on natural resources, greatly attached to their occupation, and typically reside in areas of high exposure to natural hazards and climate change. In this study, we examined the perceptions of Puerto Rican fishers to changes in the marine environment and climate. Our findings show that fishers perceive the local environment and climate to have undergone significant changes in the past couple of decades and they believe these changes have been affecting the fishery and consequentially leading them to adapt. Changes and adaptations mentioned by fishers support research that shows that the Puerto Rico marine environment and fisheries are at risk from pollution, climate change, and overfishing. Fishers' adaptations to these impacts, which consist mostly of seeking new fishing grounds, have also led them to increase their exposure to risks, particularly among SCUBA divers fishing in deeper waters and farther away from the coast, thus presenting implications for their safety and health.

Our findings show that having more negative perceptions of the status of fishery resources is related to higher levels of environmental awareness and awareness of anthropogenic impacts which in turn lead to increased concern about climate change. Education as well as age and fishing experience, an indicator of LEK, were both important factors influencing fishers' perceptions of the status of fishery resources. These findings have significant implications for the development of policy and educational strategies aimed at increasing sustainability and well-being in fishing communities. Strategies that take into account fishers' perceptions and factors influencing these perceptions are not only more context appropriate but are more likely to be met with higher acceptance and compliance levels, thus resulting in more effective and just outcomes.

Acknowledgment This work was supported by the NOAA Coral Reef Conservation Program (Award number: NA16NOS4820051).

References

Adger WN, Barnett J, Brown K, Marshall N, O'Brien K (2013) Cultural dimensions of climate change impacts and adaptation. Nat Clim Chang 3:112–117

Adger WN (2003) Social capital, collective action, and adaptation to climate change. Econ Geogr 79(4):387–404

Albins MA, Hixon MA (2013) Worst case scenario: potential long-term effects of invasive predatory lionfish (*Pterois volitans*) on Atlantic and Caribbean coral-reef communities. Environ Biol Fish 96(10–11):1151–1157

Allison EH, Perry AL, Badjeck MC, Adger WN, Brown K, Conway D, Halls AS, Pilling GM, Reynolds JD, Andrew NL, Dulvy NK (2009) Vulnerability of national economies to the impacts of climate change on fisheries. Fish Fish 10:173–196

Alongi DM (2015) The impact of climate change on mangrove forests. Curr Clim Chang Rep 1(1):30–39

Apostle RL, Kasdan L, Hanson A (1985) Work satisfaction and community attachment among fishermen in Southwest Nova Scotia. Can J Fish Aquat Sci 42(2):256–267

Appeldoorn RS (2008) Transforming reef fisheries management: application of an ecosystem-based approach in the USA Caribbean. Environ Conserv 35(3):232–241

Appeldoorn R, Beets J, Bohnsack J, Bolden S, Matos D, Meyers S, Rosario A, Sadovy Y, Tobias W (1992) Shallow water reef fish stock assessment for the U.S. Caribbean. NOAA Technical Memorandum NMFS/SEFSC/304. 70 p

Appeldoorn R, Yoshioka P, Ballantine D (2009) Coral reef ecosystem studies: integrating science and management in the Caribbean. Caribb J Sci 45(2–3):134–137

Aswani S, Vaccaro I, Abernethy K, Albert S, de Pablo JF (2015) Can perceptions of environmental and climate change in island communities assist in adaptation planning locally? Environ Manag 56(6):1487–1501. https://doi.org/10.1007/s00267-015-0572-3

Baker AC (2014) Climate change: many ways to beat the heat for reef corals. Curr Biol 24(24):R1166–R1168

Betancur RR, Hines A, AAcero P, Ortı G, Wilbur AE, Freshwater DW (2011) Reconstructing the lionfish invasion: insights into greater Caribbean biogeography. J Biogeogr 38:1281–1293

Binkley M (1995) Risks, dangers and rewards in the Nova Scotia offshore fishery. McGill-Queen's University Press, Montreal, p 192

Browne MW (2009) Path analysis (RAMONA). In: SYSTAT Software (ed) SYSTAT 13 statistics III. SYSTAT Software, Inc., Chicago

Bryan DR, Smith SG, Ault JS, Feeley MW, Menza CW (2016) Feasibility of a Regionwide probability survey for coral reef fish in Puerto Rico and the U.S. Virgin Islands. Mar Coast Fish 8(1):135–146. https://doi.org/10.1080/19425120.2015.1082520

Cinner JE, Folke C, Daw T, Hicks CC (2011) Responding to change: using scenarios to understand how socioeconomic factors may influence amplifying or dampening exploitation feedbacks among Tanzanian fishers. Glob Environ Chang 21:7–12

Cinner JE, Adger WN, Allison EH, Barnes ML, Brown K, Cohen PJ, Gelcich S, Hicks CC, Hughes TP, Lau J, Marshall NA, Morrison TH (2018) Building adaptive capacity to climate change in tropical coastal communities. Nat Clim Chang 8:117–123. https://doi.org/10.1038/s41558-017-0065-x

Davis ME (2012) Perceptions of occupational risk by us commercial fishermen. Mar Policy 36(1):28–33

Daw T, Adger WN, Brown K, Badjeck M-C (2009) Climate change and capture fisheries: potential impacts, adaptation and mitigation. In: Cochrane K, De Young C, Soto D, Bahri T (eds) Climate change implications for fisheries and aquaculture: overview of current scientific knowledge. FAO fisheries and aquaculture technical paper. No. 530. FAO, Rome, pp 107–150

Donner SD, Knutson TR, Oppenheimer M (2007) Model-based assessment of the role of human-induced climate change in the 2005 Caribbean coral bleaching event. Proc Natl Acad Sci 104(13):5483–5488

García-Quijano CG (2009) Managing complexity: ecological knowledge and success in Puerto Rican small-scale fisheries. Hum Organ 68(1):1–17

García-Quijano C, Poggie JJ, Pitchon A, Del Pozo M (2015) Coastal resource foraging, life satisfaction, and Well-being in southeastern Puerto Rico. J Anthropol Res 71(2):145–167

García-Sais JR, Castro R, Sabater-Clavell J, Esteves R, Carlo M (2006) Monitoring of coral reef communities from natural reserves in Puerto Rico, 2006: Isla Desecheo, Rincon, Mayaguez Bay, Guanica, Ponce and Isla Caja de Muerto. Final Report submitted to the Department of Natural and Environmental Resources of Puerto Rico, San Juan, P.R. pp 151

García-Sais J, Appeldoorn R, Battista T, Bauer L, Bruckner A, Caldow C, Carrubba L, Corredor J, Diaz E, Lilyestrom C, García-Moliner G, Hernández-Delgado E, Menza C, Morell J, Pait A, Sabater J, Weil E, Williams E, Williams S (2008) The state of coral reef ecosystems of Puerto Rico. Pp. 75-116. In: Waddell JE, Clarke AM (eds) The state of coral reef ecosystems of the United States and Pacific freely associated states: 2008. NOAA technical memorandum NOS NCCOS 73. NOAA/NCCOS Center for Coastal Monitoring and Assessment's Biogeography Team, Silver Spring. 569 pp

Griffith D, Valdés-Pizzini M (eds) (2002) Fishers at work, workers at sea – a Puerto Rican journey through labor and refuge. Temple University Press, Philadelphia. 265 pp

Griffith D, Valdés-Pizzini M, García-Quijano C (2007) In: Agar JJ, Stoffle B (eds) Entangled Communities: socioeconomic profiles of fishers, their communities, and their responses to marine protective measures in Puerto Rico, NOAA Series on U.S. Caribbean Fishing Communities. NOAA Technical Memorandum NMFS-SEFSC-556. , 524 p

Grothmann T, Patt A (2005) Adaptive capacity and human cognition: the process of individual adaptation to climate change. Glob Environ Chang 15:199–213

Hernández-Delgado EA, Shivlani M, Sabat AM (2014) Ecosystem-based and community-based model integration to designate coral reef no-take marine protected areas: a case study from Puerto Rico. Nat Resour 5:538–560

Holding S, Allen DM (2015) From days to decades: numerical modelling of freshwater lens response to climate change stressors on small low-lying islands. Hydrol Earth Syst Sci 19:933–949

Huchim-Lara O, Salas S, Fraga J, Méndez-Domínguez N, Chin W (2016) Fishermen's perceptions and attitudes toward risk diving and management issues in small-scale fisheries. Am J Human Ecol 5(1):1–10. https://doi.org/10.11634/216796221605760

Huchim-Lara O, Salas S, Chin W, Montero J, Fraga J (2015) Diving behavior and fishing performance: the case of lobster artisanal fishermen of the Yucatan coast, Mexico. Undersea Hyperb Med Soc 42(4):285–296

Juda L (1999) Considerations in developing a functional approach to the governance of large marine ecosystems. Ocean Develop Int Law 30(2):89–125

Limuwa MM, Sitaula BK, Njaya F, Storebakken T (2018) Evaluation of small-scale fishers' perceptions on climate change and their coping strategies: insights from Lake Malawi. Climate 6(34). https://doi.org/10.3390/cli6020034

Loh T, McMurray SE, Henkel TP, Vicente J, Pawlik JR (2015) Indirect effects of overfishing on Caribbean reefs: sponges overgrow reef-building corals. Peer J 3:e901. https://doi.org/10.7717/peerj.901.

Marshall NA, Fenton DM, Marshall PA, Suton SG (2007) How resource dependency can influence social resilience within a primary resource industry. Rural Ecol 72(3):359–390

Matos-Caraballo D (2009) Lessons learned from the Puerto Rico's commercial fishery, 1988 – 2008. Proc Gulf Caribb Fish Inst 61:123–129

Matos-Caraballo D, Agar JJ (2011) Census of active commercial fishermen in Puerto Rico: 2008. Mar Fish Rev 73(1):13–27

McArdle JJ, McDonald RP (1984) Some algebraic properties of the reticular action model for moment structures. Br J Math Stat Psychol 37:234–251

McCay BJ, Brandt S, Creed CF (2011) Human dimensions of climate change and fisheries in a coupled system: the Atlantic surfclam case. ICES J Mar Sci 68(6):1354–1367

Miller J, Waara R, Muller E, Rogers CS (2006) Coral bleaching and disease combine to cause extensive mortality on reefs in U.S. Virgin Islands. Coral Reefs 25(3):418

Mora C (2008) A clear human footprint in the coral reefs of the Caribbean. Proc R Soc B 275(1636):767–775

Mumby PJ, Edwards AJ, Arias-González JE, Lindeman KC, Blackwell PG, Gall A, Gorczynska MI, Harborne AR, Pescod CL, Renken H, Wabnitz CCC, Llewellyn G (2004) Mangroves enhance the biomass of coral reef fish communities in the Caribbean. Nature 427:533–536

Nellemann C, Hain S, Alder J (2008) In dead water: merging of climate change with pollution, over-harvest, and infestations in the World's fishing grounds. United Nations Environment Programme, GRID-Arendal, Norway

Nicholls RJ, Cazenave A (2010) Sea-level rise and its impact on coastal zones. Science 328(5985):1517–1520

NOAA (2018a) Hurricane Season – 2017. Retrieved from: https://sos.noaa.gov/datasets/hurricane-season-2017/

NOAA (2018b) Teams study and restore corals following Hurricanes Irma and Maria. Retrieved from: https://coralreef.noaa.gov/aboutcrcp/news/featuredstories/mar18/welcome.html

Perry AL, Low PJ, Ellis JR, Reynolds JD (2005) Climate change and distribution shifts in marine fishes. Science 308(5730):1912–1915. https://doi.org/10.1126/science.1111322

Pfeiffer L, Gratz T (2016) The effect of rights-based fisheries management on risk taking and fishing safety. Proc Natl Acad Sci 113(10):2615–2620

Pinsky ML, Mantua NJ (2014) Emerging adaptation approaches for climate-ready fisheries management. Oceanography 27(4):146–159

Pittman J, Armitage D, Alexander S, Campbell D, Alleyne M (2015) Governance fit for climate change in a Caribbean coastal-marine context. Mar Policy 51:486–498

Poggie JJ, Pollnac RB, Van Dusen C (1996) Intracultural variability in the cognition of danger among southern New England fishers. Mar Resour Econ 11(1):23–30

Poggie JJ, Pollnac RB, Jones S (1995) Perceptions of vessel safety regulations: a southern New England fishery. Mar Policy 19(5):411–418

Pollnac RB, Poggie JJ, Cabral SL (1998) Thresholds of danger: perceived risk in a New England fishery. Hum Organ 57(1):53–59

Pollnac RB, Poggie JJ (1988) The structure of job satisfaction among New England fishermen and its application to fisheries management policy. Am Anthropol 90(4):888–901

Pollnac RB, Poggie JJ (2008) Happiness, Well-being and psychocultural adaptation to the stresses associated with marine fishing. Hum Ecol Rev 15(2):194–200

Pollnac RB, Abbott-Jamieson S, Smith C, Miller ML, Clay PM, Oles B (2008) Toward a model for fisheries social impact assessment. Mar Fish Rev 68(1–4):1–18

Pollnac RB, Seara T, Colburn LL (2015) Aspects of fishery management, job satisfaction, and Well-being among commercial fishermen in the northeast region of the United States. Soc Nat Resour 28(1):75–92

Ramos-Scharrón CE, Torres-Pulliza D, Hernández-Delgado EA (2015) Watershed- and island wide-scale land cover changes in Puerto Rico (1930s–2004) and their potential effects on coral reef ecosystems. Sci Total Environ 506-507:241–251

Randall CJ, van Woesik R (2015) Contemporary white-band disease in Caribbean corals driven by climate change. Nat Clim Chang 5(4):375–379. https://doi.org/10.1038/nclimate2530

Rhein M, Rintoul SR, Aoki S, Campos E, Chambers D, Feely RA, Gulev S, Johnson GC, Josey SA, Kostianoy A, Mauritzen C, Roemmich D, Talley LD, Wang F (2013) Observations: ocean. In: Stocker TF, Qin D, Plattner G-K, Tignor M, Allen SK, Boschung J, Nauels A, Xia Y, Bex V, Midgley PM (eds) Climate change 2013: the physical science basis. Contribution of working group I to the fifth assessment report of the intergovernmental panel on climate change. Cambridge University Press, Cambridge, UK/New York

Rogers C (2009) Coral bleaching and disease should not be underestimated as causes of Caribbean coral reef decline. Proc R Soc B 276(1655):197–199

Seara TP, Clay M, Colburn LL (2016) Perceived adaptive capacity and natural disasters: a fisheries case study. Glob Environ Chang 38:49–57

Seara T, Pollnac R, Poggie J (2017a) Changes in job satisfaction through time in two Major new England fishing ports. J Happiness Stud 18(6):1625–1640

Seara T, Pollnac R, Poggie J, Garcia-Quijano C, Monnereau I, Ruiz V (2017b) Fishing as therapy: impacts on job satisfaction and implications for fishery management. Ocean Coast Manag 141:1–9

Short FT, Kostenb S, Morgan PA, Malone S, Moore GE (2016) Impacts of climate change on submerged and emergent wetland plants. Aquat Bot 135:3–17

Smith C, Clay PM (2010) Measuring subjective and objective Well-being: analyses from five marine commercial fisheries. Hum Organ 69(2):158–168

Sumaila UR, Cheung WWL, Lam VWY, Pauly D, Herrick S (2011) Climate change impacts on the biophysics and economics of world fisheries. Nat Clim Chang 1:449–456

Taylor JG, Stewart TR, Downton M (1988) Perceptions of drought in the Ogallala Aquifer region. Environ Behav 20:150–175

Tonioli FC, Agar JJ (2011) Synopsis of Puerto Rican Commercial Fisheries. NOAA Technical Memorandum NMFS-SEFSC-622, pp 69

USDA (2012) 2012 Census Publications. Retrieved from: https://www.agcensus.usda.gov/Publications/2012/Online_Resources/County_Profiles/Puerto_Rico/

Valdés-Pizzini M, García-Quijano CG, Schärer-Umpierre MT (2012) Connecting humans and ecosystems in tropical fisheries: social sciences and the ecosystem-based fisheries Management in Puerto Rico and the Caribbean. Caribb Stud 40(2):95–128

Valdés-Pizzini M, Schärer-Umpierre M (2014) People, habitats, species, and governance: an assessment of the social-ecological system of La Parguera. Interdisciplinary Center for Coastal Studies, University of Puerto Rico, Mayagüez

Weatherdon LV, Magnan AK, Rogers AD, Sumaila UR, Cheung WWL (2016) Observed and projected impacts of climate change on marine fisheries, aquaculture, coastal tourism, and human health: an update. Front Mar Sci 3:48. https://doi.org/10.3389/fmars.2016.00048

Weeratunge N, Béné C, Siriwardane R, Charles A, Johnson D, Allison EH, Nayak PK, Badjeck M (2014) Small-scale fisheries through the wellbeing lens. Fish Fish 15:255–279

Whitfield PE, Hare JA, David AW, Harter SL, Muñoz RC, Addison CM (2007) Abundance estimates of the Indo-Pacific lionfish *Pterois volitans/miles* complex in the Western North Atlantic. Biol Invasions 9(1):53–64

Wilkinson C, Souter D (2008) Status of Caribbean coral reefs after bleaching and hurricanes in 2005. Global Coral Reef Monitoring Network, and Reef and Rainforest Research Centre, Townsville. 152 pp

Woody K, Atkinson A, Clark R, Jeffrey C, Lundgren I, MIller J, Monaco M, Muller E, Patterson M, Rogers C, Smith T, Spitzak T, Waara R, Whelan K, Witcher B, Wright A (2008) Coral bleaching in the U.S. Virgin Islands in 2005 and 2006. In: Wilkinson C, Souter D (eds) Status of Caribbean coral reefs after bleaching and hurricanes in 2005. Global Coral Reef Monitoring Network and Reef and Rainforest Research Centre, Townsville, pp 68–72. 152 pp

Chapter 3
"We Used to go Asking for the Rains": Local Interpretations of Environmental Changes and Implications for Natural Resource Management in Hwange District, Zimbabwe

Zénaïde Dervieux and Melody Belgherbi

Abstract In Hwange District, Zimbabwe, people living in the vicinity of the largest protected area of the country are facing rapid climate and environmental changes. Adopting an ethnoecological perspective, we sought to understand the way changes are understood locally in an area where people have interacted with their environment for centuries. In this chapter, we examine local people's knowledge, expertise, and interpretative diagnoses about the environmental and climate changes they perceive around them. Qualitative fieldwork, including participant observation, semi-structured interviews, and structured free-listing interviews, was carried out over a three-month period in the communal lands of the district. Among changes related to wildlife interactions and landscape transformations, people mainly mobilize knowledge of trees and birds to predict rainfall and explain climate variability (related to seasons, precipitation, and temperatures). The most important findings of this research lie in people's descriptions of ecological changes and their interpretations and explanations for these changes, which focus on arguments that are cultural (abandonment of ritual practices, access to ancestral sites), demographic (population growth), and political (wildlife management). For example, the disturbances in precipitation patterns are understood as a manifestation of the anger of ancestral spirits. We argue that these interpretive frameworks reflect the strong marginalization of the communities of the district from the national program of community-based natural resource management, CAMPFIRE, and that these discourses allow silenced voices to express themselves about sociopolitical concerns in an authoritarian context.

Z. Dervieux (✉)
Geography, Sorbonne Université, Paris, France
e-mail: zenaide.dervieux@gmail.com

M. Belgherbi
Anthropology, Université Paris Nanterre, Nanterre, France

M. Welch-Devine et al. (eds.), *Changing Climate, Changing Worlds*,
Ethnobiology, https://doi.org/10.1007/978-3-030-37312-2_3

Keywords Local knowledge · Environmental and climate change · Natural resource management · Hwange

3.1 Introduction

Over the past two decades, anthropological descriptions of climate change, its impacts, and its potential impacts on ecosystems and biodiversity have increased in number, documenting local knowledge and interpretations (Barnes et al. 2013; Diemberger et al. 2012; Dounias 2010; Fiske 2009; Hastrup 2013; Hastrup and Skrydstrup 2013; Marino and Schweitzer 2009; Orlove et al. 2010). From centuries-long interactions with their environment, the most natural resource-dependent societies have managed to adapt their resource management practices to a wide range of changes, including climate variability as a normal phenomenon (Nakashima et al. 2012). In this paper, we focus on the Southern African region, where aridification has become an alarming problem in the last fifty years (New et al. 2006) and where people have developed specific knowledge and adaptation strategies (Nyong et al. 2007). Rainfall analysis has shown that drought severity increased in Northwestern Zimbabwe in the course of the twentieth century (Chamaillé-Jammes et al. 2007a). In this uncertain context, local communities of the Hwange District also live on the borders of the largest protected area of the country and are affected by increasingly common interactions with wildlife, especially African elephants (*Loxodonta africana*), which cause substantial damage to fields during the growing period and late wet season (Guerbois et al. 2013; de Garine-Wichatitsky et al. 2013). Thus, inhabitants of the communal lands of the district live in an environment that is gradually deteriorating and subject to the consequences of human growth. These contrasted climate and environmental transformations have a powerful impact on local perceptions, knowledge dynamics, and traditional systems of natural resource use in the region.

This study uses qualitative research carried out over a three-month period in 12 villages of Hwange District to analyze local perceptions of climate and environmental changes from an ethnoecological perspective.[1] The results show that people discern a wide range of environmental and climate changes. We will see, in the first section of this paper, that communities' modes of natural resource use - which rely on fruit gathering and agropastoral activities - changed significantly in the course of the last century. This is especially marked by policy changes during two distinct eras: the colonial period, with the regulation of living places and natural resources through centralized protectionism, and, after independence, notably with the

---

[1]This reflection has been carried out within the framework of the PIAF program (interdisciplinary program on indigenous indicators of fauna and flora), which aims at comparing indigenous diagnoses of environmental changes in different continents and seeks to build an international and interdisciplinary network of young researchers around the question of human-nature interactions and issues of biodiversity conservation (ANR-13-JSsouthernH1–0005, http://www.anr-piaf.org/)

community-based natural resource management model. In a second section, we emphasize people's direct and indirect observations related to climate to show the existence of local knowledge and associated practices of natural resource conservation that could be encompassed in national policies for natural resource management. And lastly, we investigate the diagnoses formulated by our informants to interpret the changes they detect in their environment.

An important finding of our research lies in the interpretations of changes, which focus on cultural (abandonment of ritual practices, access to ancestral sites), demographic (population growth), and political (wildlife management) arguments. As such, the disturbances of precipitation patterns are understood as a manifestation of the anger of ancestral spirits. Given that local communities' involvement in the actual management of natural resources is very low, we argue that these ways of representing environmental change reflect the fact that the communities of the district are significantly marginalized from the national community-based natural resource management program, CAMPFIRE, and that we cannot apprehend local environmental and climate knowledge without taking cultural systems into consideration.

3.2 Studying Perceptions of Environmental Changes in a Context of Rapid Transformations

Within the framework of this study, we visited inhabitants living on the communal lands of Hwange District near the northeastern boundary of Hwange National Park. This unfenced protected area of 14,651 km^2, created in 1928, is a product of a land management plan implemented by the Southern Rhodesian authorities (Davison 1967). In this area, the villagers are affected by the proximity of the park, which has major impacts on their subsistence. Furthermore, interactions between humans and their environment in the communal lands are influenced by the broader context of international conservation dynamics and the noteworthy case of Zimbabwe. To better contextualize interactions between these communities and their environment, we will see that the country's colonial past generated an unequal land tenure distribution and centralized modes of natural resource management. The territorial organization of Hwange District impacts the way people interact with their surroundings and respond to its rapid changes, in a context of poor involvement in environmental management.

3.2.1 Land Distribution and Conservation in Hwange: Strong Constraints for Subsistence

In Zimbabwe, the colonial era durably impacted land tenure distribution. In 1930, the Land Apportionment Act divided the national territory into three categories: state land, land reserved for Whites, and tribal lands, later renamed "communal" lands (Herbst 1990). This resulted in a rural population confined to the less fertile lands of the country (Cling 2001), three-quarters of whom live in areas with less than 650 mm of rain per year (Cavendish 2000). This unequal land distribution was maintained after independence (Compagnon 2003). The emergence of centralized protectionism, which remains a characteristic of wildlife legislation in much of Africa and beyond, contributed to the marginalization of local populations from the fauna and flora of their local environment (Child 1996).

The violent history of forced mobility in the area had started with the Ndebele invasions in the mid-nineteenth century (Ncube 2004). Nearly a century ago, the process of creating what is known today as Hwange National Park began when inhabitants of the northwestern region of the district were displaced outside of the protected area (Andersson and Cumming 2013). These resettlements continued with the creation of the Gwaai and Shangani Native Reserves in 1896 and the creation of the colonial farms (Alexander et al. 2000; Kwashirai 2008). Later on, the delimitation of the Sikumi Forest in 1968 fueled further displacement. The national park now has one of the highest worldwide densities of African elephants (Blanc et al. 2005).

As a result of this migration and displacement, Hwange District has a complex social composition including various ethnolinguistic groups with contrasting backgrounds. A recent study recorded the following groups: Ndebele, Tonga, Nambya, Chewa, Chikunda, Nyanja, Doma, Zezuru, Kalanga, Mlubale, Karanga, and Shangwe (Le Bel et al. 2011). The actual tourism enhancement of the park, built on the representation of the area as "wilderness," is perceived by certain groups - such as the Nambya people - as a denial of their history (McGregor 2005). Indeed, culturally important sites located within Hwange National Park are no longer accessible, such as the Bumbusi and Mtoa ruins that mark previous Nambya settlements (Fig. 3.1). We will see that whereas these sacred places, specially used for rainmaking ceremonies, are neglected by the park managers, and this is not the case for contiguous populations.

3.2.2 Community-Based Natural Resource Management

In Zimbabwe, rapid population growth, the introduction of modern technologies, and the development of agriculture (Child 1996) led to the "deliberate elimination" of approximately 680,000 mammals between 1919 and 1960, encouraged by the then Department of Conservation and Extension, which considered wildlife as competing with food production and livestock (Muboko and Murindagomo 2014: 206). While centralized conservation management resulted in the creation of protected areas, however, it did not promote preventive measures and benefits related to wildlife management in the communal lands that had to support the economic costs of living alongside wildlife (Child 1996). The concomitant awareness resulting from this drastic decrease of wildlife in the country enabled the rise of a pioneering model of community-based management in the communal lands, the CAMPFIRE (*Communal Areas Management Programme for Indigenous Resources*) program, in the late 1980s (Taylor 2009).

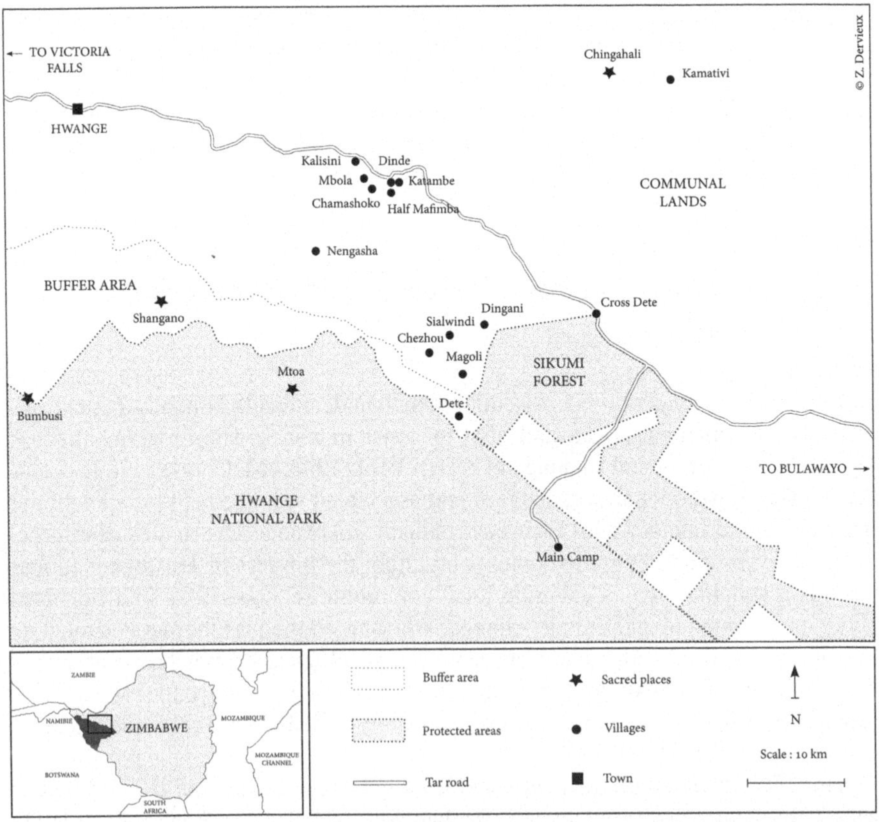

Fig. 3.1 Map of the villages visited in Hwange District. (Sources: © OpenStreetMap contributors)

This model, adapted in most of sub-Saharan Africa, can be described as a "concept of managing wildlife and wildlife habitat in the communal lands of Zimbabwe for the benefit of the people living in these areas" (Frost and Bond 2008: 777). This form of decentralized conservation is based on a drive toward natural resource collectivization and defining a "productive community," i.e., a management unit that receives direct profits from wildlife revenues. Its success lies in the devolution of ownership to the local level, which is required to enable effective management of natural resources (Murphree 1993).

However, this environmental management scheme, in which community participation is formalized, does not appear to be effective. Many districts face implementation challenges in devolving ownership and defining participatory institutions for natural resource management (Murombedzi 1999; Dzingirai 1999). Moreover, we observed that local management practices and systems do not seem to be recognized in Hwange District despite the fact that they are rooted in long-term observation and knowledge of the environment. Local subsistence systems are based on agropastoralism and fruit gathering (Guerbois 2012; Perrotton 2015). Their centuries-old interactions with the natural resources of their environment, through activities of collecting and rearing (firewood, fruits, grazing, firewood), have enabled them to develop in-depth local knowledge.[2] In this context, the protected area hosts biodiversity of high value on which local communities depend. The conservation framework of the study area allows us to interrogate the actual implementation of community-based natural resource management in Hwange District.

3.2.3 *Local Knowledge and Perceptions of Environmental Changes*

Integrating local perceptions is essential both to understand the impacts of environmental and climate changes and also to assist in the decision-making process regarding environmental management (DeWalt 1994; McCorkle 1989; Sabar 2012). The jurisdictional organization and associated management policies frame the way people interact with their environment in Hwange District. Furthermore, the conservation and climatic contexts highlight the borders of Hwange National Park as a suitable area to examine local communities' knowledge and practices about environmental and climate change. Working on the park borders enabled us to embrace a diversity of subjects, ranging from ecological knowledge to strategies for adaptation to climate and environmental changes. Work on park borders also provides lessons for the Zimbabwean community-based natural resource management model, CAMPFIRE.

This study is based on data collected from fieldwork carried out from March to May 2015. It was conducted among 70 inhabitants of different villages of wards 11, 13, and 15 of the communal lands of the Hwange Rural District of Matabeleland North Province in Zimbabwe (Fig. 3.1). Our informants have diverse livelihood portfolios encompassing a range of occupations: small farming, herding, fruit gathering and selling, commercial gardening, house building, work at the mine, domestic work, and "traditional" crafting and weaving. We interviewed both men and women from different ethnolinguistic groups. As elderly and mature individuals were identified by the traditional leaders we worked with as custodians of local environmental knowledge, age range of the sample is 50–60 years. However, young adults were also included in the protocol and demonstrated that knowledge is being transmitted between grandparents and their children.

The corpus of qualitative data combines semi-structured interviews, participant observation, and a free-listing protocol. The semi-structured interviews did not specifically target climate change, in order to put this specific phenomenon, which we expected to be mentioned, into perspective. We also specifically asked whether respondents knew the causes of those phenomena. The free-listing exercise was

[2] Within the framework of the free lists, we gathered 48 different tree species and 34 different mammal species.

conducted to encourage the informants to cite all the items they know within three semantic fields: "birds," "mammals," and "trees" (Brewer 2002; Quinlan 2005). We then examined whether, by asking them to describe and list the markers of environmental change around them, it could be possible to access indigenous theories related to climate and environmental changes. Questions and responses were translated from English to Nambya, Ndebele, Tonga, Dombe, and Leya dialects (and back) with the assistance of interpreters.

3.3 Climate Change: Local Knowledge and Related Practices of Natural Resource Management

In this study, we apprehended the environment understood in the broad sense, as all the elements that constitute the surroundings of the villagers. From a methodological point of view, this approach was built on semi-structured interviews focused on people's life histories and the complex ways that natural resource interactions related to these life histories. In this section, we will focus on the direct and indirect observations related to climate, which has been a theme discussed in most of the interviews conducted. Local knowledge of fauna and flora diversity and the most important changes people perceive around them appeared to be a relevant source for the local comprehension of climate and environmental changes. We first explain how people, in their own terms, describe these changes. Second, we show that the descriptions of climate and its changes rely, to a great extent, on the observation of bird and tree species used to predict rains. Finally, we emphasize the role of cultural and traditional practices in the protection of certain tree species.

3.3.1 Climate Knowledge and the Unpredictability of the Rain

During the interviews, certain changes emerged with markedly greater local salience than others, such as those related to weather and climate variability (seasons, rainfall, and temperatures), increasingly frequent interactions with wildlife (mostly elephants and lions), landscape transformations ("it was like a bush," "the land is now too open and bare, this area used to be a thick forest," "it's been cleared now"), and changes in the abundance of bird and tree species. As expected, a majority of informants of the various villages visited (53 inhabitants) describe changes affecting seasons, rainfall, and temperatures that are occurring in their direct environment (fields, homesteads, villages).

Discourses converge on the "unpredictability of the rain," both over the course of a year and from one year to the next, as 11 villagers report an abnormal frequency of droughts and floods and 7 a shift in the start of the rainy season. With regard to this last observation, the respondents' narratives are consistent on the "season coming later," a process that started approximately 45 years ago: "A long time ago, the rain used to come around October and, by December, the crop used to be 1 meter high. Now the rains come with an unusual delay" (villager from Magoli, 2015). The terms "lower," "less," and "poor" are most often used to describe the changes linked to the rainfall and are mentioned by 45 respondents. Sometimes, people directly describe associated consequences for food production – this is "bad for the crops" – and water access due to boreholes and rivers that are drying up. Respondents estimated that the decrease in rainfall started 7–10 years ago, a period that coincided with the 2008 drought, mentioned by 10 respondents.[3] Temperatures are also viewed as rising. Within the collected accounts, extreme climate events such as droughts and floods[4] are mentioned, especially those of 1947 and 2008, which caused famines.

Some of these observations are corroborated by scientific analyses of annual rainfall inside Hwange National Park.[5] These data relate to the rainy season, which concentrates 98% of the rains between October and April (mean annual rainfall of 606 mm per year). First of all, the scientific results confirm both the shift in the rainy season and the high interannual fluctuations of the rainfall for the period 1982–2002, with an interannual coefficient of variation of annual precipitation of 25% (Chamaillé-Jammes et al. 2006). In the second place, Chamaillé-Jammes et al. (2007a) highlighted that droughts worsened within the park during the course of the twentieth century, dry years became between 20% and 50% drier, and the park experienced drier conditions during the 1982–2002 period - which includes two of the worst droughts - than during 1928–1981. The qualifying terms used by the respondents (lower, less) are compatible with this study but do not go back as far as 1928.

They furthermore highlighted that droughts worsened within the park during the course of the twentieth century and that dry years became between 20% and 50% drier (Chamaillé-Jammes et al. 2007a). They revealed that the park experienced drier conditions during the 1982–2002 period - which includes two of the worst droughts (in 1982 and 1995) - than during 1928–1981 and that 1982, 1984, 1987, 1995, 1998, and 2002 were the driest years of the 1982–2002 period (Chamaillé-Jammes et al. 2006). 1928, 1947, and 1951 were also drought years (Zone Atelier, unpublished data). Some of these years were mentioned by the informants or with a time lag of one year (such as the ones of 1952, 1982, 1987, and 1996), and the qualifying terms used by the respondents (lower, less) are compatible with this study but do not go back as far as 1928.

[3] 2008 is remembered well most probably because it is more recent and because the extreme situation was reinforced by the economic and politic crisis.

[4] Years of drought mentioned are the following: 1952, 1957, 1958, 1978 (a year of poor harvests, attacks by baboons reported in Magoli), 1981, 1982, between 1986 and 1988, 1992, and 1996. Years of floods noticed are those of 1958, 1984, 2000, and 2002, with the year 2000 posting very high temperatures.

[5] See Chamaillé-Jammes et al. (2006) and Chamaillé-Jammes et al. (2007a) for a more detailed analysis of the rainfall patterns of the area of Hwange National Park.

The most reliable evidence reported by the respondents was related to the unpredictability of the rains and the lag of the rainy season. It is interesting to note that while most of the years cited by the respondents are only approximative, they refer to particular economic and historical events. For instance, the respondents mentioned a drought in 1996, during which soil erosion and land degradation grew worse. This can be explained by the respondents' feeling of water stress resulting from the drought recorded in 1995. It can also be explained by the fact that people remember this particular year better because they were in negotiations with the Forestry Commission at Sikumi Forest in 1996 that resulted in the possibility of accessing pastures within the first 3 km of forests.[6]

3.3.2 Forecasting the Rain: The Main Role of Birds and Trees

The local perceptions presented in the previous section are based on diagnoses respondents built using their own sensory experiences of long-term changes in weather elements (the wind, the rains, the sky) over decades. These direct sources of knowledge of climatic irregularities are combined with indirect sources such as the observation of certain species of birds, trees, and insects to forewarn of the beginning of the rainy season and the amount of rain people might expect, as practiced in many other natural resource-dependent societies (Roncoli et al. 2009; Orlove et al. 2010). Both the bird and tree species indicate the arrival and abundance of rainfall or may even "call" the rain (Table 3.1). But whereas the seasonal pattern is estimated according to the birds' behavior, the indicators for trees relate to the different stages of their vegetative cycle, notably leaf growth and fruit formation.

This empirical knowledge shows that the inhabitants of the area have developed a reliable diagnostic system of the elements that surround them, especially with regard to forecasting the rains. These methods of predicting seasonal rainfall and related quantities have forewarned of the onset of the rainy season and driven local decision-making on the agricultural calendar for generations (Perrotton 2015).

However, some of the seasonal indicators seem to be affected by climate changes. This represents a high social risk linked to a break in the transmission of knowledge. These environmental elements are now subjected to transformations that affect their reliability, and some of the species referred to most frequently, such as the *inkonjani*/Ndebele (*Hirundo rustica*) or *inkanku*/Ndebele (*Jacobin cuckoo*), are observed in the sky less often: "Normally when there is rain, they are flying in the sky and now there are no more of them" (a villager of Magoli). As a result, the use of these indicators appears to be gradually eroding. If the knowledge about these indicators is threatened, it is difficult to predict which new sources of knowledge people will turn to in the future as early warning systems about trends in seasonal rainfall (Roncoli et al. 2002).

---

[6] Some breeders explained that demographic pressure and poor rainfall had forced them to change pastoral techniques. As a result, they destroyed veterinarian fences to lead their herds inside the Sikumi Forest to access grazing areas and water points in order to limit livestock mortality.

Table 3.1 Bird, tree, and insect species used to forewarn of rains[a]. Interviews inventory 5 species of birds, 6 species of trees, and 1 species of insects[b]

Latin name	Local name	Behavior of the species	Announced forecast
Species of birds			
Dicrurus adsimilis	Katengwe/Ndebele	"Cry a lot"	"Rain is coming"
European swallow	Inkonjane/Ndebele	"Fly in the sky"	"It will rain"
Ground hornbill	Momba/Nambya	Presence	"Too much rain," "floods"
Jacobin cuckoo	Inkanku/Ndebele	Presence	"Too much rains," "floods," "indicates and calls the rain"
Red bishop/lesser masked weaver	Intaka/Ndebele	"Coming"	"More food in the fields"
Species of trees[c]			
Acacia galpinii	Acacia/English	"Abundant flowers"	"Good rains"
Adansonia digitata	Baobab/English	"Abundant fruits"	"No rainfall, low rainfall"
Afzelia quanzensis	Unkamba/Nambya	"First to bloom" "No leaves" "Green leaves"	"Onset of the rains" "Rains won't come" "Rains are coming, prepare to go to the fields"
Berchemia discolor	Umnyi/Ndebele	"Abundant fruit" "Starting to have flowers and leaves"	"A drought is about to occur" "Onset of the rains"
Kirkia acuminata	Umvimila/Ndebele	"Losing leaves"	"The rainy season is about to come, ploughing season is starting"
Lannea discolor	Chingangatsha/Nambya	"Abundant fruits"	"The rainy season is about to start"
Species of insects			
Gonimbrasia belina	Mopane worms/English	Presence	"Means more rains"

[a] Perrotton (2015) recorded complementary atmospheric elements (thunder, lightning, and wind) and three different tree species in Hwange District: umfuthi/Ndebele (*Brachystegia boehmii*), igonde/Ndebele (*Brachystegia spiciformis*), and unkotonga/Nambya (*Pterolobium exosum*)

[b]Other signs, based on clouds and moon observations, also enable prediction of the onset of the rains and expected quantities: "And also, the moon. It would show two circles, a smaller one in a bigger one, something orange, that will inform us that the rain will come" (villager from Dinde, April 2015). The clouds *makole vula*/Nambya are the "raining clouds" and can be recognized by their darker color at the base, whereas the yellow, red, and white clouds, the *makole chilimo*/Nambya, which are seen during the winter, do not "bring the rain"

[c]The following tree species were also described as "bringing the rains": umkusu/Ndebele (*Baikiaea plurijuga*), umtshibi/Ndebele (*Guibourtia coleosperma*), iphane/Ndebele (*Colophospermum mopane*), umkhaya/Ndebele (*Acacia nigrescens*), uxakuxaku/Ndebele (*Azanza garckeana*), and mnyelenyele/Ndebele (*Ochna pulchra*)

3.3.3 Protecting Sacred Trees by Perpetuating Rainmaking Ceremonies

The previous sections show that inhabitants of the communal lands of Hwange District have developed detailed expertise about changes in their environment. Embracing this knowledge, culture represents a core prism whereby people detect climatic and environmental disturbances. It also frames the interactions with and uses of certain species viewed as sacred among the studied communities. Across Africa, interactions between people and their changing environments involve cultural understandings (Croll and Parkin 1992). Rusinga and Maposa stated about the Ndau people of southeastern Zimbabwe that "the success in managing nature is inextricably linked to proper management and control of society" (2010: 204). For instance, the taboos, established generations ago, limit the use of some species through "emotional and ethical norms designed to govern behavior and the context of resource utilization" (Mandondo 1997: 353). In many regions of Zimbabwe, the "violation of taboos is thought to invoke the anger of the spiritual world" (Chemhuru and Masaka 2010: 122). They play a special role for conservation as they contribute to the protection of symbolic species that are sometimes facing extinction such as pythons, pangolins, fish, and rhinoceros (ibid.).

In Hwange, several interviewees attest to the continuing importance of natural resource conservation through informal governance by elders and traditional leaders. These modes of conservation are based on the attribution of symbolic values to certain tree species and expressed by associated restrictions on utilization of and interaction with those species. Among the interviews, respondents specified that the taboos mainly concern species of trees which are involved in rainmaking ceremonies and have a function in the rituals. As such, it is strictly forbidden to cut down the trees that are used for firewood during rituals,[7] reserved for making dishes to be used during the ceremonies,[8] or used to forewarn of the rains or known to "bring the rain" (Table 3.1). If these prohibitions are transgressed, the inhabitants fear reprisals from the ancestral spirits: disturbances in the rains for cutting down the *umkamba*/Ndebele or the *umnyi*/Ndebele (an action committed by the whole community) or marital breakdown for the use of the *ichithamuzi*/Ndebele (*Psorospermum febrifugum*) (consequences on an individual level).

Moreover, the species subject to taboos play a key role for the community: ancestor worship, consumption (mostly fruit gathering), firewood, and construction for both domestic and commercial uses. For the village chiefs, their elders made relevant choices in the establishment of the laws:

> The ancestors are so clever, they didn't want us to cut down this tree [*the umnyi*] because of the fruits. It's a tree that takes a long time to bear fruits so they established that law. There are more [*umnyi*] now, because of that law: it's a good sign. (Village head of Half Mafimba, April 2015)

[7] The following trees are used as firewood in the course of rituals and rainmaking ceremonies: *umbuyu*/Nambya (*Adansonia digitata*), *umkamba*/Ndebele (*Afzelia quanzensis*), *mukwakwa*/Nambya (*Strychnos madagascariensis*), and *umkhaya*/Ndebele (*Acacia nigrescens*).

[8] Such as the *umkamba*/Ndebele (*Afzelia quanzensis*)

We therefore used free lists to investigate changes in the abundance of the tree species mentioned in the interviews, as well as changes in their associated uses. Admittedly, the abundance of the *umnyi*/Ndebele and the *umbuyu*/Nambya is increasing according to several sources. This is most likely thanks to the controlled restrictions enforced by the traditional authorities. However, 60% of the species listed are reported to be declining in abundance, and these traditional rules appear to be less and less respected.

Among the studied communities, climate knowledge encompasses more than the atmospheric conditions and their direct impacts; it also includes modes of resource protection through traditional control. To some extent, these regulations are part of natural resource management but in ways that are not recognized by environmental management programs, such as CAMPFIRE. Within the CAMPFIRE institutions, decision-making at the local level, while still functional, concerns more benefit sharing of wildlife revenues rather than management systems. The study of local systems of natural resource management and its integration into the framework of the various environmental committees which coexists on the ground would warrant further investigations in the region.

3.4 What Do These Representations Say About Inequalities in Environmental Management?

The previous section emphasized the value of local knowledge and practices and the role of culture in the mediation and management of climatic and environmental phenomena through taboos on certain tree species. In this section, we explore how individuals explain the abovementioned phenomena by studying more closely their emic theory about environmental changes and the causal linkages they establish. Analytically, this is based on the view that the appearance of a phenomenon is integrated into an explanatory system that can only be understood through thorough consideration of the social and cultural representations of the communities themselves (Evans-Pritchard 1976). We will see that a first step of interpretation - characterized in the coming discussion as "proximal causes" - shifts toward a critique of management politics. We then show that there is broad consensus that ancestors' anger is the ultimate cause for the rapid changes in our interviewees' direct environment and that this reflects the fact that local communities lack proper involvement in the actual management of the natural resources that surround them.

3.4.1 Mobilizing Management Policies as Proximal Causes

The analysis of our interviews shows that people's discourses mobilize proximal and ultimate causes to explain the appearance of these abnormal phenomena, identifying multiple causes. Whereas these changes are often ecological or climatic in nature, their interpretations mobilize religious, cultural, and political explanations (Fig. 3.2).

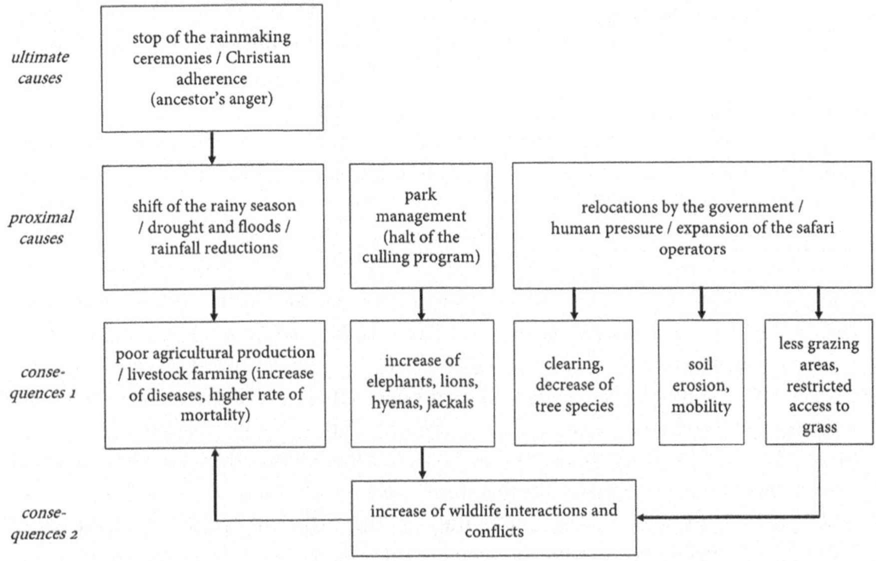

Fig. 3.2 Diagram of the main causal linkages noted in the discourses

For instance, transformations in the landscape are noticed through soil erosion and decreases in tree species and in the availability of grazing areas: "it was like a bush," "the land is now too open and bare, this area used to be a thick forest," and "it's clear now." These changes are explained by demographic pressure and anthropic activities, such as land clearing for housing and agriculture, understood as the combined product of natural growth and forced and spontaneous mobility: government-ordered relocations and the attractiveness of the protected area for finding employment. The following quotation confirms the interconnectedness of these changes:

> In the 1950s, people used to stay in Hwange National Park but were told to move out in the 1960s. Those who used to stay in the Forestry Area were asked to move in 1972. There was an increase in population in the new resettlements, and people started to cut a lot of trees, causing deforestation and soil erosion. This increase in population led to even higher levels of unemployment in this area and by cutting trees, people will be trying to make a living. (Villager of Magoli, March 2017)

Furthermore, many villagers attribute the increased number of elephants going outside the boundaries of Hwange National Park to the fact that an elephant culling program was halted in 1986.[9] This change in the management of the park, due to unethical purposes (Chamaillé-Jammes et al. 2007b), marks a decisive shift in the respondents' memory. Indeed, the culling program was used to regulate elephant densities in the park, to limit the wildlife dispersal into agricultural production areas, and to improve food security for the inhabitants by creating employment and

[9] "A culling program was initiated in 1966, and in 1974, the park adopted a policy of maintaining a population of 13,000 elephants" (Cumming 1981, cited by Chamaillé-Jammes et al. 2007b: 627).

producing considerable quantities of elephant meat. Although the inhabitants only mention the significant time of culling, the increase in wildlife movements outside the park appears to be determined by multiple factors, among them the veterinary fence between protected areas and the communal lands used to control the movement of elephants (Chamaillé-Jammes et al. 2009). However, people do not point out this event as a fact that may have a potential impact on the increasing frequency of interactions with wildlife.

Furthermore, while the dynamics are complex (Murwira et al. 2013), wildlife intrusions often depend on the proximity to the protected areas (Anthony et al. 2010). Surprisingly, villagers of Nengasha, Mbola, and Chimboma seem overwhelmed by these intrusions, despite a greater distance from the park, whereas they are less mentioned in the closest village. In Nengasha, political changes in park management, government policies, and private actors are cited as key elements behind the growing interactions with elephants. Villagers mention that they face both silence and incomprehension from park managers and the councilor, who no longer come to stop elephants in the framework of Problem Animal Control[10] or CAMPFIRE. This shows that, depending on the different wards of the district, inhabitants do not benefit from the same level of care and the villagers who live closer to the park might have maintained better relations with park managers, giving them access to quicker support and interventions. Thus, perceptions of environmental changes are directly linked to overlapping political responsibilities, mismatches, and conflicts. Furthermore, interventions no longer solve these issues in a long-term perspective, as the culling used to do. Other compensations, previously implemented by CAMPFIRE in certain villages, such as jobs or collective equipment, could make the pressure of wildlife conflicts more acceptable.

3.4.2 *"We used to go asking for the rain"*

The diagram of causal linkages highlights the significance of frustrations in territorial and resource management decisions (Fig. 3.2). It also points out that the most widespread interpretations of climate changes are the dissolution of ritual practices and the appeal of Christianity (a correlation mentioned in 33 interviews). The abandonment of rainmaking practices is a subject of high concern in a context of restricted access to privileged sites such as the Bumbusi and Matoa ruins located inside Hwange National Park. "Climatic observations," "cultural abandonment," and "agricultural production" are interconnected in the large majority of local discourses:

---

[10] Problem Animal Control (PAC) is a management service offered by the Zimbabwe Parks and Wildlife Management Authority (ZPWMA) to control animals that cause damage to fields. It is more a way to provide meat for people than to control their displacements or to limit human-wildlife interactions (Guerbois 2012).

> The changing of seasons is due to the fact that people don't follow the culture. Long ago, people asked ancestors for the rain. Now, people are going to church. (Villager of Magoli, June 2015)

> There is a huge change in the coming of the rain. People are asking each other questions: 'Why is the rain not coming? Why is it so little?' And they can't find the answers. For the other years, even though it was scarce, it was coming at nighttime and it was enough for the crops. […] Now, it's difficult because people who used to ask for rain have gone to church. There is a time when the elders have suggested that they go back to that culture of asking for rain, and people have refused. They said: 'No, we are now in churches, we can't do that. (Village head of Mbola, April 2015)

Some respondents talked about "cultural abandonment," and others cited the "banning of tradition" and the "interruption of the ceremonies." These quotations show that local perceptions of climate change are linked to cultural and religious changes. Indeed, adherence to Christianity and the abandonment of ritual practices of ancestor worship are held directly accountable for ancestors' anger and are a very sensitive topic. The villagers express regrets over the disappearance of these practices, formerly performed by the spirit mediums, traditional healers, and rainmakers (*iwosana*/Ndebele or *mhande*/Nambya). This loss of traditional values is also perceived in daily practices, by using forbidden kitchen utensils, for instance: "this is because people don't want to follow traditions and customs: people use black pots to fetch water." Furthermore, attending church is sometimes presented as incompatible with worshipping ancestors, and these debates about returning to ancestral practices are controversial in the villages, especially because of the massive influence of the churches.

We noticed that in the villages where inhabitants continue to practice rainmaking ceremonies (Nengasha) at different shrines and sacred places, such as Shangano and Chingahali (Fig. 3.1), respondents rarely discuss climate variability. This shows that the topics of climate change and cultural abandonment are less salient in areas where the rainmaking ceremonies are still practiced.

3.4.3 People's Marginalization from Their Rights Over the Land and the Natural Resources

The process of dispossession, exclusion, and poverty, resulting from the creation of protected areas, is becoming better described in the bordering areas of Southern Africa (Brockington and Igoe 2006; Cernea and Schmidt-Soltau 2006; West et al. 2006). In Hwange District, CAMPFIRE aimed at addressing the previous issues by compensating inhabitants of the communal lands for the high costs of living with wildlife, due to their proximity with conservation areas (protected areas, hunting concessions). Political ecology contributions to CAMPFIRE analysis helped in understanding implementation challenges of the program, including in identifying the marginalization of some users. These studies also illustrate how different discourses are used to voice grievances. Among these studies, Matema and Andersson (2015) have shown that the upsurge of lion attacks in 2010 in Mbire District, Zimbabwe, is locally interpreted as lions' guardian spirits – also known as *mhondoro*/Shona[11] – communicating with the inhabitants. Local interpretations of the lion attacks vary but generally highlight social issues involving immigrants and safari hunting (designated as responsible for driving lions into human settlements, as observed as well in our study). Thus, the *mhondoro* provides a voice of protest to villagers and traditional leaders to raise sociopolitical concerns: "while the *mhondoro* discourse appears as apolitical, it strongly resonates with Zimbabwean political discourse at national level" (Matema and Andersson 2015: 116).

We observed the same mechanism in the communal lands of Hwange District. Climate and environmental changes are partly understood through the prism of cultural and political systems. Local perceptions show, in different ways, the influence of the protected area on people's daily lives, either through the way they relate to the changes in park management (to express the increasing pressure from wildlife), the spatial expansion of the safari operators, or to their limited access to natural resources and sacred sites that are located within the protected area. According to our interviewees, the rains are disturbed because of the anger of ancestors, who are avenging themselves for the notorious lack of interest of the inhabitants who no longer worship their forebears. However, if these rituals are practiced less frequently, this is also due to the difficulties people face in gaining access to the main sacred sites and, therefore, to park management.

These discourses allow us to introduce a reinterpretation of the discourses relative to climate and environmental changes in the region: while the whole ecological system seems to be affected, the decision-making power of local people with regard to the actual state of their resources appears to be very weak. The government, through the Zimbabwe Parks and Wildlife Management Authority (ZPWMA) and the Rural District Council (RDC), as well as the private sector, through safari operators, controls access and management over land and natural resources. Despite efforts begun with CAMPFIRE in the 1980s, the real hold of communal lands' inhabitants on their surroundings and its resources remains quite limited. Therefore, the fact that people's interpretation of changes in climate turns to the anger of ancestral spirits can be viewed as an expression of their own deprivation: inhabitants of the area have been dispossessed of their rights over the land and the natural resources of their environment for more than a century. In other terms, it reflects the inefficiency of a community-based management program that had been promising.

3.5 Conclusion

Individuals, within their own memories, amass certain remarkable facts that have affected their lives and which they assimilate to the notion of "change." These changes date their experience, alter their "affect," and participate in the construction of a reference guide of individual behaviors, in other terms, to their episodic memory (Craik 1994). Indeed, the most reliable datings are the ones individuals connect

[11] As "deceased chiefs are believed to continue to guide the current inhabitants of their chiefdoms as mhondoro or 'lion spirits'" (Matema and Andersson, 2015: 100).

to political and economic events such as the creation of the park, the halting of the elephant culling program, or the economic and politic crisis of 2008. Although they are less precise, they are also less altered by the fading of respondents' old memories. This local knowledge and this memory provide a strong and reliable basis for understanding local dynamics, as argued in other contexts (Nazarea 2006).

Studying local knowledge and practices gives a voice to those who are most affected by climate and environmental uncertainty (Batterbury 2008). This study conducted on discourses related to environmental changes in the Hwange region reveals a complex interconnection of causal links. Among the different discourses, these changes appear to be part of the daily conversations and the social lives of the inhabitants. The linkages are a translation of global transformations that are embedded in the individuals' environment as a system, linked not only to the environmental sphere but also to the sociopolitical and cultural ones. These observations bear similarities with the Ndau people who have a "complex system that fuses environmental management into social life" which are presented as "traditional arrangements that promote sustainable utilization of natural resources" (Rusinga and Maposa 2010: 205).

Mutekwa (2010) showed that spirit possession has been used by politicians and traditional leaders as a metaphor for oppression in Zimbabwean society. In our case study, these metaphoric utilizations come from the villagers. We interpret the central place of the ancestors' anger in the discourses as a translation of a significant frustration of the inhabitants themselves, which emerges from tangible realities. We adhere to the argument that these discourses allow silent voices to express themselves about sociopolitical issues in an authoritarian context (Matema and Andersson 2015). Inhabitants of Hwange District observe and interpret significant environmental changes but remain disregarded in the actual development of environmental management policies. So, if the ancestral spirits do not express the frustration of the inhabitants of this highly marginalized territory, then who will?

Acknowledgements "We would like to thank our colleagues in the PIAF project, the field assistants who contributed to data collection, as well as the interviewed families for agreeing to participate in the study. This chapter has been produced from data collected thanks to the French Agence Nationale de la Recherche for funding support through the SAVARID project (ANR-11- CEPL-0003) and the PIAF project (ANR-13-JSH1-0005). This work was also made possible through funding and technical support from the Zone Atelier Hwange (Hwange LTSER) and the Research Platform "Production and Conservation in Partnership" (www.rp-pcp.org)."

References

Alexander J, McGregor J, Ranger TO (2000) Violence and memory: one hundred years in the 'dark forests' of Matabeleland, Zimbabwe. Heinemann and James Currey, Portsmouth/Oxford

Andersson JA, Cumming DHM (2013) Defining the edge: boundary formation and TFCAs in Southern Africa. In: Andersson JA, de Garine-Wichatitsky M, Cumming D, Dzingirai V, Giller K (eds) Transfrontier conservation areas: people living on the edge. Earthscan Publications, London, pp 25–61

Anthony BP, Scott P, Antypas A (2010) Sitting on the fence? Policies and practices in managing human-wildlife conflict in Limpopo Province, South Africa. Conserv Soc 8(3):225–240. https://doi.org/10.4103/0972-4923.73812

Barnes J, Dove M, Lahsen M, Mathews A, McElwee P, McIntosh R, Moore F, O'Reilly, Orlove B, Puri R, Weiss H, Yager K (2013) Contribution of anthropology to the study of climate change. Nat Clim Chang 3(6):541–544. https://doi.org/10.1038/nclimate1775

Batterbury S (2008) Anthropology and global warming: the need for environmental engagement. Aust J Anthropol 19(1):62–68. https://doi.org/10.1111/j.1835-9310.2008.tb00108.x

Blanc JJ, Barnes RFW, Craig GC, Douglas-Hamilton I, Dublin HT, Hart JA, Thouless CR (2005) Changes in elephant numbers in major savanna populations in eastern and southern Africa. Pachyderm 38:19–28

Brewer DD (2002) Supplementary interviewing techniques to maximize output in free listing tasks. Field Methods 14(1):108–118. https://doi.org/10.1177/1525822X02014001007

Brockington D, Igoe J (2006) Eviction for conservation: a global overview. Conserv Soc 4(3):424

Cavendish W (2000) Empirical regularities in the poverty-environment relationship of rural households: evidence from Zimbabwe. World Dev 28:1979–2003. https://doi.org/10.1016/S0305-750X(00)00066-8

Cernea MM, Schmidt-Soltau K (2006) Poverty risks and national parks: policy issues in conservation and resettlement. World Dev 34(10):1808–1830. https://doi.org/10.1016/j.worlddev.2006.02.008

Chamaillé-Jammes S, Fritz H, Murindagomo F (2006) Spatial patterns of the NDVI–rainfall relationship at the seasonal and interannual time scales in an African savanna. Int J Remote Sens 27(23):5185–5200. https://doi.org/10.1080/01431160600702392

Chamaillé-Jammes S, Fritz H, Murindagomo F (2007a) Detecting climate changes of concern in highly variable environments: quantile regressions reveal that droughts worsen in Hwange National Park, Zimbabwe. J Arid Environ 71:321–326. https://doi.org/10.1016/j.jaridenv.2007.05.005

Chamaillé-Jammes S, Valeix M, Fritz H (2007b) Managing heterogeneity in elephant distribution: interactions between elephant population density and surface-water availability. J Appl Ecol 44(3):625–633. https://doi.org/10.1111/j.1365-2664.2007.01300.x

Chamaillé-Jammes S, Valeix M, Bourgarel M, Murindagomo F, Fritz H (2009) Seasonal density estimates of common large herbivores in Hwange National Park, Zimbabwe. Afr J Ecol 47(4):804–808. https://doi.org/10.1111/j.1365-2028.2009.01077.x

Chemhuru M, Masaka D (2010) Taboos as sources of Shona people's environmental ethics. J Sustain Develop Afr 12(7):121–133

Child B (1996) The practice and principles of community-based wildlife management in Zimbabwe: the CAMPFIRE programme. Biodivers Conserv 5(3):369–398. https://doi.org/10.1007/BF00051780

Cling J (2001) Réforme agraire et démagogie électorale au Zimbabwe. Afrique contemporaine 197:64–75

Compagnon D (2003) La prétendue « réforme agraire » au Zimbabwe : à qui profite le crime ? Etude 397:297–307

Craik FI (1994) Memory changes in normal aging. Curr Dir Psychol Sci 3(5):155–158. https://doi.org/10.1111/1467-8721.ep10770653

Croll E, Parkin D (eds) (1992) Bush base, forest farm: culture, environment, and development. Routledge, London and New York

Cumming DHM (1981) The management of elephant and other large mammals in Zimbabwe. In: Jewel PJ, Holt S, Hart D (eds) Problems in management of locally abundant wild animals. Academic Press, New York, pp 91–118

Davison T (1967) Wankie: the story of a great game reserve. Books of Africa, Cape town

de Garine-Wichatitsky M, Fritz H, Chaminuka P, Caron A, Guerbois C, Pfukeni D, Matema C, Jori F, Murwira A (2013) Consequences of animal crossing the edges of transfrontier Parks. In: Andersson JA, de Garine-Wichatitsky M, Cumming D, Dzingirai V, Giller K (eds) Transfrontier conservation areas: people living on the edge. Earthscan Publications, London, pp 137–162

DeWalt B (1994) Using indigenous knowledge to improve agriculture and natural resource management. Hum Organ 53(2):123–131. https://doi.org/10.17730/humo.53.2.ku60563817m03n73
Diemberger H, Hastrup K, Schaffer S, Kennel CF, Sneath D, Bravo M, Graf HF, Hobbs J, Davis J, Nodari ML, Vassena G, Irvine R, Evans C, Strathern M, Hulme M, Kaser G, Bodenhorn B (2012) Communicating climate knowledge. Curr Anthropol 53(2):226–244. https://doi.org/10.1086/665033
Dounias E (2010) Perception du changement climatique par les peuples des forêts tropicales. In: Barbault R, Foucault A (eds) Changements climatiques & biodiversité. Vuibert-AFAS, Paris, pp 243–255
Dzingirai E (1999) Big men and CAMPFIRE: a comparative study of the role of external actors in conflicts over local resources. Zambezia 26(1):77–92. https://hdl.handle.net/10520/AJA03790622_415
Evans-Pritchard EE (1976) The notion of witchcraft explains unfortunate events. In: Evans-Pritchard EE (ed) Witchcraft, oracles and magic among the Azande. Clarendon Press, Oxford University Press, Oxford, pp 18–32
Fiske SJ (2009) Global change policy making from inside the beltway: engaging anthropology. In: Crate SA, Nuttall M (eds) Anthropology and climate change: from encounters to actions. Routledge, London and New York, pp 87–115
Frost PG, Bond I (2008) The CAMPFIRE programme in Zimbabwe: payments for wildlife services. Ecol Econ 65(4):776–787. https://doi.org/10.1016/j.ecolecon.2007.09.018
Guerbois C (2012) Considérer les aires protégées dans la dynamique des systèmes socio-écologiques pour une conservation intégrée et durable de la faune sauvage africaine. Ph.D. dissertation. Muséum National d'Histoire Naturelle, Paris
Guerbois C, Dufour AB, Mtare G, Fritz H (2013) Insights for integrated conservation from attitudes of people toward protected areas near Hwange National Park, Zimbabwe. Conserv Biol 27(4):844–855. https://doi.org/10.1111/cobi.12108
Hastrup K (2013) Anthropological contributions to the study of climate: past, present, future. Wiley Interdiscip Rev Clim Chang 4(4):269–281. https://doi.org/10.1002/wcc.219
Hastrup K, Skrydstrup M (eds) (2013) The social life of climate change models: anticipating nature. Routledge, London and New York
Herbst J (1990) State politics in Zimbabwe. University of California Press. (Perspectives on Southern Africa, 45)
Kwashirai VC (2008) Poverty in the Gwai Forest Reserve, Zimbabwe: 1880-1953. Glob Environ 1(1):146–175. https://doi.org/10.3197/ge.2008.010106
Le Bel S, Murwira A, Mukamuri B, La Grange M, Czudek R, Taylor R (2011) Human wildlife conflicts in southern Africa: riding the whirl wind in Mozambique and in Zimbabwe. In: Pujol JL (ed) The importance of biological interactions in the study of biodiversity. InTech, Rijeka, pp 283–322
Mandondo A (1997) Trees and spaces as emotion and norm laden components of local ecosystems in Nyamaropa communal land, Nyanga District, Zimbabwe. Agric Hum Values 14(4):353–372. https://doi.org/10.1023/A:1007498110104
Marino E, Schweitzer P (2009) Talking and not talking about climate change in Northwestern Alaska. In: Crate SA, Nuttall M (eds) Anthropology and climate change: from encounters to actions. Routledge, London and New York, pp 87–115
Matema S, Andersson JA (2015) Why are lions killing us? Human–wildlife conflict and social discontent in Mbire District, northern Zimbabwe. J Mod Afr Stud 53(1):93–120. https://doi.org/10.1017/S0022278X14000664
McCorkle CM (1989) Toward a knowledge of local knowledge and its importance for agricultural RD&E. Agric Hum Values 6(3):4–12. https://doi.org/10.1007/BF02217664
McGregor J (2005) The social life of ruins: sites of memory and the politics of a Zimbabwean periphery. J Hist Geogr 31(2):316–337. https://doi.org/10.1016/j.jhg.2004.12.026
Muboko N, Murindagomo F (2014) Wildlife control, access and utilisation: lessons from legislation, policy evolution and implementation in Zimbabwe. J Nat Conserv 22(3):206–211. https://doi.org/10.1016/j.jnc.2013.12.002

Murombedzi JC (1999) Policy arena: devolution and stewardship in Zimbabwe's CAMPFIRE programme. J Int Dev 11(2):287–293. https://doi.org/10.1002/(SICI)1099-1328(199903/04)11:2<287::AID-JID584>3.0.CO;2-M

Murphree MW (1993) Communities as resource management institutions. IIED, London. (Gatekeeper Series, 36)

Murwira A, de Garine-Wichatitsky M, Zengeya F, Poshiwa X, Matema S, Caron A, Guerbois C, Hellard E, Fritz H (2013) Resource gradients and movements across the edge of transfrontier parks. In: Andersson JA, de Garine-Wichatitsky M, Cumming D, Dzingirai V, Giller K (eds) Transfrontier conservation areas: people living on the edge. Earthscan Publications, London, pp 123–161

Mutekwa A (2010) The avenging spirit: mapping an ambivalent spirituality in Zimbabwean literature in English. Afr Stud 69(1):161–176. https://doi.org/10.1080/00020181003647264

Nakashima DJ, McLean KG, Thulstrup HD, Castillo AR, Rubis JT (2012) Weathering uncertainty: traditional knowledge for climate change assessment and adaptation. UNESCO and UNU, Paris/Darwin

Nazarea VD (2006) Local knowledge and memory in biodiversity conservation. Annu Rev Anthropol 35:317–335. https://doi.org/10.1146/annurev.anthro.35.081705.123252

Ncube GT (2004) A history of northwestern Zimbabwe, 1850–1960. Mond Books, Kadoma

New M, Hewitson MB, Stephenson DB, Tsiga A, Kruger A, Manhique A, Gomez B et al (2006) Evidence of trends in daily climate extremes over southern and West Africa. J Geophys Res Atmos 111(D14). https://doi.org/10.1029/2005JD006289

Nyong A, Adesina F, Elasha BO (2007) The value of indigenous knowledge in climate change mitigation and adaptation strategies in the African Sahel. Mitig Adapt Strateg Glob Chang 12(5):787–797. https://doi.org/10.1007/s11027-007-9099-0

Orlove B, Roncoli C, Kabugo M, Majugu A (2010) Indigenous climate knowledge in southern Uganda: the multiple components of a dynamic regional system. Clim Chang 100(2):243–265. https://doi.org/10.1007/s10584-009-9586-2

Perrotton A (2015) Cattle herding and coexistence between protected areas and their periphery: a participatory approach to a wicked problem. Ph.D. dissertation. Montpellier: Université de Montpellier

Quinlan M (2005) Considerations for collecting freelists in the field: examples from ethobotany. Field Methods 17(3):219–234. https://doi.org/10.1177/1525822X05277460

Roncoli C, Ingram K, Kirshen P (2002) Reading the rains: local knowledge and rainfall forecasting among farmers of Burkina Faso. Soc Nat Resour 15:411–430. https://doi.org/10.1080/08941920252866774.

Roncoli C, Crane T, Orlove B (2009) Fielding climate change in cultural anthropology. In: Crate SA, Nuttall M (eds) Anthropology and climate change: from encounters to actions. Routledge, London and New York, pp 87–115

Rusinga O, Maposa R (2010) Traditional religion and natural resources: a reflection on the significance of indigenous knowledge systems on the utilisation of natural resources among the Ndau people in South-eastern Zimbabwe. J Ecol Nat Environ 2(9):201–206

Sabar B (2012) Bringing culture back: traditional agricultural knowledge, food production and sustainable livelihood among Chuktia Bhunjia of Orissa. Soc Chang 42(2):203–227. https://doi.org/10.1177/004908571204200204

Taylor R (2009) Community based natural resource Management in Zimbabwe: the experience of CAMPFIRE. Biodivers Conserv 18(10):2563–2583. https://doi.org/10.1007/s10531-009-9612-8

West P, Igoe J, Brockington D (2006) Parks and peoples: the social impact of protected areas. Annu Rev Anthropol 35:251–277. https://doi.org/10.1146/annurev.anthro.35.081705.123308

Chapter 4
Indigenous Knowledge and Dynamics Among Himalayan Peoples, Vegetation, and Climate Change

Jan Salick, Ben Staver, and Robbie Hart

Abstract Based on two decades of our ethnobotanical and alpine plant ecological research on Himalayan climate change, we use system dynamics modeling to integrate our results for insights into long-term trends. Himalayan climate change is perceived by traditional peoples, who adapt, mitigate, and integrate these changes into their cosmologies. Himalayan alpine vegetation is also responding quickly to climatic changes: mountain vegetation is proliferating with lower elevational species moving into alpine areas, while high alpine areas – at the elevational limit of plant life – are being colonized by endemic species. At lower elevations, the model predicts reduced pasturing and increased agroforestry and tree and crop introductions. With the invasion of non-alpine plants and agriculture, there will likely be reduced habitat for alpine species. At high elevations, the model predicts displacement of alpine meadows and yak grazing to yet higher elevations. Medicinal herb collection, a major economic activity, will likely expand and subsequently experience overharvesting. Yak grazing at increasingly high elevations may stress the animals, the pastures, and the herders. Tourism may put yet further pressure on highest elevational habitats. Indigenous peoples will be forced to adapt rapidly by both traditional and contemporary means including the use of indigenous knowledge, agricultural innovation, and development of tourist economies, which can also be threatened by climate change as mountains lose their snowy and glacial beauty. Abandonment of previously adaptive indigenous cultures and out-migration may be strategies of last resort.

J. Salick · R. Hart (✉)
Missouri Botanical Garden, St. Louis, MO, USA
e-mail: robbie.hart@mobot.org

B. Staver
ICF, Atlanta, GA, USA

M. Welch-Devine et al. (eds.), *Changing Climate, Changing Worlds*,
Ethnobiology, https://doi.org/10.1007/978-3-030-37312-2_4

Keywords Himalaya · Climate change · Ethnobotany · Indigenous knowledge · Traditional ecological knowledge · Alpine plant ecology · Medicinal plants · Mountain agriculture · System dynamics modeling

Himalayan peoples perceive (Salick and Byg 2007; Byg and Salick 2009; Salick et al. 2009; Salick and Moseley 2012; Salick 2013), adapt to (Konchar et al. 2015; Salick et al. 2018), mitigate (Salick et al. 2014), and integrate into their cosmologies (Salick et al. 2013, 2018) the severe climate change taking place in the Himalaya (Salick et al. 2014). Himalayan alpine vegetation is also responding quickly to these climatic changes (Salick et al. 2009, 2014, 2019); specifically, mountain vegetation is proliferating, with lower elevational species moving into alpine areas, while high alpine areas – at the elevational limit of plant life – are being colonized by endemic species (Salick et al. 2019). But how do these processes of climate change, vegetation change, and human adaptation interact and influence each other? How do people and plants interact in response to climate change? What roles do indigenous knowledge and traditional practices play in these processes? Here, based on our two decades of data and analyses, we integrate our previous Himalayan work through system dynamics modeling to suggest processes by which human responses to climate change may affect vegetation responses to climate change and vice versa.

We need to introduce caveats immediately and be clear that climate and vegetation changes along with human adaptations are not the only transformations taking place in the Himalaya. Social, cultural, political, and economic changes are extensive and interact meaningfully with climate change vulnerability (Nagoda 2015; Nagoda et al. 2017). However as botanists, our research, which concentrates on the ethnobotany of climate change, has not addressed these factors directly. Our model incorporates the data we have collected, touching on broader issues such as politics and economics only indirectly. Unlike the lower Himalaya, other issues such as deforestation and population growth are not relevant to alpine areas (and may be misconstrued elsewhere as discussed by Ives 2005). The alpine by definition (Nagy and Grabherr 2009) is above tree line and forests and thus the pressures of deforestation. Human populations in the alpine are often declining due to out-migration because of both economics and politics (Craig 2002; Childs et al. 2014).

4.1 Himalayan Climate Change

The eastern Himalayas are critical for understanding the consequences of climate change to biodiversity. At the nexus of the temperate and subtropical Asian floras and the dry alpine flora of the Tibetan plateau, the eastern Himalayas are among the most biologically diverse temperate regions (Mutke and Barthlott 2005; Mittermeier et al. 2004) and the most affected by climate change outside polar regions (Williams et al. 2007). Himalayan temperatures have increased faster than the global mean, with increasing rates in this century (1.5–1.75 times global mean increase; Shrestha et al. 2012; IPCC 2013). In an area already famous for torrential monsoons, Himalayan precipitation is predicted to increase and to become yet more variable and extreme (Kohler and Maselli 2009; Mittal et al. 2014). As ice and snow melt and rains intensify, the "water tower" (Immerzeel et al. 2010) services of this region – providing a steady and reliable source of water to over a billion people downstream – threaten to become ever more unstable (Fig. 4.1). Instability may result from lack of water during droughts, which had previously been provided by slow, steady glacial melt, and flooding during increasing monsoons and snowmelts (Xu et al. 2009). The billion people downstream, dependent on the great rivers that originate in the Himalaya, are thereby made vulnerable to climate change (Füssel 2007).

To review our results on the effects of Himalayan climate change, we organize two decades of research around the most general feedback model (Fig. 4.2). Himalayan climate change affects both people and vegetation, which in turn affect each other. Later, we break down these processes and detail a system dynamics model to analyze how indigenous knowledge and traditional practices can play a role in understanding and adapting to climate change.

Fig. 4.1 Glacier melting above Manang, Nepal, has formed a precarious glacial lake. Glacial lake outbreak floods (GLOFs) are threatening many populations in the Himalaya and downstream, as well as the "water tower" services provided by glaciers. This picture was taken by a Manangi villager who participated in a photovoice project to document effects of climate change, of which villagers are well aware and actively discussing. (Previously published in Konchar et al. 2015 and reprinted with permission)

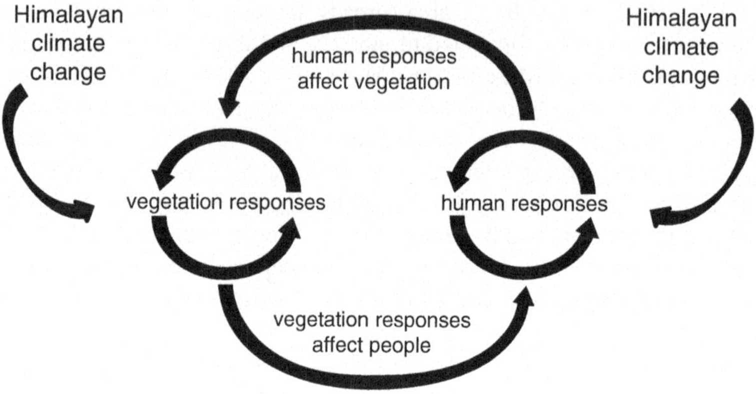

Fig. 4.2 A general feedback diagram with both vegetation and humans affected by and responding to climate change. In turn, these affect and respond to each other

4.2 Himalayan Research Sites, Peoples, and Methods

We have conducted two decades of botanical and ethnobotanical research in the eastern Himalayas including the Hengduan Mountains and Tibetan Plateau of Southwestern China, western and central high-elevation Bhutan, and eastern, central, and western Himalayan Nepal. The Himalayan peoples of these regions with whom we worked included Tibetan peoples of NW Yunnan and Lhasa regions; Naxi of NW Yunnan; high-elevation Bhutanese; and Sherpa, Tamang, and Manangi of Nepal. These peoples are far from culturally monolithic. Not only are their histories, languages, and cultures distinct, their economic and political contexts within China, Bhutan, and Nepal are very different. However, there are recognizable similarities in the ways these high-altitude Tibeto-Burman peoples have responded to climate change.

Briefly, our ecological methods included longitudinal alpine plot sampling following the GLobal Organization for Research In Alpine areas (GLORIA, Pauli 2015; Salick et al. 2009, 2014, 2019), elevational transects (Hart and Salick 2018), and geographical information systems (Salick et al. 2005; Anderson et al. 2005), with statistical analyses in the R framework in all our publications. Participatory methods included mapping, calendar construction (Salick et al. 2005, 2018), photovoice (Salick and Moseley 2012; Konchar et al. 2015), pile sorts (Hart and Salick 2017), and accompanied field visits ("walk in the woods," Salick et al. 2009, 2014, 2018), along with many, many semi-structured interviews.

Overall, the eastern Himalayas are biologically and culturally extremely diverse with among the highest diversity of languages (Turin 2007; Gorenflo et al. 2012) and of plants (Mutke and Barthlott 2005; Mittermeier et al. 2004) in the world. The extreme topography isolates both plant and human populations (Zurick and Pacheco 2006). These same people and plants are facing among the fastest climate change in the world (IPCC 2013). We have seen recognizable trends in plant and human adaptations, which we address in this paper.

4.3 Vegetation Responses to Climate Change

Throughout the world's mountain ranges, shifts in vegetation zones along elevational gradients show a consistent response to warming temperatures (Parmesan and Yohe 2003; Kelly and Goulden 2008; Chen et al. 2011; Morueta-Holme et al. 2015). Within plant communities, the vegetation gradually becomes dominated by species adapted to warmer temperatures (Gottfried et al. 2012). At the landscape level, this results in upward shifts of tree lines, shrublines, and of the lower limit of alpine meadows. Our high alpine permanent plots across the eastern Himalayas (Salick et al. 2009, 2014, 2019) show some of these same responses along with unique characteristics due to the dominance of rhododendrons (especially ~4000 masl) and of endemic species at very high elevations (~5000 masl). With increasing temperatures, vegetation measures – number of plant species, frequency, diversity, and endemic and useful plant species – increase significantly. On lower elevation summits, especially, plants adapted to warmer temperatures increase in relative abundance and in elevational range. These responses are statistically significant over as little as 7 years (our resampling period). Similar changes are reported elsewhere in the Himalaya, as woody vegetation encroaches into alpine meadows (Brandt et al. 2013) and plants are found at increasingly high elevations never before reported (Dolezal et al. 2016).

Shifts have also been seen in phenology, the timing of plant life histories (Parmesan and Yohe 2003; Parmesan 2006; Wolkovich and Cleland 2014). Across the globe, spring life history events such as emergence, leaf out, and flowering have become earlier (Cleland et al. 2007; Wolkovich et al. 2012; Ellwood et al. 2013), and fall life history events often occur later, expanding the season (Fridley 2012; Gallinat et al. 2015). However, the specific trajectories of change are constrained by species biology (Cook et al. 2012; Keller and Körner 2003). In the Himalaya, the direction and degree of phenological change varies with geography, life history event, measurement method, and phylogenetic and physiological differences (e.g., Yu et al. 2010; Zhang et al. 2013; Hart et al. 2014, 2016; Hart and Salick 2018). For example, Himalayan *Rhododendron* species (Fig. 4.3) usually respond to warming with earlier spring flowering; however, in especially warm winters, "chilling requirements" are not met, which results in *delayed* spring flowering (Hart et al. 2014).

4.4 Human Responses to Climate Change

Himalayan peoples clearly perceive climate change. In Diqing Tibetan Autonomous Prefecture (Byg and Salick 2009; Salick et al. 2018), Tibetans living in small, dispersed villages with little external information about climate change still had many observations on, adaptations to, and interpretations of climate change. There were some consensus and much variability among their observations, adaptations, and interpretations. Most Tibetans consistently observed warming temperatures, less snow, and glacial retreat. Consequences of and adaptations to these phenomena

Fig. 4.3 Rhododendrons shrubs are expanding their ranges up mountains and are changing their phenology in response to climate change. (Photo by Robbie Hart)

included longer yak grazing in the highlands, earlier sowing and harvesting of crops, lower crop yields, and increased insect and disease attacks (Salick et al. 2005, 2018). Perceptions of climate change were more variable regarding river levels and landslides frequency (Byg and Salick 2009), which were themselves extremely variable and unpredictable, rendering adaptations problematic.

For these Tibetans, not connected to international news at the time of this research (before the Chinese government had widely acknowledged climate change), the causes of climate change were often obscure or variously attributed. Cosmology (Salick et al. 2013) prominently informed their interpretations of change, as well as their responses. Among many rural Tibetans in this area, there were beliefs that climate is changing and that bad deeds have caused this and good deeds will mitigate it (connected to Buddhist beliefs), that fickle gods must be supplicated and appeased (connected to shamanic beliefs), and/or that there are material causes and solutions (connected to scientific/materialist beliefs). These various interpretations could often be held simultaneously. In the extreme, some Tibetans feel that their traditional culture – food, clothing, and livelihoods – is no longer adaptive and that, along with their political woes, Tibetan culture is further doomed by climate change.

Those most closely associated with the natural environment may be those who perceive the most change. In Southwestern China, for example, we found that Naxi and Yi participants from rural villages – in contrast to those from urban Lijiang only 25 km distant – reported more knowledge about the elevational and seasonal distributions of rhododendrons and were the most likely to report changes to those distributions, including unpredictable phenology and species disappearances (Hart and Salick 2017).

Fig. 4.4 Proud Manangi woman in front of her new vegetable garden and tourist guest house. Agriculture is changing due to climate change and economics is changing with an influx of tourists. (Photo by Ben Staver. Previously published in Konchar et al. 2015 and reprinted with permission)

In Manang, Nepal (Konchar et al. 2015), perceptions were more informed by media and open access to information about climate change. Personal perceptions of rapid change in temperature, precipitation, permanent snow, and glacial retreat were commonly held. Villagers knew and were immediately concerned about the potential for a glacial lake outburst flood (GLOF) in Manang (Fig. 4.1). Adaptation strategies included a shift from traditional agropastoralism (yak and limited farming). Recently, fruit tree and vegetable crops were newly introduced into alpine areas, as well as agroforestry practices (Fig. 4.4). Yak grazing had expanded into (was relegated to) the highest alpine areas, stressing both animals and herders. Tourism from the Annapurna trekking circuit was the major economic strategy in Manang and the new agricultural products were grown with tourist penchants in mind.

4.5 Vegetation Responses Affect People

The dependence of human livelihoods (farming, grazing, and tourism) and well-being (health, culture, and economics) upon the environment is clear, as are changes in vegetation driven by climate such as elevational shifts and phenology. However, the indirect effects upon people of environmental responses to climate change are less well studied (Pecl et al. 2017). For example, how are vegetation changes brought on by climate change affecting Himalayan peoples? Our work has shown diverse examples of how Himalayan peoples have noticed and responded to these vegetation responses.

Elevational shifts of vegetation affect agropastoralism. In general, agriculture most directly affects Himalayan people (Salick et al. 2005, 2018). New crops are grown, traditional crops are grown at higher elevations, and yak grazing on alpine meadows is moving ever higher, while shrubs and trees are encroaching on alpine meadows at lower elevations. Complex changes are also observed. In Nepal, herders report that deteriorating pasture grass quality caused declines in blue sheep populations, which in turn led to more snow leopard predation on yak (Bhimtang Herder, personal communication, 2016).

Changes in plant communities and populations are also perceived. Tibetan doctors notice or are concerned that native medicinal plants may be affected by climate change (Salick and Moseley 2012) at the same time that commercial harvesting is threatening these plant populations (Law and Salick 2005, 2006). They also notice that native plant phenology – flowering and fruiting – and distributions are changing, which can affect traditionally prescribed times when and places where they can collect their medicines.

Himalayan people also report changes in plant phenology – emergence, flowering, and fruiting – in ways that defy expectation and disrupt traditional ecological calendars, seasonal indicators, and connections between plants and religious holidays (Hart and Salick 2017). For example, for Yi informants in Northwest Yunnan, *Rhododendron racemosum* flowering traditionally indicated when to plant buckwheat, and the species was named accordingly "shouxma mgeqy" – "buckwheat rhododendron" (Fig. 4.5). However, this indicator is now desynchronized and unpredictable (Hart and Salick 2017). Similar climate-driven changes to ecological calendars are noted across the region (Klein et al. 2014; Ingty 2017) and the globe (Lantz and Turner 2003; Armatas et al. 2016).

Tibetans have shifted their seasonal crop planting and harvesting timelines to adapt to a changing climate. Effects of climate change are being incorporated into traditional Tibetan agropastoral calendars, which have been produced for more than a millennium, now with earlier recommended planting and harvesting dates (Salick and Moseley 2012). Monks who formulate the Tibetan calendar remark on their struggle to interpret so many recent changes in the environment. Traditionally, these monks use Tibetan astrology in concert with pragmatic knowledge about agriculture and local ecology. Shifts in planting dates have been earlier and of three planting dates traditionally designated by the Tibetan calendar, recently, the earliest recommended dates have been consistently the best (Salick et al. 2018).

4.6 Human Responses Affect Vegetation and Climate Change

As people are affected by vegetation responses to climate change, so reciprocally do people affect vegetation. As agriculture, both cultivation and grazing, moves to higher elevations, natural vegetation is displaced in the lower alpine regions and experiences more grazing pressure at higher elevations. Additionally, as grazing is pushed higher and to more remote areas (Fig. 4.6.), it becomes less integrated with

Fig. 4.5 Indicator species no longer flower with their associated resources. Traditionally, (**a**) *Rhododendron racemosum* flowering indicated (**b**) when to plant buckwheat and the species was named accordingly "shouxma mgeqy" or "buckwheat rhododendron." With climate change, this association has broken down. (Photos by Robbie Hart)

on farm management and herding becomes a more independent and specialized occupation, with remote summer herder villages putting pressures on scarce high alpine resources (Yeh 2007). These remote summer villages also specialize in commercial collection of medicinal plants (e.g., chong cao (*Ophiocordyceps sinensis*), snow lotus (*Saussurea laniceps*), and bei mu (*Fritillaria cirrhosa*)), which greatly affects alpine plant populations, resulting in overharvest of traditional medicinal plants (Law and Salick 2005; Salick and Moseley 2012).

Himalayan mitigation of climate change is affected by traditional practices of incorporating organic matter into mineral and unstable soils on steep hillsides (Salick and Moseley 2012). Worldwide, organic matter in soils contains more carbon than does vegetation (Raich and Schlesinger 1992). New Himalayan practices of growing tree crops and agroforestry in lower alpine areas also capture carbon (Konchar et al. 2015).

Fig. 4.6 Himalayan climate change is pushing yak grazing to higher and higher elevations. Shrubs and expanding agricultural production are encroaching on yak pastures at lower elevations. (Photo by Katie Konchar)

4.7 A Dynamic Whole

In order to integrate and clarify interactions and responses among Himalayan climate change, alpine plants, and indigenous peoples, we have constructed a system dynamics model (Hovmand 2014). Building from the general feedback diagram (Fig. 4.1), we modeled the dynamics of this complex system of climate-vegetation-human interactions informed by our field research over the last two decades (Fig. 4.2). Our model distinguishes between short- and long-term responses to climate change. However, there is a caveat in that system dynamics models cannot deal with unremitting change (a very real prognosis for climate change) because models loop endlessly without resolution, so interactions and results are never resolved. Thus, in order to evaluate results, we are forced to impose an overly optimistic end to long-term climate change.

Results of our model are illustrated in Fig. 4.7. This model concentrates on observed responses of lower alpine woody vegetation (~3–4000 m) to both climate change and human management and on observed responses of high alpine nonwoody vegetation (~4–5000 m) to human/yak impacts. At lower elevations, the model predicts reduced pasturing and increased agroforestry and tree and crop

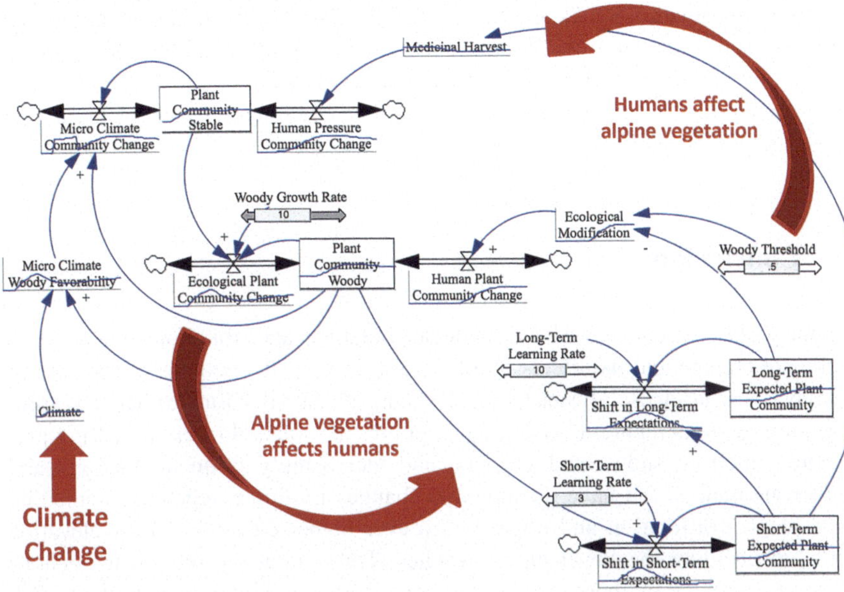

Fig. 4.7 A system dynamics model based on two decades of climate change research and locally observed changes. *Red arrows and labels* relate to the general feedback model (Fig. 4.1). *Blue arrows* illustrate connections between variables that are based on underlying calculations of the variables. *Blue lines* graph values of variables changing over time. *Boxes with black labels* are stocks; *black axes with black labels* are variables. *Black arrows with triangles* are flows. *Clear arrows with black outline* are adjustable variables. *Boxes within arrows including numbers* are assigned values of variables (in years or a 50% threshold for woody vegetation). Positive or negative relationships among variables are indicated by + or –. (Model by Ben Staver)

introductions. At high elevations, the model predicts expansion of alpine meadows and yak grazing, along with simultaneous expansion and overharvesting of medicinal herbs, followed by abandonment of medicinal harvesting because of overharvesting.

This model also suggests that the Himalaya may face severe depletion or fluctuations in native plant communities due to overharvesting of useful plants and increased agricultural activities at ever higher elevations. Human management of alpine pastures in lower elevation alpine environments either will need extensive woody vegetation control (e.g., traditional cutting and burning, now often prevented by governmental policies) to maintain alpine pastures or people will convert them to cash crops, agroforestry, and tourism (e.g., contemporary adaptations). At high elevations, yak herding and medicinal plant harvest will likely move yet higher (to 5000 masl and beyond for some medicinal plants) putting great stress on plants, animals, and people. Adventure tourism (Fig. 4.4) may also extend to these great elevations, further stressing these fragile, highly erodible environments. Adaptations based on traditional practices informed by indigenous knowledge tend to be milder and potentially more sustainable (Salick and Moseley 2012). These may build from traditional livelihood activities, such as burning and grazing alpine meadows. In contrast, contemporary adaptations include more drastic responses with the introgression of market economies (e.g., tourism, cash cropping, commercial sale of medicinal plants, and out-migration).

4.8 Conclusion

Over the past two decades, we have conducted botanical and ethnobotanical research on climate change and its effects on plants and indigenous peoples of the eastern Himalayas, including Southwest China, Bhutan, and Nepal. Climate change is similar in these regions including rising temperatures, erratic and often increasing precipitation, melting snows and glaciers, and increasingly frequent disasters and weather anomalies. We have documented changes in alpine vegetation including alpine plant proliferation and migration to ever higher elevations. Low-elevation plants are invading the lower alpine reaches. Throughout the region, indigenous peoples are stressed by these changes but are adapting and innovating both through indigenous knowledge of traditional practices and through contemporary applications of new agricultural crops and techniques and new economies such as tourism, as well as yet more drastic out-migration. When we use a system dynamics model to synthesize our data and to project into the future, there are worrying predictions including overexploitation of medicinal plant populations. Yak grazing at increasingly high elevations may stress the animals, the pastures, and the herders. Tourism may put yet further pressure on highest elevational habitats. Lower in the alpine, non-alpine plants and agriculture may invade and reduce habitat for alpine species. Indigenous peoples will be forced to adapt rapidly by both traditional and contemporary means including the use of indigenous knowledge, agricultural innovation, and development of tourist economies (which can be threatened by darkening mountains left without snow or glaciers). Abandonment of previously adaptive indigenous cultures and out-migration may be strategies of last resort.

Acknowledgements International research on this scale can only be done with the collaboration of many people, institutions, and governments. Collaborating scientists and workers (not directly responsible for this publication) include Dr. Wayne Law, Dr. Anja Byg, Dr. Nanci Ross, Katie Konchar, Fang Zhendong, and his team in the Tibetan Autonomous Prefecture, Dr. Suresh Ghimire and his graduate students in Nepal, and Sangay Dema and her team in Bhutan. In addition, we are thankful for the support of Drs. Xu Jianchu and Yang Yongping (Kunming Institute of Botany – Chinese Academy of Sciences) and Drs. Georg Grabherr and Harold Pauli (GLORIA Coordination, University of Vienna, Austria). Chinese, Bhutanese, and Nepalese government permits for this research were issued by national and prefectural governments, without which we could not have done the work. Funding was provided by National Geographic Society, The Nature Conservancy, Ford Foundation, and National Science Foundation ("Biodiversity Conservation and Sustainable Development in Southwest China" IGERT DGE#0549369) along with salaries of the authors from their home institutions. For all of these services, we are profoundly grateful.

Bibliography

Anderson DM, Salick J, Moseley B, Ou Xiaokun X-K (2005) Conserving the sacred medicine mountains: a vegetation analysis of Tibetan sacred sites in Northwest Yunnan. Biodivers Conserv 14:3065–3091

Armatas CA, Venn TJ, McBride BB, Watson AE, Carver SJ (2016) Opportunities to utilize traditional phenological knowledge to support adaptive management of social-ecological systems vulnerable to changes in climate and fire regimes. Ecol Soc 21:16

Brandt JS, Haynes MA, Kuemmerle T, Waller DM, Radeloff VC (2013) Regime shift on the roof of the world: alpine meadows converting to shrublands in the Southern Himalayas. Biol Conserv 158:116–127

Byg A, Salick J (2009) Local perspectives on a global phenomenon—climate change in Eastern Tibetan villages. Glob Environ Chang 19:156–166

Chen H, Yin KP, Wang HY, Zhong SX, Wu N, Shi FS, Zhu D et al (2011) Detecting one-hundred-year environmental changes in Western China using seven-year repeat photography. PLoS One 6:e25008

Childs G, Craig S, Beall CM, Basnyat B (2014) Depopulating the Himalayan highlands: education and outmigration from ethnically Tibetan communities of Nepal. Mt Res Dev 34:85–94

Cleland EE, Chuine I, Menzel A, Mooney HA, Schwartz MD (2007) Shifting plant phenology in response to global change. Trends Ecol Evol 22:357–365

Cook BI, Wolkovich EM, Parmesan C (2012) Divergent responses to spring and winter warming drive community level flowering trends. PNAS 109:9000–9005

Craig S (2002) Place, work, and identity between Mustang, City. Stud Nepali Hist Soc 7:355–403

Dolezal J, Dvorsky M, Kopecky M, Liancourt P, Hiiesalu I, Macek M, Altman J et al (2016) Vegetation dynamics at the upper elevational limit of vascular plants in Himalaya. Sci Rep 6:24881. https://doi.org/10.1038/srep24881

Ellwood ER, Temple SA, Primack RB, Bradley NL, Davis CC (2013) Record-breaking early flowering in the eastern United States. PLoS One 8:e53788

Fridley JD (2012) Extended leaf phenology and the autumn niche in deciduous forest invasions. Nature 485:359–362

Füssel H-M (2007) Vulnerability: a generally applicable conceptual framework for climate change research. Glob Environ Chang 17:155–167

Gallinat AS, Primack RB, Wagner DL (2015) Autumn, the neglected season in climate change research. Trends Ecol Evol 30:169–176

Gorenflo LJ, Romaine S, Mittermeier RA, Walker-Painemilla K (2012) Co-occurrence of linguistic and biological diversity in biodiversity hotspots and high biodiversity wilderness areas. PNAS 109:8032–8037

Gottfried M, Pauli H, Futschik A, Akhalkatsi M, Barančok P, Benito Alonso JL, Coldea G et al (2012) Continent-wide response of mountain vegetation to climate change. Nat Clim Chang 2:1–5

Hart R, Salick J (2017) Dynamic ecological knowledge systems amid changing place and climate: Mt. Yulong rhododendrons. J Ethnobiol 37:21–36

Hart R, Salick J (2018) Vulnerability of phenological progressions over season and elevation to climate change: Rhododendrons of Mt. Yulong. Perspect Plant Ecol Evol Syst 34:129–139

Hart R, Salick J, Ranjitkar S, Xu JC (2014) Herbarium specimens show contrasting phenological responses to Himalayan climate. PNAS 111:10615–10619

Hart R, Georgian EM, Salick J (2016) Fast and cheap in the fall: phylogenetic determinants of late flowering phenologies in Himalayan Rhododendron. Am J Bot 103:198–206

Hovmand PS (2014) Community based system dynamics. Springer, New York

Immerzeel W, van Beek L, Bierkens M (2010) Climate change will affect the Asian water towers. Science 328:1382–1387

Ingty T (2017) High mountain communities and climate change: adaptation, traditional ecological knowledge, and institutions. Clim Chang 145:41–55

IPCC (2013) Climate change 2013: the physical science basis. Cambridge University Press, Cambridge
Ives J (2005) Himalayan misconceptions and distortions: what are the facts? Himal J Sci 3:15–24
Keller F, Körner C (2003) The role of photoperiodism in alpine plant development. Arct Antarct Alp Res 35:361–368
Kelly AE, Goulden ML (2008) Rapid shifts in plant distribution with recent climate change. PNAS 105:11823–11826
Klein JA, Hopping KA, Yeh ET, Nyima Y, Boone RB, Galvin KA (2014) Unexpected climate impacts on the Tibetan Plateau: local and scientific knowledge in findings of delayed summer. Glob Environ Chang 28:141–152
Kohler T, Maselli D (2009) Mountains and climate change: from understanding to action. Geographica Bernensia, Bern
Konchar KM, Staver B, Salick J, Chapagain A, Joshi L, Karki S, Lo S, Paudel A, Subedi P, Ghimire SK (2015) Adapting in the shadow of Annapurna: a climate tipping point. J Ethnobiol 35:449–471
Lantz TC, Turner NJ (2003) Traditional phenological knowledge of aboriginal peoples in British Columbia. J Ethnobiol 23:263–286
Law W, Salick J (2005) Human induced dwarfing of Himalayan snow lotus (Saussurea laniceps (Asteraceae)). PNAS 102:10218–10220
Law W, Salick J (2006) Comparing conservation priorities for useful plants among botanists and Tibetan doctors. Biodivers Conserv 16:1747–1759
Mittal N, Mishra A, Singh R, Kumar P (2014) Assessing future changes in seasonal climatic extremes in the Ganges river basin using an ensemble of regional climate models. Clim Chang 123:273–286
Mittermeier RA, Robles GP, Hoffmann M, Pilgrim J, Brooks T, Mittermeier CG, Lamoreux J, da Fonseca GAB (2004) Hotspots revisited: earth's biologically richest and most endangered ecoregions. CEMEX, Mexico City
Morueta-Holme N, Engemann K, Sandoval-Acuña P, Jonas JD, Segnitz RM, Svenning J-C (2015) Strong upslope shifts in Chimborazo's vegetation over two centuries since Humboldt. PNAS 112:12741–12745
Mutke J, Barthlott W (2005) Patterns of vascular plant diversity at continental to global scales. Biologiske Skrifter 55:521–531
Nagoda S (2015) New discourses but same old development approaches? Climate change adaptation policies, chronic food insecurity and development interventions in northwestern Nepal. Glob Environ Chang 35:570–579
Nagoda S, Eriksen S, Hetland O (2017) What does climate change adaptation mean for humanitarian assistance? Guiding principles for policymakers and practitioners. IDS Bull 48:125–137
Nagy L, Grabherr G (2009) The biology of alpine habitats. Oxford University Press, Oxford
Parmesan C (2006) Ecological and evolutionary responses to recent climate change. Annu Rev Ecol 37:637–669
Parmesan C, Yohe G (2003) A globally coherent fingerprint of climate change impacts across natural systems. Nature 421:37–42
Pauli H, Gottfried M, Lamprecht A, Nießner S, Rumpf S, Winkler M, Steinbauer K, Grabherr G (2015) The GLORIA Field Manual: standard multi-summit approach, supplementary methods and extra approaches. http://www.gloria.ac.at/downloads/GLORIA-FIELDMANUAL_5thEd_2015_ONLINE.pdf
Pecl GT, Araujo B, Bell JD, Blanchard J, Bonebrake TC, Chen I-C, Clark TD et al (2017) Biodiversity redistribution under climate change: impacts on ecosystems and human well-being. Science 355:eaai9214
Raich JW, Schlesinger WH (1992) The global carbon dioxide flux in soil respiration and its relationship to vegetation and climate. Tellus B 44:81–99
Salick J (2013) Indigenous knowledge integrated with long term ecological monitoring: Himalaya. In: Thaman R, Lyver P, Mpande R, Perez E, Cariño J, Takeuchi K (eds) The contribution of

indigenous and local knowledge systems to IPBES: building synergies with science. UNESCO/UNU, Paris
Salick J, Byg A (2007) Indigenous peoples and climate change. Tyndall Centre, Oxford, UK. http://tinyurl.com/salickbyg2007
Salick J, Moseley R (2012) Khawa Karpo: Tibetan tradition knowledge and biodiversity conservation. Monogr Syst Bot 121:273pp
Salick J, Yang YP, Amend A (2005) Tibetan land use and change in NW Yunnan. Econ Bot 59:312–325
Salick J, Fang ZD, Byg A (2009) Tibetan Ethnobotany and climate change in the eastern Himalayas. Global Environ Change 19:147–155
Salick J, Byg A, Bauer K (2013) Contemporary Tibetan cosmology of climate change. J Study Relig Nat Cul 6:552–577
Salick J, Ghimire S, Fang ZD, Dema S, Konchar K (2014) Himalayan alpine vegetation, climate change and mitigation. In climate change special volume. J Ethnobiol 34:276–293
Salick J, Byg A, Konchar K (2018) Innovation and adaptation in Tibetan land use and agriculture coping with climate change. In: Indigenous peoples, marginalized populations and climate change. Cambridge University Press, Cambridge, UK
Salick J, ZD Fang, Hart R (2019) Rapid changes in eastern Himalayan alpine flora with climate change. Am J Bot 106:520–530
Shaoliang Y, Ning W, Peng L, Qian W, Fusun S, Geng S, Jianzhong M (2007) Changes in livestock migration patterns in a Tibetan-style agropastoral system. Mt Res Dev 27:138–145
Shrestha UB, Gautam S, Bawa KS (2012) Widespread climate change in the Himalayas and associated changes in local ecosystems. PLoS One 7:e36741
Turin M (2007) A multitude of mountain voices. Sustain Mt Dev 52:11–13
Williams JW, Jackson ST, Kutzbach JE (2007) Projected distributions of novel and disappearing climates by 2100 AD. PNAS 104:5738–5742
Wolkovich EM, Cleland EE (2014) Phenological niches and the future of invaded ecosystems with climate change. AoB Plants 6:plu013
Wolkovich EM, Cook BI, Allen JM, Crimmins TM, Betancourt JL, Travers SE, Pau S, Regetz J, Davies TJ, Kraft NJB, Ault TR, Bolmgren K, Mazer SJ, McCabe GJ, McGill BJ, Parmesan C, Salamin N, Schwartz MD, Cleland EE (2012) Warming experiments underpredict plant phenological responses to climate change. Nature 485:494–497
Xu J, Grumbine RE, Shrestha A, Eriksson M, Yang X, Wang Y, Wilkes A (2009) The melting Himalayas: cascading effects of climate change on water, biodiversity, and livelihoods. Conserv Biol 23:520–530
Yu H, Luedeling E, Xu J (2010) Winter and spring warming result in delayed spring phenology on the Tibetan plateau. PNAS 107:22151–22156
Zhang G, Zhang Y, Dong J, Xiao X (2013) Green-up dates in the Tibetan Plateau have continuously advanced from 1982 to 2011. PNAS 110:4309–4314
Zurick D, Pacheco J (2006) Illustrated atlas of the Himalaya. University of Kentucky Press, Lexington

Chapter 5
Observing "Weeds" to Understand Local Perceptions of Environmental Change in a Temperate Rural Area of Southwestern France

Anne Sourdril, Emilie Andrieu, Cécile Barnaud, Louise Clochey, and Marc Deconchat

Abstract Rural areas of the temperate European countries are affected by climate changes that are not always perceived by local communities. We focus on how local discourses on biodiversity, in our case wild flora, provide insight into what people see as changing in their environment. We conducted ethnographic research, including interviews and participant observation, on perceptions of biodiversity change in Bas-Comminges, a rural area of France where agriculture consists primarily of extensive mixed farming. Wild flora management there is shaped by traditional agricultural practices, rural and agricultural policies, and warmer temperatures and other climatic changes. We will show that (1) wild flora is seen as growing and expanding due to changes in local institutions in charge of green spaces, changes in agriculture, and warmer temperatures; (2) discourses on those impacts reveal different types of knowledge and uses of local flora; and (3) social conflicts are emerging around local flora management, and these conflicts reveal tensions between different objectives for the land within a changing community. We will demonstrate that warmer temperatures are not always linked to global climate changes by local residents and that environmental and social changes cannot be apprehended separately from climatic ones. More broadly, we want to understand how rural populations are facing and adapting to the major transformations of their land and society, and we show that conflicts can be used by different types of local residents to take back control of their land and maintain their communities. This

A. Sourdril (✉) · L. Clochey
CNRS, UMR 7533 Ladyss, Univerity of Nanterre, Nanterre, France
e-mail: asourdril@parisnanterre.fr

LTSER Zone Atelier « PYRÉNÉES GARONNE », 31320 Auzeville-Tolosane, France

E. Andrieu · C. Barnaud · M. Deconchat
DYNAFOR, Université de Toulouse, INRA, Castanet-Tolosan, France

LTSER Zone Atelier « PYRÉNÉES GARONNE », 31320 Auzeville-Tolosane, France

M. Welch-Devine et al. (eds.), *Changing Climate, Changing Worlds*, Ethnobiology, https://doi.org/10.1007/978-3-030-37312-2_5

research is part of a larger interdisciplinary and comparative program on local perceptions of environmental change funded by the French Agence Nationale de la Recherche (ANR).

Keywords Weeds and wild flora as revealers of environmental change · Social and climate change · Ethnography · Southwestern France

5.1 Introduction: "*More and More Weeds*": How to Get Rid of Weeds in the Cemetery?

Our interest in the perception of and knowledge about "weeds" as a local revealer of change began with the observation of a land-use conflict about weed management in a local community of Southwestern France. Local residents are experiencing increasing growth of wild flora which seems to be due both to climate variations and to changes in public infrastructure management. This increase has led to a land-use conflict about the origin of the problem and how to get rid of what are considered "weeds" in community-managed public spaces. The conflict, which is especially important in local cemeteries, has pitted two groups of people against one another. On one side are those who want to use herbicides; these are mainly native residents that voice arguments such as: "Anyway, why not use herbicides? Everyone there is already dead." On the opposing side are those who favor manual weeding; these are mostly newcomers who argue, "We need to respect life and biodiversity even within the cemetery."

To refer to what Mary Douglas (1966) has written about disorder, weeds appear to be mostly composed of wild flora that grow in the wrong place. The way local residents perceive weeds therefore reveals how those residents think their environment, land, and society should be. Following this conflict, local people (both natives and newcomers) managed to overcome the tensions and organize collectively to take care of what they see as a problem on communal land. We describe this action and its causes in our chapter. This study allows us to investigate various types of knowledge and uses of wild flora, as well as different interpretations of environmental issues (including climate change) and social changes (including the reconfiguration of local society).

Temperate rural areas in Europe are currently facing numerous environmental and social changes, ranging from rural exodus, territorial reform, and land-use transformations linked to new agricultural systems and the EU Common Agricultural Policies to a range of global changes that are prompting biodiversity decline (IPBES 2018). These changes and their impacts are increasingly documented on a global scale in an expert or scientific perspective that mainly focuses on the natural sciences (Fiske et al. 2014). Knowledge of their local and regional consequences is growing, but more needs to be done to understand local communities' perceptions of those impacts, both in Europe and more broadly (Crate, 2011). Social scientists endeavor to understand how communities perceive these changes and how different kinds of knowledge or cultural contexts can favor or hinder the adaptation of societies to new climate conditions (Hastrup 2016; Roncoli et al. 2002; Strauss and Orlove 2003). Anthropologists have asserted that environmental and social transformations and their impacts are generally noticed simultaneously at a local scale and that relevant and successful adaptation attempts will come from this level (Vedwan and Rhoades 2001). Describing these local ecological perceptions and knowledge of climate and environmental changes is an essential supplement to formal scientific knowledge and can improve our understanding of the effects of global changes (Crate and Nutall 2009; Roncoli et al. 2009; Shaffer 2014).

Numerous studies have been carried out in extreme areas of the world to understand how vulnerable communities are experiencing and facing such changes. Researchers have shown how the consequences of climate change cannot be understood in isolation from other changes. Joana Roque de Pinho (this book), for example, shows how Kenyan pastoralists' perceptions of climate change effects are strongly linked to transformations affecting traditional social organization and land use. Privatization of community land and tourism, combined with increasing periods of severe droughts, make herds increasingly vulnerable due to a lack of grazing areas and water. The pastoralist community is gradually disappearing, leading to other future social and land-use changes (Roque de Pinho 2019, 2020). In northern Alaska, Elizabeth Marino has shown how local environmental expertise on weather, as well as on wildlife behavior, is essential to understand the effects of climate change. She has noted that adaptation attempts, such as relocation proposals following destruction of habitat due to rising sea levels or melting permafrost, can only come from collaboration with the affected communities and that this collaboration must respect traditional forms of social organization and cultural backgrounds (Marino 2015). In temperate areas, a number of studies have examined adaptation following extreme events such as major storms linked to climate change. Camille Hochedez and Benoit Leroux (2018) show, for example, how the floods following Cyclone Xynthia in France in 2010 devastated the vineyards on the Île de Ré. They demonstrate how conflicts among farmers were overcome so that communities could deal with the consequences of the storm and help the most affected producers recover by sharing a portion of harvests, lending tools, and providing mutual aid. Our case study provides additional support for the finding that collaborative responses to climate impacts are possible, as well as recommendations for how this might be promoted.

Many studies in anthropology and human ecology have also shown that local societies do not monitor the climate only by observing the sky or meteorological conditions (Strauss and Orlove 2003). They also observe their environments in general, including changes in flora and fauna (Dounias 2009; Sourdril et al. 2017). The observation of indicators or biological markers of seasonality (such as plant phenology, behavior or migration of animals, or even biological invasions) allows people to perceive and anticipate the transformations of their environments (Berkes and Jolly 2001; Vedwan and Rhoades 2001; Veteto and Carlson 2014). This research thus shows how local ecological knowledge – including knowledge of climate and weather, but extending to broader knowledge about environmental change – is essential to understand how climate changes are actually unfolding in the field.

Climate change, especially paired with land-use and policy changes, has a direct impact on the phenology and growth of cultivated plants and wild flora at a local scale (Root et al. 2003). People within our study site experience this impact using local wild flora, including weeds, as an indicator and revealer[1] of changes. Unusual weather variations, exacerbated by a worsening shortage in the workforce needed to deal with such unusual weather patterns, put the future of farming at stake. The rural exodus, for example, leads to a decrease in the rural population and a shortage of labor for dealing with land-use issues at a local scale. Territorial reforms, such as changes in administrative divisions, notably in public services, threaten local institutions. As a result, management of the public green areas of villages is currently in jeopardy, reflecting the changes in the local community and its environment.

The place, role, and representations of wild flora in urban public spaces have been widely studied lately by geographers, anthropologists, and sociologists (Barthélémy et al. 2016; Lefebvre et al. 2013; Lizet et al. 1997; Pellegrini and Baudry 2014). Researchers investigate the place and benefits of green areas in cities for local communities' health and well-being (Carrus et al. 2015); the potential role of green areas for mitigation in places facing climate changes (Bowler et al. 2010); their role as habitat for biodiversity, contributing to the building of green networks (Blanc et al. 2012); and the process through which various stakeholders become committed to these areas (Baudry 2014). Indeed, some research shows that revaluing urban green spaces is also a way to revalue the land and community (Lizet 2010). Wild flora has been a focus of the renewal of urban nature studies for the past decade, either from the standpoint of invasive, proliferating plants and the increase of vacant land or in relation to the renaturalization of urban public spaces (Ambrosino and Andres 2008; Groth and Corijn 2005; Menozzi 2007). In rural areas in Europe, some literature in social sciences on wild flora or weeds focuses on traditional ecological knowledge and the related medicinal, artisanal, or culinary uses in the indigenous or past traditional rural communities (Molina et al. 2014; Nedecheal et al. 2007; Pieroni et al. 2002). Other studies deal with the proliferation of species encroaching on private or community agricultural land (Cavailhès and Normandin 1993). Encroachment, proliferating weeds, and uncontrolled growth of wild flora have been associated with environmental and social changes, often reflecting social disorder in the perspective both of local people and of researchers (Claeys and Sirost 2010; Dupré 2005; Luginbühl 1989). Disorder has been associated with important reorganizations of local society following major traumas (such as World War II) or social changes (Coughlan and Gragson 2016; MacDonald et al. 2000), as well as with the decline of certain agricultural systems whose faltering affects the landscape and biodiversity (Gibon et al. 2010).

Ecological and agronomic studies have highlighted the potential benefits of wild flora not only for local biodiversity but also for agriculture, notably through the increase of pollinators or crop pests' natural enemies (Alignier et al. 2014). An increasing number of studies also attempt to link the growth of some weeds to climate change (Peters

[1] We make a distinction between "indicator" and "revealer." We consider that an indicator helps people make sense of an unknown change, whereas a revealer helps people understand already known mutations of land or society. In this paper, we focus mainly on revealers.

et al. 2014). However, few studies in ecology or agronomy have studied links between social changes, weeds, and their combined impacts on wildlife.[2] Understanding how local people connect wild flora growth and phenology to climatic or seasonal variations is a useful addition for developing an in-depth and integrative understanding of these changes and the way they affect local ecosystems and communities.

In this chapter, we illustrate how observing people dealing with weeds and wild flora in public spaces can reveal local perceptions of the changes affecting their communities and the attempts to adapt to these changes. Another question we want to ask is what is considered a weed: how do some wild flora species come to be regarded as weeds in certain places, and how can weeds become indicators of change? More generally, we ask how everyday biodiversity is viewed by local communities as an indicator of change, how are local diagnostics of environmental changes built at a local scale, and how can these indicators and diagnostics help local communities adapt to changes – notably to climate change.

We begin this chapter by describing the context and methodology of our study, then we look at what constitutes a "weed" for local residents, how weeds and wild flora are revealers of particular social and ecological changes, and how these changes are seen as interconnected. We will then demonstrate that knowledge competition is at stake in the weed conflict and that the conflict is used by both native residents and newcomers to regain control of the land and maintain their communities.

5.2 Study Area and Methods: Ethnography of Rain, Weeds, and Rural Exodus in a House-Centered Society in Rural Southwestern France

This research is part of an interdisciplinary and comparative program (ANR Piaf) to understand how local stakeholders in four different countries, and along a gradient of urban, protected, and rural field sites, perceive and make sense of socioecological changes through observation of their everyday biodiversity (Sourdril et al. 2017).[3] The site we focus on is located in Southwestern France, in a rural area called Bas-Comminges,

[2] But see the innovative paper by Gaüzere et al. whose research endeavors to link the impacts of climate and land-use changes to dynamics not of flora but avifauna (Gaüzère et al. 2019).

[3] PIAF (Programme interdisciplinaire sur les Indicateurs Autochtones de la Flore et de la faune) is a project funded by the ANR Young Researcher Program #ANR-13-JSH1–0005-01 from 2014 to 2018: http://www.anr-piaf.org. PIAF brings together a research team from six research institutes and ten research laboratories and is coordinated by Anne Sourdril (CNRS, UMR7533 Ladyss). The team is composed of (by laboratory and alphabetical order) A. Sourdril, Émilie Andrieu, Cécile Barnaud, Marc Deconchat, Wilfried Heinz, and Sylvie Ladet (Inra UMR1201 Dynafor); Nadia Bélaïdi (MNHN UMR7206 Éco-anthropologie et Ethnobiologie); Éric Garine, Émilie Guitard, and Jean Wencélius (Université Paris Nanterre, UMR7206 LESC); Christine Raimond (CNRS, UMR8586 Prodig); Michel de Garine-Wichatitsky (Cirad); Sylvain Aoudou Doua (Université de N'Gaoundéré, Cameroon); Brian J. Burke (Appalachian State University, USA); Ted Gragson, Meredith L. Welch-Devine (University of Georgia, USA); and Hervé Fritz and Chloé Guerbois (CNRS, UMR5558 LBBE).

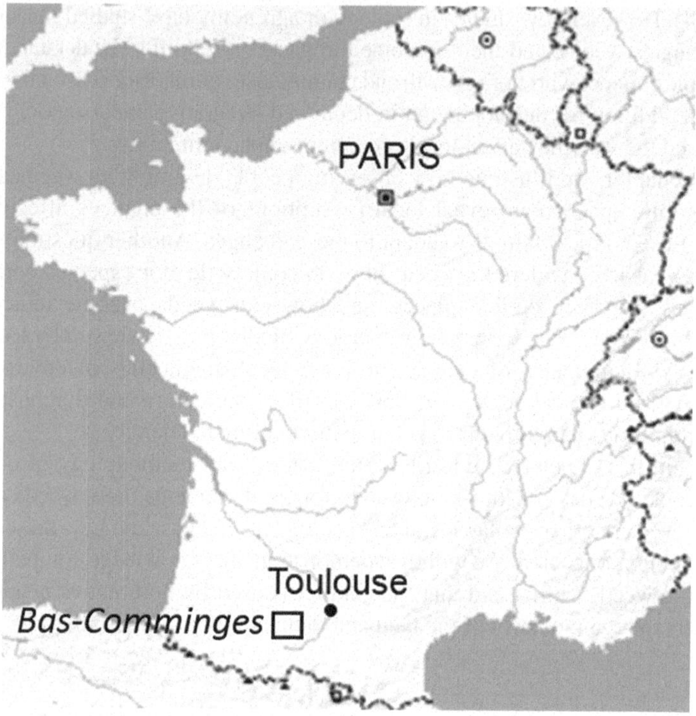

Fig. 5.1 Map of the studied site

within the Zone Atelier Pyrénées-Gascogne (ZA PYGAR),[4] approximately 80 km southwest of Toulouse (see Map/Fig. 5.1). The area and the dynamics of its socioecological systems are well-known and have been studied by ecologists and social scientists since the 1980s (Deconchat et al. 2007; Sourdril et al. 2012). The whole area is characterized by the presence of a specific social system known as the "house-centered system," which is based on a social entity, "the house."[5] House societies are characterized by a real-estate holding transmission system with a single heir who is also successor to

[4]The Zone Ateliers are CNRS platforms for long-term research on environmental issues. There are 14 official ZAs. The ZA Pygar was officially recognized in 2017, but long-term and interdisciplinary research has been conducted in the area since the 1980s; the ZA Pygar is specialized in the study of spatial and temporal dynamics of socioecological systems. See: http://www.za-inee.org/.

[5]"House" is defined as a "moral person, keeper of a domain composed altogether of material and immaterial property, which perpetuates itself by the transmission of its name, of its fortune and of its titles in a real or fictive line held as legitimate on the sole condition that this continuity can express itself in the language of kinship or of alliance, and most often, of both together" (Lévi-Strauss 1979, translated by Gillespie 2007: 33). House-centered approaches are used by anthropologists in kinship studies throughout the world, e.g., France, Algeria, Indonesia, and South America, to look at the reproduction of social organizations as domestic groups or local societies (Augustins 1989; Cunningham 1964; Bourdieu 1973; Rogers 1991). Researchers are trying to renew house-centered approaches by looking at the house in its material and spatial dimensions (Carsten and Hugh-Jones 1995; Gillespie 2000; Hamberger 2010).

the house as an agricultural undertaking. This leads to stability of social organization, of real-estate and farm holdings, and of the landscape in general. While "houses" reproduce, relations between neighboring "houses" are also perpetuated and are generally expressed in the form of strong mutual aid (for agricultural work, daily help, or during ritual events such as burials). The system has been threatened for a few decades by major social and agricultural changes affecting the area, such as the rural exodus and the arrival of new inhabitants from Northern Europe. All of this has an impact on the transmission model, the shape of farms, and also the neighborhood networks and the community itself. We will see in this chapter that the weed situation reveals the local community's fear of seeing its social system threatened by environmental and social changes.

The Bas-Comminges area is a hilly region with a temperate climate with oceanic and slight Mediterranean influences. Winters are moderately cold, summers are hot, springs are rainy, and autumns are quite dry and sunny. The whole chain of the Pyrenees spreads out on the horizon and forms the background of the hills. Hills, valleys, rivers, and streams, together with hedges and small private forests, form a fragmented landscape. The region is not densely populated (23 inhabitants/km^2 according to the general census of 2010) and is still largely agricultural despite a significant rural exodus. Many farm units are abandoned following the retirement of older farmers. Their lands are being rented by the remaining farmers. At the township level (*canton* in French), farms decreased from 460 in 1979 to 650 in 2010, while the number of farmers declined from 225 to 200 in those same years (AGRESTE data). Farms – based on a mixed farming system – are mostly tended by a "domestic" unit. In the past, this was typically the farmer, his wife, and his retired parents; today it is mostly only the farmer and his retired father since the women generally have work outside the farms.

Farmers have traditionally had a multicrop-livestock breeding model (veal calves from grazing beef breeds). Most of the harvest of wheat, corn, and sorghum was kept as feed for the cattle with a view to farm self-sufficiency, but a portion of these crops is now sold to the local farmers' cooperative. The area has some dairy and free-range chicken farms, with the latter being tended by the few remaining women farmers (mostly the wives of dairy farmers). Residences are generally dispersed, with houses and barns built on hilltops along roads that descend to the valleys; small villages are also usually built on hilltops and are home to the local institutions of the villages (town council, schools, and shops), which are also declining.

At the same time, the arrival of new populations has been changing the social composition of local society. In the late 1990s, a high number of retirees or people with multiple jobs from Northern Europe (United Kingdom, Netherlands, Germany) settled in the area, and in the 2010s, active newcomers from nearby urban areas were building houses on lots recently opened for development. The population has increased since the late 1990s after a constant decline since the middle of the twentieth century (at the scale of the township, the population was 4158 in 1975, 3906 in 1990, and 4334 in 2012, according to the national census).

Following the overall decline in France's rural population, a national "réforme territoriale," voted into law in 2014, impacts the communal and political institutions in charge of village upkeep. In our study area, the villages were grouped together into a "Communauté de Communes"[6] that shares governmental services, notably a green area management team that takes care of public land. The latest reforms and notably the NOTRE law[7] have broadened the limits of this region's Communauté de Communes from 19 villages in 2013 (and 4355 inhabitants) to 104 (and more than 44,000 inhabitants) in 2015. This has reorganized the local institutions and the sharing of some services, among them the management of village green areas. At the same time, a "zero phyto" law was passed at the national level to ban the use of pesticides in public green areas in villages and towns (including sidewalks and green areas around schools or parks). However, the use of such products can be permitted in spaces that are not reserved for walking or hiking, such as stadiums or cemeteries. All these reforms have an impact on the management of village green areas.

This chapter is based on ethnographic work carried out in 2015 in three villages of Bas-Comminges within the ANR-Piaf framework. A common research protocol was built and applied to the nine field sites of the project, based on a common sampling strategy and similar interview guidelines, transect walks, and free list elicitation methods. This shared protocol formalized both data collection (sampling and interview guidelines) and data management (timelines, tables to be filled). The inquiries in Bas-Comminges were compared with long-term investigations by social scientists in the area since 2003 to understand how changes are affecting the local landscape and social composition (Sourdril 2008). Investigations in 2015 consisted of long-term ethnography, participant observation during collective work on public space, and semi-directed interviews on land-use and socio-environmental changes with 60 people: 40 native residents (i.e., born there, ranging in age from 25 to 80 years old) and 20 newcomers (having arrived in the 2000s, ranging in age from 50 to 65 years old). Moreover, following the 2015 investigation, additional interviews and participant observations were carried out in 2016 and 2017.

While the ANR PIAF project broadly considered various kinds of biodiversity indicators (mammals, birds, trees, and plants), as revealers of different social and environmental changes, our specific interest in weeds emerged from the fieldwork conducted in this rural French study site. In this site, people talked about the changing phenology or behavior of trees, birds, or mammals, but the way they talked about "weeds" made it a particularly salient and unexpected revealer of both environmental change (especially, indirectly, climate change) and social change. It revealed in particular a conflict between native and new residents and the existence of different representations and links between observed changes that we did not anticipate. The generic term "weeds" we are using here ("mauvaises herbes" in French) comes from the interviews, as people referred to certain species of wild flora using this term. We will refine the identified species in the next part of this chapter.

[6]A *Communauté des Communes* is a federation of communes sharing a framework within which local tasks are carried out together. It is "a public institution of intercommunal cooperation bringing together several municipalities in one piece and without enclaves. Its purpose is to associate municipalities within a solidarity area, with a view to developing a common project for development and spatial planning" (Article L 5214–1 of the General Code of Territorial Communities).

[7]Loi pour la Nouvelle Organisation Territoriale de la République, adopted in 2015, in order to reduce the number of intercommunal institutions and to make them more efficient and economical.

5.3 Results and Discussion – When Weed Management Becomes an Issue in a Community Facing Climate and Social Changes: How to Deal with a Growing Problem with Fewer People?

When interviewed about changes in the community and its immediate environment, an important concern for the local people in 2015 was to keep the village cleared of weeds: “The village is no longer taken care of. It’s like a jungle everywhere and it’s kind of dirty in the cemetery, whereas you want the cemetery to be clean, especially for those who come to visit the dead. Weeds grow faster here than ever, and there is nobody anymore to take care of them” (55-year-old woman, native resident, shopkeeper) (Fig. 5.2a and b: pictures of the cemetery and green sidewalks). Before explaining how weeds are managed by the local community and how they reveal

Fig. 5.2 (**a**) Dirty cemetery: weeds as revealer of social, land-use, and climate changes

Fig. 5.2 (**b**) Greening, weeding, or spraying the sidewalks?

perceptions of both environmental and social changes, we need to clarify what plants the local residents consider to be weeds.

5.4 What Is a Weed? Knowing and Using Wild Flora: Archaic or Valued Knowledge?

There is a general consensus about what a "weed" is among the different types of informants (newcomers or native residents), but as we are going to see in this section, there are differences in the *types of knowledge* associated with weeds and disagreement about their *potential uses and management*. As a farmer told us: "a weed, I mean a grass that we are going to consider a weed on a sidewalk, like a rye-grass, is not a weed in our meadow. On the contrary, it's a very good

Name in French	English translation	Latin name	Who mentioned it as a weed?	Where is it typically found?	Current uses
Ambroise	Ragweed	*Ambrosia artemisiifolia* L.	Native residents/ farmers	Fields, meadows, trails, pits	–
Ronces	Blackberry bush	*Rubus* sp. .	Native residents/ newcomers	Meadows, encroached areas, trails, ditches	Fruits, jam, pies, liquors
Brome	Bromegrass	*Bromus* sp.	Native residents/ farmers	Fields, meadows	–
Pâquerettes	Daisies	*Bellis perennis* L.	Native residents/ newcomers	Trails, ditches, pathways, cemetery	Ornamental
Pissenlit	Dandelions	*Taraxacum* sp.	Native residents/ newcomers	Trails, ditches, pathways, cemetery	Salad, jam
Datura	Datura	*Datura stramonium* L.	Native residents/ farmers	Fields, trails, ditches	–
Herbes	Grasses	Poaceae	Native residents/ newcomers	Fields, meadows, trails, ditches, encroached areas, pathways	–
Orties	Nettles	*Urtica dioica* L.	Native residents/ newcomers	Meadows, trails, ditches, pathways, cemetery	Soup
Ray-grass	Ryegrass	*Lolium perenne* L.	Native residents/ farmers	Fields, meadows, trails, pits, encroached areas, pathways, cemetery	–
Chardons	Thistles	*Carduus* sp., *Cirsium* sp., *Silybum marianum* (L.) Gaertn., *Eryngium campestre* L.	Native residents/ newcomers	Fields, meadows, trails, ditches, encroached areas, pathways, cemetery	–
Folle avoine	Wild oats	*Avena* sp.	Native residents	Fields, meadows, trails, ditches, encroached areas, pathways, tree beds, cemetery	–
Sauge sauvage	Wild sage	*Salvia pratensis* L.	Native residents/ newcomers	Trails, ditches, encroached areas, pathways, tree beds, cemetery	Ornamental
Soucis	Marigold	*Calendula officinalis* L.	Native residents/ newcomers	Pathways, tree beds, cemetery	Ornamental
Xanthium	Xanthium	*Xanthium strumarium* L.	Native residents/ farmers	Fields, meadows, trails, ditches	–

Fig. 5.3 Table of plants mentioned as weeds during the interviews

grass for what we want to use it for, but on a sidewalk it's dirty" (50-year-old man, native resident, farmer). A weed is a weed when it grows in the wrong place. The general definition of a weed is common among informants: weeds are usually defined as a group of species sharing common characteristics such as long roots, rapid growth, high dispersal capacities, no conspicuous flowers or thorns, and sometimes toxic to humans or cattle. "Weeds, usually, are plants that grow fast, it's hard to pull them up and usually they scratch, you just want to get rid of them!" (66-year-old man, native resident, retiree). "Weeds are plants that are not pretty, you can have pretty wild plants with flowers that you want to keep on the sidewalk, but nettles, for example, are not pretty" (55-year-old woman, newcomer, retiree). The word "weeds" remains general, most of the residents use the terms herbs or grasses ("herbes") to refer to an undetermined mix of weeds along a pathway or a meadow, and more than half of the informants were unable to distinguish more than a few species. They generally mentioned common species of wild flora, such as nettles, wild oats, dandelions, thistles, or blackberry bushes. Native residents and newcomers have basic knowledge of the names of weed species and are only familiar with some generic terms (Fig. 5.3: table of the mentioned weeds).

We nevertheless note some specific features among the plants that native residents and newcomers see as weeds, knowledge about them, and how they are used.

Among the native residents, the older ones know some vernacular terms in local patois associated with local wild flora and past uses. "Nettles, we called them *ortiga* in patois, my mother used to make soup with the young sprouts, and when you have *ortiga* in a field, you know that the soil is very rich" (85-year-old man, native resident, retired farmer). The current native farmers, as well as the retired ones, are familiar with wild plants in the natural meadows and fields. They mentioned some new exotic species such as datura, xanthium, or ragweed that can be dangerous for cattle or for humans: "We have had unwanted plants with the increase in corn crops, they come with the seeds or with the reapers. There is xanthium, datura, and ragweed coming with the sunflowers, coming from elsewhere, and it can grow in a field, but it is toxic and it can kill a cow, so we have to weed them. [As for] ambrosia, some people are allergic to it" (55-year-old man, native resident, farmer).

These unwanted plants can be found everywhere in the fields, as they are "transported by the wind, the rain, or the birds, they re-seed themselves everywhere" (same informant). Farmers say they "know the names of plants thanks to their agricultural advisers" (50-year-old woman, native resident, farmer). "The weeds are becoming resistant to the products we used to use, and there are new ones coming with the seeds from the cooperative, so it's difficult.... It is necessary to know how to recognize the weeds and grasses to know what to do with them. The technician of the farmers' cooperative, when he comes, he knows them. When he comes, we go around in the fields so we learn them. Then we can buy the right products.... They sell us [products for] what we have to fight, so we must know what it is" (38-year-old man, native resident, farmer). Nowadays, native residents' knowledge about weeds is a mix of local ecological knowledge related to past uses of wild flora and expert knowledge emphasizing the destruction of weeds.

For native residents, some wild flora species are acceptable in the public space "when they are pretty and add something to the beauty of the village" (45-year-old man, native resident, employee) and when they remind the population of a past when the community was vibrant, such as species of wild sage or wildflowers – daisies, poppies, or cornflowers. These species are not eradicated and are even appreciated at the foot of trees or in flowerbeds. They are sometimes seen as signs of the revival of the land: "We no longer have cornflowers in the crop fields, while 50 years ago they were everywhere with butterflies, but with the pesticides and herbicides, everything has disappeared. Now it's good to see some growing again in the ditches" (52-year-old woman, native resident, employee).

Newcomers speak of weeds but mainly use the term "wild flora," following the spread of environmentally friendly discourses on biodiversity. Most of the newcomers have an aesthetic view of the wild flora, without naming the species precisely: "We like the green spaces, the wild flora, the little flowers in the tree beds. Maybe it's because we are English, but it would be beautiful to have green pathways and cemeteries. Here people are not used to that [...] green pathways, you have to get used to them, but in the end it's less work and fewer products, so it would be beneficial for the village" (55-year-old woman, newcomer, retired). A few newcomers come to the area to farm or garden and have great interest in wild flora. Some practice organic farming and permaculture. They have specific botanical knowledge, mentioning Latin names, orders, and species, associating the species together, and presenting this knowledge as essential to understand how the ecosystem works. Newcomers have acquired this knowledge during their formal education or in situ observations in the field and are actually familiar with and making use of the weeds: "In my garden, I let what they call 'weeds' grow. For me they are not weeds, they are just part of the nature from here, even if I weed sometimes when my vegetables are covered... But you can do a lot with wild flora, and it's protecting the balance of the flora and the soil. In the area, we have medicinal plants like marshmallow or willow, a lot of aromatics which can be medicinal too, horsetail for the richness of the soil. I gather plants in the nearby hedges or in the forest, you just need to know which plants to gather, and you can do a lot with wild plants" (40-year-old man, newcomer, organic farmer).

Many newcomers consider this (ethno)botanical knowledge to be crucial for ensuring the future of the environment, and they want to promote it through everyday uses, changes in the management of weeds, community workshops, and botanical hikes. "You need to be aware of what is surrounding you, the environment, and that you are part of everything. When you see someone spraying Roundup next to the school, it's shocking. You should teach the kids the names of the plants and how they can use them, not how they can destroy them. There is this guy who is doing botanical hikes, and the schoolteacher here is really into this stuff, and that's good. With the village Beautification Committee's actions… that's what we want to tell people, too: respect the environment" (55-year-old woman, newcomer, employee).

All in all, knowledge about weeds and wild flora was more specific, even if it was not based on the same classifications, for native farmers, older native people, and the ecologically friendly newcomers who might be practicing organic farming or permaculture.

Uses of weeds and wild flora are also changing, showing breaks and gaps relating to ecological knowledge among the residents. Newcomers favor the use of wild flora and "weeds," gathering and using nettles to make soup or natural fertilizer, dandelions for salads, and blackberries for jams, whereas those uses have mostly disappeared among native residents. Newcomers also favor manual weeding in vegetable plots or public areas, while most native residents chose to use herbicides.[8] Several weeds and wild flora species were used for consumption or agricultural matters until the late 1970s by the native residents, notably by the older generations. Those uses and even the related ethnobotanical knowledge were depreciated in favor of technical knowledge fostered through the modernization of agriculture. "My parents used to gather nettles to make soup or dandelions for salad, but we did not do that, we did not have time, and we had everything in the vegetable plots. We no longer needed to gather wild plants" (60-year-old woman, native resident, employee).

This knowledge of wild flora was gradually lost by the younger generations of farmers or local inhabitants, and the current younger residents are no longer able to name wild flora. "My grandparents could name a lot more plants than my siblings or I. They knew. We don't. They never taught us. They thought it was better for us to learn our lessons in school" (55-year-old man, native resident, farmer). For the older residents (born in the 1930s and 1940s), naming and using wild flora was not useful for the younger generations: "Why bother them with weeds? We had Roundup and they had other things to deal with" (82-year-old woman, native resident, retired farmer). These were considered old-fashioned and useless skills, whereas the older generations wanted their descendants to learn technical tools to "do better than they had" (same woman), either by getting a job in the cities and leaving the countryside, which they considered to be secluded, or by acquiring new skills to improve the productivity of the farms. Nowadays, among the native population, it seems that there has been a shift from knowing species for their uses to knowing them so that they can be eradicated. This is not the view promoted by the newcomers, and this situation led to the conflict within the community as described above. In the next section, we will see how the community is currently succeeding in overcoming the tensions to work collectively to tend for the land and take care of the community.

---

[8] Newcomers are also gathering fruit from wild fruit trees in hedges or on woodlots, such as nuts, apples, or quinces, when those trees were no longer used by the local residents. Native residents were gathering and using fruit in the hedges until the 1970s–1980s, but with the intensification of the agricultural work and the lifting of the hedges, those uses have been abandoned during the last decades. Native residents remain really interested in mushroom gathering (Sourdril et al. 2012; Blanco et al. 2018).

5.5 Managing Weeds to Adapt to Changes, to Be Accepted Within the Community, and to Revive Old Institutions?

5.5.1 An Emerging Conflict: A Matter of Growing Weeds, Perceptions of Changes and of Nature?

Different ideas about what counts as useful knowledge and what is the proper way to deal with weeds led to conflicts among the people in our study site. In 2016, an incident occurred between two residents of the community (a newcomer and a native resident) related to the management of weeds in public spaces. This incident reveals a broader disagreement about the management of land and different perceptions of the environment, and it was the impetus for creating the village Beautification Committee, as described in the next paragraph. The newcomer told us his side of the story: "He was spraying Roundup or something similar on the sidewalks. He was not even wearing anything to protect himself. I came up to him to tell him this was really dangerous for him and for everybody, those products go into the soil and the water we drink, this has to stop!" The native resident described the event in different terms: "I was taking care of the village and he came up to me saying I should not use these products. What should I do? I am the only one taking care of the village right now, and I have been doing that for years now, spraying, weeding, planting, or watering the plants. We are trying organic products to deal with the weeds, but they are very expensive, and I do not know if the village can afford that and we don't even know if those products are really working."

At stake is the legitimacy of knowledge and practices. Native residents have been contributing to the management of the land for generations, and they claim to be responsible for the landscape and the diversity of the environment. They cannot stand criticism from newcomers, who they see as lacking the legitimacy to manage the land or even to be there. "They are criticizing us because we use pesticides to kill the weeds, but what else can we do? They should come and try to weed our fields or borders! The researchers who are working here are doing ecological stuff and are saying that the environment is very rich here…. And yet they [the newcomers] are coming for the beauty of the landscape and nature, but who made this landscape if not us farmers?" (50-year-old woman, native resident, farmer).

The "weed situation" shows a gap between different types of perceptions of nature that must now coexist due to the reorganization of local society. What is common between all the residents is the diagnostic of the growing weeds. When we asked why the weeds were growing faster, most people mentioned several causes. "[As for the] weeds, we had a really wet season and then it was hotter than usual, and plants – especially weeds – like that, and they were everywhere after that and there is nobody to take care of them. The Communauté des Communes comes less often to weed than before" (40-year-old man, native resident, manager). Public areas were taken care of mainly by the Communauté des Communes and by a municipal employee, but as the Communauté des Communes grew, the availability of the public and green area management team decreased. The employee retired in 2006 and was not replaced. This created a feeling of relative abandonment of the public land, reinforced by meteorological conditions, which were gradually considered to be more favorable for wild flora growth. The "weed situation" tells us about different types of changes that people think may have had an impact: changes in weather, as well as changes in the social composition of the local community, and changes in land-use practices in farming and public lands. Taking care of the weeds collectively now appears to be an attempt to adapt to these changes all together.

5.5.2 *From Conflict to Adaptation: Reorganizing Local Society to Cope with Environmental and Social Changes*

The weed situation in the village we studied created tensions, conflicts, and then an adaptation attempt at the community level with the creation of a local work group or committee in charge of the management of the village public land. In order to deal with the spread of weeds and to overcome the lack of workforce at the institutional level, some people at the Communauté des Communes level in the surrounding villages – mainly residents who are part of the town councils – had organized themselves into informal groups to maintain community land: "I know that in [the neighboring village] they are doing the weeding in the cemetery twice a year. It's a group of 5–6 people from the town council weeding and spreading herbicides, it's quite informal. Here in Lussignac, we wanted to make things more official, even if it is official only in the village, but we created what we called the 'village Beautification Committee' with the agreement of the town council" (55-year-old woman, newcomer, retiree).

What makes this movement unique is that the committee was created at the initiative of a newcomer, who arrived in the area in the early 2000s, in order to overcome the conflicts arising within the community. The committee has four members: two native residents and two newcomers from the town council and outside. "We were the ones initiating the committee, but we wanted to make sure to have people originally from the village and young people as well – dynamic young people that are the future of the village… and also farmers and children of local farmers from the old families to get them involved as well. With the future threat to local institutions, we need to create alternative institutions to take over and maintain an idea of the community" (same woman). The idea was mainly to take care of public spaces but also to find a way to revive the community and to "gather people together, clean up some dirty spots, make the village more cheerful, and favor future recreational activities. For example, if we take care of the *pétanque* field, we could play again, like a few years ago, to restart the Friday evening events there and even organize barbecues during the summer" (55-year-old man, native resident, farmer).

The name of the committee was chosen to convey a positive spirit and not to frighten people with the work to be done: "We thought about 'Cleaning the Village' or 'Clean Village' for the name of the committee, but it did not sound right and people do not want to clean or weed, so we chose 'beautification' because it's more positive and it can bring people together" (same woman). Since 2016, the Beautification Committee has organized a "citizens' afternoon" every month during the spring and autumn. Specific tasks are decided on in advance with the community members, for instance, weeding the cemetery, planting fruit trees and ornamental plants in front of the town hall, or pruning hedges in front of the public school. The afternoon always ends with a get-together to share a drink, talk about the tasks achieved and the future works to be done, or the life of the village and the community itself (see Fig. 5.4, description of the third citizens' day, and Fig. 5.5, citizens' day photos).

The first citizens' days were a great success, leading to massive participation and a lot of work done (see Fig. 5.6: article in a local newspaper). The following ones attracted fewer people but remained efficient, and people were loyal to the cause. All generations, origins, and genders were represented. These citizens' days are moments of knowledge sharing around uses of public land and wild flora, land management, and changes occurring within the community and its environment. For example, "We always share a drink and a snack at the end of the day. Everybody brings something. It's a time to meet people and it's a nice time. We talk about the future work to be done for the next days and the news of the village" (65-year-old man, native resident, retired worker). Agreements and disagreements arise between the community members, contributing to the dynamism of the exchanges within the community. A newcomer told us: "During the days of the Beautification Committee, we work together, sometimes we fight together [laughter], people do not agree on what to do with the weeds, for example. Some just want to spray Roundup everywhere because it's easier and it lasts longer, whereas with manual weeding, you have to come back every 2–3 weeks or so, especially during the period when they grow so fast, and yes you can't come that often... Yes, there were some disagreements between us, but they tried organic products instead of Roundup the other day, so it's good to talk anyway, it makes people think and try some new things. I hope the mentality will change" (40-year-old man, newcomer, unemployed). At the same time, a native resident was saying: "They want us to weed, [but] we don't have time for that. Roundup is bad, they say. What else should we do? [...] He [the newcomer] wanted to weed and to put the weeds in the compost. You can't do that, you have to burn them, otherwise they will spread everywhere" (70-year-old man, native resident, retired farmer).

It is interesting to note how newcomers and native residents describe these days differently and use them for different types of social projects. Newcomers and some articles in the local newspapers refer to them as "citizens' days," whereas native residents refer to "mutual aid days." For newcomers, they are an attempt to build an integrative community where they have their place and legitimacy. For the native residents, these days are used as a means to maintain local institutions and to revive traditional social networks. The fact that most of the village neighborhoods were

At 10 am on a Saturday of April 2016, members of the Beautification Committeemet to organize the day's operations. The objectives of the day were to clean the cemetery and the pétanque field, to prepare the land for a garden at the entrance of the village, and to weed some sidewalks. No one knew for sure who would come and what they would achieve. It would depend on people's mood, but also their work obligations, notably for the farmers who were dependent on the weather to do their own work in the fields. As we left the meeting with the committee's chairperson, we met a woman in the village who told us "I'm not sure if I will come in the afternoon because I cannot stand weeding and I have enough work to do in my own garden. You should just spray and that would do!" After this encounter, the chairperson told me: "We have to be careful if we want people to participate. We don't want to force people to do what they don't want to, such as weeding as they hate that, so we let people decide what they want to do as well. They choose their tasks. Some activities are viewed favorably, such as pruning the hedges, because it's technical and it involves tools and it's work that you notice and that lasts, whereas weeding is an unappreciated task."

At 2 pm the same day, around 20 people (men, women, children, long-time residents, and newcomers) met in front of the cemetery and split into several groups. Every neighborhood was represented, as in the old days for the collective agricultural work. Not many farmers were present as the weather was good for working in the fields. Members of the committee seemed satisfied with the crowd. Men brought tools such as chainsaws, spades, pitchforks, and trailers to haul the weeds away. Native resident men worked mostly at getting the soil ready for the garden or pruning trees in the cemetery. Weeding seemed to be a woman's job in the cemetery. Discussions arise between native residents and newcomers about what to weed, about some wild flora that could be left next to the stone wall: "This little sage plant, you could let it grow, it will make nice little flowers and could maintain the stone wall." However, some native residents were "not ready to see wild plants invade the cemetery, which needs to be clean for visits by people from outside the village who have their dead here." A newcomer brought a trailer and put all the weeds in it to put in his compost pile at home, whereas a native resident said they should burn everything so as not to spread the weeds. In the end, the weeds were left on the side of the road because the green area management team from the Communauté des Communes was supposed to come to clean some of the public spaces. As a native resident said, "We can leave the weeds there. The Communauté des Communes will pick them up and take care of them. It's the least they can do now that we have to do their job." Then the residents went to the pétanque field to weed it.

At 4 pm, the weather had deteriorated and rain was coming. People encouraged the others to work fast in the pétanque field. People were laughing and children playing. Weeds have strong roots and are hard to pull out, and when they come out, they leave large holes in the ground, which have to be filled with dirt again "if we want to ever play here again." The rain finally came and only half the field had been cleaned: "We'll finish next time." People gathered under the nearby elementary school porch with drinks, homemade cakes, and snacks. They talked about the heavy rain, which was unusual at that time of the year, the weeds that would grow faster, about the temperatures that were hotter and climate change that was happening (or not?), about the cemetery that should be expanded to take over the adjoining pétanque field one day, and the fact that it could be good to revive the Friday evening games that had lasted until around 2005. Conversations continued about the future of the village school that had been saved thanks to the mayor's efforts, and fears for the institutional future of the village: "Are we going to lose the town hall after the territorial reforms? What will become of the village?" The community's future was at stake.

Fig. 5.4 Description of the third citizens' day – spring 2016 (fieldwork notes)

Fig. 5.5 Working together to magnify the communal land (and strengthen the local society)

represented during the citizens' days is highly significant of the social organization and system. The native residents are using these moments to revive traditional neighbor networks and a mutual aid organization based on the traditional house-centered social system. "These citizens' days are a good thing because it gives us an opportunity to meet the other people of the village. Otherwise we never see them. You can get news from them and do something for the village... The other day, most of the neighborhoods were represented at the citizens' day, just as during the mutual aid in the old days" (23-year-old woman, native resident, student).

"In the old days," social organization, mutual aid, and sociability within this area were based on house-centered neighborhood networks that are currently threatened by the rural exodus and arrival of a new population that is unaware of the social system and its obligations. Mutual aid, notably for traditional agricultural work or institutional obligations with the town council, was organized at the neighborhood or village levels, involving specific houses. For some collective agricultural work, it was said that every neighborhood had to be represented by

Publié le 18/03/2016 à 03:49, Mis à jour le 18/03/2016 à 09:01

Embellissement : 2e journée plantations

Partager Tweeter G+ Partager Commenter

En octobre dernier, près de quarante personnes étaient venues participer à la 1re journée consacrée à la taille des végétaux./Photo DDM

Samedi, une journée plantations est organisée à En effet, la commission pour l'entretien, l'amélioration du fleurissement et la décoration du village propose de nouvelles journées pour continuer à rendre encore plus joli le village. «Cette première journée du 10 octobre, consacrée à la taille, nous a vraiment donnés à vivre un très bon moment de partage et nous en remercions encore les habitants», souligne l'un des membres de la commission, qui poursuit : «Nous poursuivrons aussi notre démarche zéro phyto en pratiquant le désherbage manuel et en paillant pour éviter une repousse rapide et maintenir l'humidité. Pour ce faire, nous avons la possibilité d'avoir de copeaux gratuitement».

À l'occasion de cette deuxième journée pour l'embellissement du village, la commission propose aux bonnes volontés de participer à la plantation des bacs, tout récemment installés derrière la mairie. Il y aussi des arbres à planter, mis gracieusement à disposition. «Et pour «varier les plaisirs», nous irons désherber un peu le cimetière, car cet endroit mérite aussi notre attention» précise la commission. Les personnes qui ont des plantes à diviser, des graines en trop, ou des boutures faites suite à la taille du mois d'octobre, peuvent les apporter, le village sera ainsi embelli avec la collaboration des habitants. Rendez-vous samedi à 14 heures à la mairie avec des gants et des outils de jardinage genre griffe, binette, petite pelle à main…

À l'issue de ce moment de jardinage, tous partageront un autre moment convivial autour d'un verre et de grignotages. Après un temps nécessaire à recenser les endroits du village pouvant être embellis par des plantations, la commission prévoit d'autres journées, mensuelles, de mars à juin, au cours desquelles il sera question de bouturer, semer et planter. La prochaine journée, prévue en avril sera dédiée à la continuation des plantations et du désherbage.

Fig. 5.6 The embellishment commission's work within the local newspaper

one house/farm. It seems that the mutual aid days organized by the Beautification Committee work on the same basis, and it is interesting that these networks were mentioned by a young woman (previous quote) who has integrated these traditional obligations even during a period, the last few decades, when they were gradually threatened and lost. In the nineteenth and early twentieth century, the village's population consisted mainly of multi-generational farm families who could stay on the same property for centuries and of sharecroppers who could move from farm to farm or village to village every 7 years. The area welcomed a lot of newcomers over the past, a process that made local society accustomed and quite open to integrating new populations within its social organization. What we want to point out here is that adaptation attempts were successful, thanks to the cultural background and history of this region, and when they were directly linked to the specific social system there.

5.6 What the "Weed Situation" Tells Us About Climate Change

People have noticed unusual meteorological or weather variations over the past two decades. They identify a major "first" event, a serious drought in 2003, which challenged farmers across France to find hay and good grasses for their cattle. Thanks to the solidarity and self-sufficiency of the farms in Bas-Comminges, farmers were able to overcome that crisis collectively. Unexpectedly, the amounts of hay and grass harvested and especially the quantities of weeds appear to be a revealer for the residents of those environmental changes, notably variations in temperature or rainfall. Links are made mostly between weeds/wild flora and weather, but we will see how this can also inform us about climate change and how adaptation attempts show us how the local community might cope with the current and future consequences of climate change. "It has been hot and it's rainy in the spring, so the plants – the wild ones – they spread everywhere. This can be pretty, but people here are used to gray sidewalks, so this is not what it should be" (50-year-old woman, newcomer, housekeeper). Changes in phenology also give information about meteorological variations: "Weeds grow faster with all the rain and the fact that winters are warmer, and they seem to grow earlier and longer during the year as well, they organize collective work to deal with the green areas of the village from the end of April to the autumn, but this year they should have begun earlier as sidewalks and the cemetery were already green" (49-year-old man, native resident, farmer).

Both native residents and newcomers notice these changes, but the native residents do not seem to have made or believed in a link between these variations in local weather and global climate changes until very recently: "There have been changes in the weather and the seasons, but who can say if it is because of climate change? We had a major drought at the beginning of the century, my grandfather said; for me it is just cycles" (45-year-old man, native resident, farmer). The link between changes in weather and changes in the global climate is more frequently made by newcomers: "Climate change has an impact here as well, it seems. I don't have the background as I just arrived a few years ago, but based on what local people have told me, winters seem to be wetter and warmer, and there have been major storms in the past few years that even the oldest residents could not recall such disasters in the past… Plants, they are acting strangely. We had tomatoes last October. The weeds are invading the vegetable plots. And this year I had to weed my path twice as much as the first year I arrived" (55-year-old woman, newcomer, employee).

Local perceptions of the link between weeds and weather variability and thereafter with climate change are changing among the native residents. It is interesting to look at the diagnostic of "weed" growth within our 15-year-old data corpus and to note how it reveals a gradual awareness of climate change even for the most reluctant people. In the 2000s, weeds were explained more as a problem of social change than of climate change. Encroachment or unwanted plants and wild flora were mentioned by native residents – without using the term "weeds" and on private land only – to talk about the rural exodus and land abandonment: "You see these hills in front of us, this is encroachment, before there were sheep grazing the meadow, but now, with the farms being abandoned and changes in production, these hills are no longer grazed and the forest will be growing. You'll see in ten years, there will be a forest there" (Man, native resident, 78 years old in 2003, retired farmer).

A few years later, some meteorological variations were noted, linked to weeds and wild flora, and were sometimes viewed positively: "We get more rainfall and have more hot summers than a few year ago, but that's sometime good for us. We made a lot of hay this year, and we did not have to open the silo until late in the year, so that was really beneficial for us. Lots of hay, lots of good grass for the cattle… We were not that happy because it grew everywhere and we had more work to clean the fences and we had to use more herbicide, but the cattle had a lot of feed" (47-year-old woman, native resident, farmer). At the time, grasses in the public spaces grew as fast and as prolifically as in the fields, but were eradicated with herbicide by the municipal employee and the employees of the Communauté des Communes. Those uses were considered normal and the proliferation of weeds did not seem as crucial a problem as nowadays.

Since 2015, the links made between meteorological and climate variations and wild flora considered to be weeds are clearer: rain and temperature are viewed as the causes of the increase in weeds on private farmland and on public land: "These past years, we had a lot of rain and the temperatures were erratic. We had very poor quality of hay and an increase in cattle diseases. The weeds were competing with the good grazing grasses and there was nothing we could do, because when you are making the hay, you cannot separate the good grass from the weeds" (55-year-old man, native resident, farmer). In the public spaces, "weeds are growing everywhere and faster than before. There is too much rain and the temperatures were warmer, and the sidewalks are all green and dirty. The cemetery is a jungle, and now you have orders not to use herbicides anymore. If you use an ounce of Roundup, you are the devil" (47-year-old woman, native resident, farmer).

These quotes suggest that the perceptions of the ongoing changes are evolving based on their actual consequences and manifestations on the land. Links are being made with long-term meteorological variations and observations, and gradually, these meteorological events are being linked to climate change and people are expressing fear for the future: "We don't know if climate change is happening here. Some say yes, some say no. But the weather has been crazy these past few years. We don't know how to predict the seasons anymore, plants are acting strangely, we can deal with the cultivated ones but not the wild ones. There are no more swallows or far fewer than before. Is it because the barns are closed or because of the climate or both? I don't know. The woodpigeons arrive more than one month later nowadays" (45-year-old man, native resident, employee).

Interview quotations and participant observation revealed intrinsic links between the various types of changes experienced at the local level (territorial reforms, changes in the local institutions, or environmental concerns). Making sense of climate change or linking actual weather variations to climate change is a hard process for the local people. A revealer such as weeds and their expansion, or long-term observations of the weather combined with the increasing familiarity with the concept of climate change – through everyday conversations, newspapers, and official discourses – make people increasingly aware of the major global environmental changes. Having talked with local residents for years now through long-term ethnography, we can see how seasonal variations are perceived in a long-term view.

To illustrate these shifts in interpretations, we include three quotes about weather and climate change from the same native farmer at three different times between 2004 and 2018. In 2004, the farmer said about the weather and the 2003 drought: "We always had bad winters or droughts. Indeed, we have less snow now in the winter. I remember going to school when I was a kid and walking in the snow to go there. And the summers are warmer than usual… They are talking about climate change. I don't know if it's true, it may just be cycles. The older people already talked about heavy droughts in summers in the past." In 2011, we spoke about the weather with the same farmer, who said, "it's true that we have droughts almost every summer now and sometimes too much rain in the spring when we do not need it. Sometimes it's hard to sow because the soil is too compact and too humid, and that's happening quite often now. Maybe it is linked to climate change, yes, because it's happening everywhere, here in the region and all over the country." In 2018, after several droughts and two major hailstorms, the same farmer said: "In 2003 we had this major drought. The weather has really been changing since then and maybe before. Winters are not cold enough, it does not kill the pests and bugs and weeds anymore, spring and summer are either too rainy or too dry, it disrupts our work in the crops, it disturbs the growth of the plants. The hailstorms we had the past few years were terrible. Yes, climate change is happening here, too."

5.7 Conclusion: "Changing While Staying the Same" and How Global Climate Change Cannot Be Perceived Separate from Local Social Change

Through our case study, we have detailed how a local community makes sense of global changes through the observation of its everyday biodiversity. Perceiving climate change at a local level can be challenging, and our informants needed long-term observations to identify and name some of its impacts on their land or even to associate the changes they were experiencing with climate change. These long-term observations were not only made by observing the weather or the climate but, more often, by observing local wildlife, its phenology, or behavior. Weeds as a revealer of changes were not what we expected at the beginning of our research – thinking that people would talk about local indicators that are obvious for us scientists, such as trends in rainfall or temperature patterns or the behavior of migratory birds. However, local perceptions and knowledge about this category of wild flora allow us to understand some diagnostics of changes made by the studied community. We realized how integrative local knowledge – and local diagnostics of climate change – are: one change is never perceived on its own but combined with the effects of other environmental or social changes. The situation showed the crucial need to understand social and environmental transformations together at the local level due to their interdependent consequences for communities.

The impacts of those related and associated changes are significant, and the community is aware of the potential consequences and can adapt proactively. In our current case study, the weed issue reveals not only the impacts of seasonal variations and climate change but also profound changes in the organization of local institutions, communities, and the French countryside. Efforts to create local institutions to overcome new social and environmental conditions, as we described in this chapter, are very interesting because they demonstrate a community's ability to successfully anticipate changes at the local level. This community's ability to overcome conflicts and maintain the social networks that keep the community alive and functional is also remarkable.

Around France and even Europe, some communities – notably in rural areas – are organizing themselves to deal with the maintenance of public areas when needed through what are broadly called "citizens' days," which give evidence of the need for larger adaptation processes to cope with changes. These days bring people together with a common objective but with different resources, tools, and knowledge to take care of the land (see, e.g., the website www.journeecitoyenne.fr). In Bas-Comminges, what we found particularly interesting is the way the citizens' days and the new local institutions were organized by individuals with different backgrounds, perceptions, and knowledge of the changes occurring, but with the willingness to make their community resilient for the future. Local residents are loyal to the traditional social organization, trying to maintain and recreate traditional networks, while integrating newcomers and new ways of managing the land.

Our chapter also shows that ethnographic work and long-term inquiries give information about the evolution of the perceptions and knowledge about changes and are essential to the necessary comprehension of local diagnostic of changes. This study was supposed to be mainly on perceptions of climate change, but in the end, we focused substantially on social changes. Anthropologists and social scientists have already shown that although meteorological variations are well documented, climate change at a local level is sometimes hardly perceived or recognized (Peterson and Broad 2009; Roque de Pinho and Galvin 2015). We then investigate the importance given to climate change compared to other types of social or environmental transformations in the local understandings of global changes.

Acknowledgements This material is based upon work supported by the ANR Young Researcher Program #ANR-13-JSH1-0005-01.

Bibliography

Alignier A, Raymond L, Deconchat M, Menozzi P, Monteil C, Sarthou JP, Vialatte A, Ouin A (2014) The effect of semi-natural habitats on aphids and their natural enemies across spatial and temporal scales. Biol Control 77:76–82

Ambrosino C, Andres L (2008) Friches en ville: du temps de veille aux politiques de l'espace. Espaces Soc 134(3):37–51. https://doi.org/10.3917/esp.134.0037

Augustins G (1989) Comment se perpétuer? Devenir des lignées et destins des patrimoines dans les paysanneries européennes. Société d'Ethnologie, Nanterre

Barthélémy C, Melin H, Lizet B, Menozzi M (2016) Introduction au mini-dossier L'ethnobotanique à l'épreuve de la nature en ville. Dev Durab Territ 7(1). https://doi.org/10.4000/developpementdurable.11193

Baudry S (2014) Plants as a tool for DIY Urbanism: between guerilla and institutionalisation. Rom J Am Stud 2:41–51

Berkes F, Jolly D (2001) Adapting to climate change: social-ecological resilience in a Canadian Western Artic Community. Conserv Ecol 5(2):18. http://www.ecologyandsociety.org/vol5/iss2/art18/

Blanc N, Glatron S, Schmitt G (2012) Trames vertes urbaines: recherches en sciences humaines et sociales. Dev Durab Territ 3(2). http://journals.openedition.org/developpementdurable/9370

Blanco J, Sourdril A, Deconchat M, Ladet S, Andrieu E (2018) Social drivers of rural forest dynamics: a multi-scale approach combining ethnography, geomatic and mental model analysis. Landsc Urban Plan 188:132–142. https://doi.org/10.1016/j.landurbplan.2018.02.005

Bourdieu P (1973) The Berber house. In: Douglas M (ed) Rules and meanings: the anthropology of everyday knowledge. Penguin, Harmondsworth, pp 98–110

Bowler D, Buyung-Ali L, Knight T, Pullin A (2010) Urban greening to cool towns and cities: a systematic review of the empirical evidence. Landsc Urban Plan 97(3):147–155

Camille Hochedez, Benoît Leroux, (2018) Développement durable et territoires (Vol. 9, n°2)

Carrus G, Scopelliti M, Lafortezza R, Colangelo G, Ferrini F, Salbitano F, Agrimi M, Portoghesi L, Semenzato P, Sanesi G (2015) Go greener, feel better? The positive effects of biodiversity on the well-being of individuals visiting urban and peri-urban green areas. Landsc Urban Plan 134:221–228. https://doi.org/10.1016/j.landurbplan.2014.10.022

Carsten J, Hugh-Jones S (eds) (1995) About the house: Lévi-Strauss and beyond. Cambridge University Press, Cambridge

Cavailhès J, Normandin D (1993) Déprise agricole et boisement: état des lieux, enjeux et perspectives dans le cadre de la réforme de la PAC. Rev For Fr 45(4):465–482
Claeys C, Sirost O (2010) Proliférantes natures. Etud Rurales 185:9–22
Coughlan M, Gragson T (2016) An event history analysis of parcel extensification and household abandonment in Pays Basque, French Pyrenees, 1830-1958 AD. Hum Ecol 44(1):65–80. https://doi.org/10.1007/s10745-016-9808-y
Crate SA (2011) Climate and culture: anthropology in the era of contemporary climate change. Annu Rev Anthropol 40(1):175–194. https://doi.org/10.1146/annurev.anthro.012809.104925
Crate SA, Nutall M (2009) Anthropology and climate change: from encounters to actions. Left Coast Press, Walnut Creek
Cunningham CE (1964) Order in the Atoni house. Bijdr Taal Land Volkenkd 120(1):34–68
Deconchat M, Gibon A, Cabanettes A, du Bus de Warnaffe G, Hewison M, Garine E, Gavaland A, Lacombe J-P, Ladet S, Monteil C, Ouin A, Sarthou J-P, Sourdril A, Balent G (2007) How to set up a research framework to analyze social–ecological interactive processes in a rural landscape. Ecol Soc 12(1):15. http://www.ecologyandsociety.org/vol12/iss1/art15/
Douglas M (1966) De la Souillure: essai sur les notions de pollution et de tabou. Maspero, Paris
Dounias E (2009) The sentinel key role of indigenous peoples in the assessment of climate change effects on tropical forests. IOP Conf Ser Earth Environ Sci 6:572008
Dupré L (2005) Des friches: le désordre social de la nature. Terrain 44:125–136. https://doi.org/10.4000/terrain.2488
Fiske S et al (2014) Anthropology and climate change. Final report of the AAA global climate change task force. American Anthropological Association, Arlington
Gaüzère P, Barbaro L, Calatayud F, Princé K, Devictor V, Raison L, Sirami C, Balent G (2019) Long-term effects of combined land use and climate changes on local bird communities in complex agricultural landscapes. Agriculture Ecosystems & Environment. 289. https://doi.org/10.1016/j.agee.2019.106722
Gibon A, Sheeren D, Monteil C et al (2010) Modelling and simulating change in reforesting mountain landscapes using a social-ecological framework. Landsc Ecol 25:267. https://doi.org/10.1007/s10980-009-9438-5
Gillespie SD (2000) Beyond kinship: an introduction. In: Gillespie SD (ed) Beyond kinship: social and material reproduction in house societies. University of Pennsylvania Press, Philadelphia, pp 1–22
Gillespie SD (2007) When is a house? In: Beck RA Jr (ed) The durable house: house society models in archaeology. Center for Archeological Investigations, Southern Illinois University, Carbondale, pp 25–50
Groth J, Corijn E (2005) Reclaiming urbanity: indeterminate spaces, informal actors and urban agenda setting. Urban Stud 42(3):503–526
Hamberger K (2010) La maison en perspective: un modèle spatial de l'alliance. L'Homme 194:7–40
Hastrup K (2016) Climate knowledge, assemblage, anticipation, action. In: Crate S, Nuttall M (eds) Anthropology and climate change. From actions to transformation. Routledge, New York/London, pp 35–57
IPBES (2018) The IPBES regional assessment report on biodiversity and ecosystem services for Europe and Central Asia. M. Rounsevell, M. Fischer, A. Torre-Marin Rando and A. Mader. Bonn, IPBES, 892p
Lefebvre J, Lizet B, Barthélémy C (2013) La Canne de Pline et le concert des disciplines. Ethnographiques.org, No. 27. http://www.ethnographiques.org/2013/Lefebvre-Lizet-Barthelemy
Lizet B (2010) Du terrain vague à la friche paysagée. Le square Juliette-Dodu, Paris, Xe. Ethnol Fr 40:597–608. https://doi.org/10.3917/ethn.104.0597
Lizet B, Wolf A-E, Celecia J (eds) (1997) Sauvages dans la ville, JATBA, revue d'ethnobiologie. MNHN, Paris
Luginbühl Y (1989) Sauvage/cultivé: l'ordre social et l'harmonie des paysages. In: Mathieu N, Jollivet M (eds) *Du rural à l'environnement, la question de nature aujourd'hui*. ARF Editions/L'Harmattan, Paris, pp 42–49

MacDonald D, Crabtree JR, Wiesinger G, Dax T, Stamou N, Fleury P, Gutierrez Lazpita J, Gibon A (2000) Agricultural abandonment in mountain areas of Europe: environmental consequences and policy response. J Environ Manag 59(1):47–69. https://doi.org/10.1006/jema.1999.0335

Marino E (2015) Fierce climate, sacred ground: an ethnography of climate change. University of Alaska Press, Fairbanks

Menozzi M-J (2007) Mauvaises herbes, qualité de l'eau et entretien des espaces. Nat Sci Soc 15:144–153

Molina M, Tardio J, Aceituno-Mata L, Morales R, Reyes-Garcia V, Pardo-de-Santayana M (2014) Weeds and food diversity: natural yield assessment and future alternatives for traditionally consumed wild vegetables. J Ethnobiol 34(1):44–47. https://doi.org/10.2993/0278-0771-34.1.44

Nedecheal A, Dogan Y, Guarrera P (2007) Plants traditionally used to make brooms in several European countries. J Ethnobiol Ethnomed 3:20. http://www.ethnobiomed.com/content/3/1/20

Pellegrini P, Baudry S (2014) Streets as new places to bring together both humans and plants: examples from Paris and Montpellier (France). Soc Cult Geogr 15(8):871–900

Peters K, Breitsameter L, Gerowitt B (2014) Impact of climate change on weeds in agriculture: a review. Agron Sustain Dev 34:707. https://doi.org/10.1007/s13593-014-0245-2

Peterson N, Broad K (2009) Climate and weather discourse in anthropology: from determinism to uncertain futures. In: Crate S, Nuttall M (eds) Anthropology and climate change. From encounters to actions. Left Coast Press, Walnut Creek, pp 70–86

Pieroni A, Nebel S, Quave C, Münz H, Heinrich M (2002) Ethnopharmacology of liakra: traditional weedy vegetables of the Arbëreshë of the vulture area in southern Italy. J Ethnopharmacol 81(2):165–185. https://doi.org/10.1016/S0378-8741(02)00052-1

Rogers SC (1991) Shaping modern times in rural France: the transformation and reproduction of an Aveyronnais community. Princeton University Press, Princeton

Roncoli C, Ingram K, Krishen P (2002) Reading the rains: local knowledge and rainfall forecasting in Burkina Faso. Soc Nat Resour 15:409–427

Roncoli C, Crane TA, Orlove B (2009) Fielding climate change in cultural anthropology. In: Crate SA, Nuttall M (eds) Anthropology and climate change: from encounters to actions. Left Coast Press, Walnut Creek

Root TL, Price JT, Hall KR, Schneider SH, Rosenzweig C, Pounds JA (2003) Fingerprints of global warming on wild animals and plants. Nature 421(6918):57–60

Roque de Pinho J, Galvin KA (2015) "Maasai Voices on Climate Change (and other changes, too)": participatory video and communication about environmental changes in the East African rangelands. In: Brites M, Jorge A, Correia Santos S (eds) Metodologias participativas: os media e educação. LabCom Books, Covilha, pp 235–248

Roque de Pinho J (2020) The Year People Helped Zebras to Stand Up: Climatic Variability and Extreme Weather Observed and Portrayed by Kenyan Maasai Pastoralists. In: Welch-Devine M, Sourdril A, Burke B (eds), Changing Climate, changing worlds. Springer editions, Collection Ethnobiology.

Shaffer LJ (2014) Making sense of local climate change in rural Tanzania through knowledge co-production. J Ethnobiol 34(3):315–334

Sourdril A (2008) Territoire et hiérarchie dans une société à maison Bas-Commingeoise: permanence et changement. Dissertation, University of Paris X, Nanterre, France

Sourdril A, Andrieu E, Cabanettes A, Elyakime B, Ladet S (2012) How to maintain domesticity of usages in small rural forests? Lessons from forest management continuity through a French case study. Ecol Soc 17(2):6. https://doi.org/10.5751/ES-04746-170206

Sourdril A, Welch-Devine M, Andrieu É, Bélaïdi N (2017) Do April showers bring May flowers? Knowledge and perceptions of local biodiversity influencing understanding of global environmental change. A presentation of the PIAF project. Nat Sci Soc 25(1):56. https://doi.org/10.1051/nss/2017009

Strauss S, Orlove B (eds) (2003) Weather, climate, culture. Berg Publishers, Oxford/New York
Vedwan N, Rhoades RE (2001) Climate change in the Western Himalayas of India: a study of local perception and response. Clim Res 19:109–117
Veteto JR, Carlson SB (2014) Climate Change and Apple Diversity: Local Perceptions from Appalachian North Carolina. Journal of Ethnobiology 34(3):359–382

Chapter 6
Whose Climate, Whose Changes? Various Views from Rural Northern Cameroon

Christine Raimond, Markus Bakaira, Sylvain Aoudou Doua, and Eric Garine

Abstract In the Upper Benoue area in Northern Cameroon, narratives about weather events and climate change are different according to one's economic and cultural profile. Formal education level, economic strategy, and autochthony versus migration are promising explanatory variables to account for the heterogeneity of discourses about climate change.

Weather conditions, especially the timing of rainfall, are highly variable from one year to another in the sahelo soudanian part of Africa. Rationales developed by local stakeholders to account for these variations are based on various sources of knowledge depending on their primary socialization within a local territory and their direct experience of it, their formal schooling and linguistic competence, their faith to monotheist religion, and their commitment to conservation and development programs. Looking across various residents of the same region, this chapter presents differing voices on local changes: global climate change is not necessarily the primary explanation.

Keywords Climate change · Biodiversity · Perceptions of changes · Rural area · Conservation · Northern Cameroon

The latest Intergovernmental Panel on Climate Change (IPCC) reports emphasize Africa's high vulnerability to weather variability and climate change, and climatologists are increasingly knowledgeable about the African climate and its variability (IPCC 2014, 2018). After the drought periods in the Sahel during the 1970s–1980s, rainfall gradually returned to the level of the 1950s (Dardel et al. 2014). In the con-

C. Raimond (✉)
UMR Prodig, CNRS, Paris, France
e-mail: christine.raimond@univ-paris1.fr

M. Bakaira
Ngaoundere University, Ngaoundere, Cameroon

S. A. Doua
Institut Supérieur du Sahel, Université de Maroua, Kousseri, Cameroon

E. Garine
UMR Lesc, Université Paris Nanterre, Nanterre, Paris, France

M. Welch-Devine et al. (eds.), *Changing Climate, Changing Worlds*, Ethnobiology, https://doi.org/10.1007/978-3-030-37312-2_6

text of climate change, results of research show a clear trend toward higher temperatures, but precipitation projections are much more uncertain and exhibit high spatial and seasonal dependence (Niang et al. 2014). In the Sudano-Sahel region of Cameroon, the climate is characterized by two seasons. Landscapes, life, and human activities are determined by this cycle of dry and rainy seasons. The study of the variability of the climate in this area shows a complex situation. As in the Sahel area, we observe high interannual variability. The droughts of the 1970s–1980s caused the southward advance of isohyets. Overall, they recently returned to the same annual rainfall after the droughts. However, trends are not very clear, and we observe sizeable local variations. Rainfall projections over the coming decades remain uncertain, but past observed trends have already had significant local impacts (Sultan et al. 2013). In the context of development policies and assistance for mitigation and/or adaptation through agricultural development projects (COP 21), is the perception of climate change by local populations changing in Sub-Saharan Africa? This point remains under-investigated.

6.1 Site Description

In this region of North Cameroon, there are two main opposing political orientations. On the one hand, a biodiversity conservation policy has been pursued since the early twentieth century to preserve Sudan's large wildlife ecosystem from logging, agriculture, and grazing by livestock. It has led to the gradual creation of three national parks and some 20 hunting zones, which in 2018 covered more than 45% of the surface area of the northern region (Fig. 6.1). This system of protected areas is a major spatial constraint and concentrates (theoretically) all human activities on the rest of the territory.

As of 2015, several types of stakeholders (note: each stakeholder category is given *in italics* below) with diverging strategies were interacting in the studied areas (Raimond and Aboubakar 2018).

Indigenous sedentary farmers (Duupa in Boumba and Wanté; Dii in Djaba and Skaje) practiced, before the arrival of the agricultural front, slash-and-burn shifting agriculture incorporating a great diversity of plants that were intercropped in the same plots and used heterogeneous forest resources for various purposes (Garine et al. 2005). Bush resources were used as additional food (hunting and plant gathering) and for cooking, basketry, and construction.

The arrival of the pioneer or settler frontier (in Djaba and Sakje in 2005; in Boumba in 2010; in Wanté in 2017) abruptly increased population densities and deeply changed practices by shifting agriculture toward a system of short fallow periods, then without fallow, with a rotation of less diversified crop plots. This trend is especially strong on the edges of the Benue National Park because a development plan strongly restricts land use by delimiting "zones of human use" beyond which inhabitants are not allowed to travel or carry out any activities. *Forest rangers* are responsible for the application of spatial and resource use standards, under the

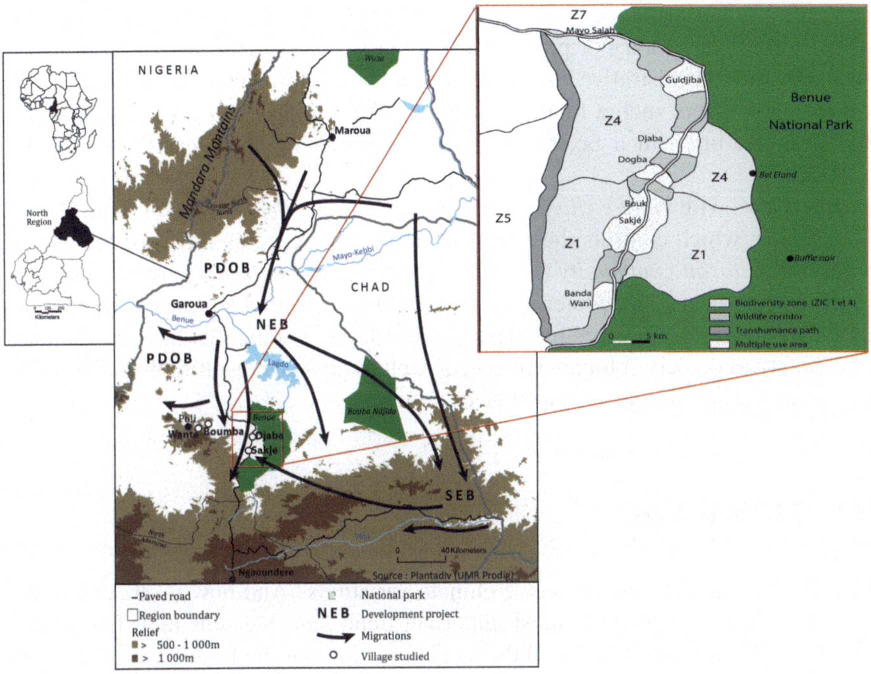

Fig. 6.1 Benue National Park location, agricultural frontier, and development plan

responsibility of the *park conservationist* and in collaboration with volunteer *community rangers* responsible for monitoring the protected areas defined in the development plan.

Migrant farmers (including the Mafa, Tupuri, and Guidar farmers interviewed in our survey) have settled in the studied areas within the last 15 years to practice intensive agriculture, often introducing commercial cotton into the rotations. They also raise livestock, with part of the herd remaining in the village and the other part entrusted to *transhumant herders* to take them to more distant pastures, especially during the crop season.

Transhumant herders are mobile throughout the region and choose their routes according to the availability of pastoral resources (water, grass, and tree foliage). Since 2010 and the rise of insecurity in the Sahel zone (Magrin and Pérouse de Montclos 2018), there has been an increasing number of them in the Sudanese zone, and they have invaded protected areas. Some herding families have settled on the outskirts of agricultural settlements and keep the farmers' herds. They were the ones interviewed in this survey, as time and financial resources did not allow us to reach remote groups in the bush.

Finally, the region has not escaped the gold fever that has plagued Sub-Saharan Africa since international prices began to rise in the late 2000s (Chevrillon-Guibert and Magrin 2018). *Gold prospectors* first entered the territories to prospect for veins abandoned since the 1970s. They then entered protected areas and reached several thousand in Benue National Park by 2015. All sedentary or traveling residents become temporary or permanent gold prospectors. The activity, considered illegal by the administrative authorities and denounced by the park conservationist, nevertheless offers such a financial windfall that a flourishing commercial sector has been set up with a taxation system controlled by the local ("traditional") authorities.

Interviews with *village chiefs*; representatives of the *Peul chiefdom* (Lamidat) of Rey Bouba, which controls the territory and people; representatives of the *decentralized territorial community* (town hall); and *sectoral delegates* are not analyzed in this paper because we prefer to give voice to the stakeholders who depend most directly on natural resources for their livelihood. We assume that they are the ones most likely to perceive climate change, as well as changes resulting from biodiversity conservation discourses and devices.

6.2 Methodology

How do local populations perceive climate variations? And how do they describe them? Is climate change the most important trend that threatens their livelihood? These are the questions we asked the local population of the Benue Valley. They fit into the PIAF[1] program, which aims to compare indigenous change indicators with scientific indicators. In order to verify the place of climate change in all locally perceived changes and therefore in the adaptation strategies implemented, we asked stakeholders about all the changes they perceive in their immediate environment. The timescale covered is therefore the lifespan of the local interviewees and seldom dates to periods prior to 1950.

We surveyed 121 people living in four villages subject to similar climatic conditions. We choose to illustrate two contrasting dynamics that characterize this area:

- The low population density enabled the creation of protected areas for wildlife (elephants, giraffes, buffaloes, and antelopes).
- Recently, population density has increased due to an inflow of migrants seeking to establish farms in the region.

We chose two villages next to the national park within the migration frontier (Djaba and Sakje, grouped below in the rural protected area or "protected" group) and two villages in non-protected countryside (Wante and Boumba, located within the migration frontier for agriculture, grouped below in the "rural" group). An opportunistic sample of informants differing in age, sex, activities (farmers,

[1] Programme Interdisciplinaire sur les Indicateurs Autochtones de la Flore et de la faune (PIAF) is a project funded by the ANR Young Researcher Program ♯ANR-13-JSH1–0005-01 from 2014 to 2018: http://www.anr-piaf.org. PIAF brings together a research team from 6 research institutes and 10 research laboratories and is coordinated by Anne Sourdril (CNRS, UMR7533 Ladyss).

Table 6.1 Sample of interviewees

Location	Rural protected			Rural			Total
Sex	Djaba	Sakdje	Total	Boumba	Wante	Total	
F	16	0	16	3	7	10	26
M	58	8	66	16	13	29	95
Total	**74**	**8**	**82**	**19**	**20**	**39**	**121**

cattle breeders, gold miners, service activities), geographic origin (indigenous communities, migrants), duration of residency, ethnolinguistic group (Duupa, Dii, Fulbe, Mafa, Tupuri, Guidar), and level of education (primary/secondary schooling/academic) was built. In 2015 and 2016, 121 interviews were conducted (Table 6.1).

In this region, two media discourses on global change intersect: climate change, which has major repercussions on ecology and agriculture, and the protection of biodiversity to preserve the large wildlife population still present in some national parks. Do these discourses have impacts on local populations? After presenting the results of the survey on the perception of environmental changes by the populations of Upper Benue and then the changes analyzed by the researchers, we will show that the perception of changes varies according to the experience of local stakeholders, with little variation according to their economic and cultural profile, and that it is not (yet?) influenced by global discourses on climate change.

The semi-structured interviews were conducted by four fieldworkers in the vernacular language of northern Cameroon (Fulfulde). They are structured around five main questions: What are your activities? Do you depend on the bush for your activities? What changes have you noticed [in the bush/your village]? How do these changes affect you? What are your reactions to these changes?

The interviews were transcribed in French by the fieldworkers in collaboration with the researchers to standardize terminology and coding and build a database.

6.2.1 *Asking Stakeholders About Changes*

When we asked the question on what has changed in the village and the bush, informants generally cited several changes (six on average with a standard deviation of 2.5), and some would specify up to 13 different changes. As the interview is semi-structured, notes were taken directly in a notebook, which preserved the verbatim statements made by the very different informants. They were recoded into 47 summary expressions (verbatim encoded) that cover similar ideas (see Table 6.2). Four informants stated that they did not observe any changes; they are grouped together in the wording “there is no change,” which may in some cases express a desire to cut the interview short.

When the same idea comes up very often in various expressions (e.g., “the bush has receded,” “the bush is destroyed,” “the bush has disappeared”), it is recoded in a more general formulation in order not to distort the ideas expressed (e.g., “the

Table 6.2 Ranking of 47 verbatim changes recoded into seven kinds of change, plus one "no change" category

Type of change	No. of occurrences/type
Recorded verbatim	No. of occurrences
Rains/hydrology	**190**
Rains have changed	91
The seasons have changed	47
Hydrological changes	32
Agricultural calendar has changed	11
Humidity changes	8
Wind has changed	1
Temperature	**41**
It's hotter	41
Land cover and diversity of flora	**154**
The bush has changed	83
There is less grass	25
Useful species are difficult to find	23
Loss of floral species diversity	12
Trees have disappeared	11
Demography/pressure on rural land	**124**
The population has grown	49
Farms have extended	31
Rising of breeding pressure	23
Gold prospectors are destroying trees and soils	10
Fires have changed	5
We can't get agricultural land anymore	2
There is no place for cattle anymore	2
Sacred places have been destroyed	1
We burn savanna to cause grass to grow again for cattle	1
Fauna	**117**
Wildlife has disappeared	73
Fewer fish in the rivers	11
Loss of faunal species diversity	10
Birds are less numerous	9
We don't eat bushmeat anymore	8
Monkeys have disappeared	3
There are fewer pests in the fields	2
Monkeys are more numerous	1
Agricultural soil fertility	**97**
Soil fertility loss	37
Farm production is lower	29
Farming technology has changed	21
Invasive plants proliferate in fields	10
Farm production is higher	1
Political/social/economic change	**43**

(continued)

Table 6.2 (continued)

Type of change	No. of occurrences/type
The village is larger	14
There is more infrastructure	7
Cost of living is higher	7
Deaths and illness are numerous	5
There are many ethnic groups	2
Now we plant corn	2
Children are vaccinated	1
People gather grass to sell *sekos* (woven mats)	1
Immigrants are invading	1
The bush is less safe	1
We no longer see any tourists	1
New products have appeared on the market	1
Nil	**4**
I don't see any changes	4
Total	771

bush has changed" or "the rains have changed" for "it no longer rains as before," "the rains stop during the season," "the rains are more variable" or "humidity changes" for "it is drier," "the drought advances"). On the other hand, very concrete expressions expressing an original idea have been retained, such as "we can't get agricultural land anymore," which expresses a land shortage; "now we plant corn," which evokes changes in agricultural practices; or "we burn savanna to cause the grass to grow again for cattle," which reflects a loss of biodiversity, pressure on the resource, a change in uses (sale), and a change in food.

6.2.1.1 Not Only the Climate Is Changing

Table 6.2 shows the diversity of perceived environmental changes, and only 15% of recoded verbatim changes are directly related to climate change (rainfall, humidity, surface water, or temperature). However, since the informants are mainly farmers, changes in the abundance and distribution of rainfall are logically the most frequently mentioned ("less rains," "rains stop during the season," "there are no longer any heavy rains as before") and gathered here in the category "rains have changed." Perceptions of the length of the rainy season ("it starts later," "it is shorter," or "it ends earlier") refine an overall perception of a decreasing trend that is consistent with the diagnosis made by climatologists (see Sect. 6.2.2.2). The consequences of these changes on agricultural calendars are also highlighted by mentioning the difficulties they have in sowing at the right time ("now we sow later," "we must sow again every year"). Changes in the river regime and more rapid well depletion are also observed and grouped under "hydrological changes." More general perceptions on the drying of soils and vegetation ("everything is drier"), as well as the decrease in dew, are gathered in "humidity changes." Winds are rarely mentioned without prompting, although informants often have a lot to say about winds when specifically asked about them.

The two types of changes, "land cover and diversity of flora" and "demography/pressure on rural land," are closely linked and widely cited by informants. They observe the decline of wild vegetation (the "bush") and the increasing scarcity of trees and grasses, especially for useful species that have to be sought further and further away. The increase in the number of humans is observed in parallel, in increasing numbers in villages, clearing vast areas of fields and raising a larger livestock population to the point where anthropic pressure leads to new practices ("fires have changed", "fires are more numerous", or "we burn savanna to get grass to regrow for cattle") and causes new situations viewed by some stakeholders as being quite harsh: "there is no place for cattle anymore," "we can't get agricultural land anymore," or "sacred places have been destroyed." We will return to these issues in Sect. 6.2.3.

Changes in wildlife are also widely discussed. "Wildlife has disappeared" includes all observations of wild mammals that are fewer in number, have moved further away from the village, or have disappeared. Some observers cite the extinction of particular species, such as elephants, rhinos, spotted hyenas, lion, and panthers, which are the most highly valued endangered species in conservation systems, or certain bird species. The latter are also perceived as less numerous, as are fish in rivers. The references to monkeys have not been grouped with the others above, as they represent an interesting indicator of pest pressure in the field. Farmers perceive their disappearance directly in frontier areas, where habitat destruction and poaching reduce their population in the vicinity of fields. Conversely, in areas not affected by the agricultural frontier, their population may increase in a sparsely populated area where food crops are particularly coveted by primates, and the pressure of these pests is acutely felt by farmers. The perception of the trend in fauna in the Upper Benue region is strongly regressive, and all the stakeholders interviewed attribute this decline to the increase in the population and its pressure on the environment and natural resources, like bushmeat, which can no longer be consumed. Interestingly, there is no explicit reference to climatic causes that would influence wildlife.

This anthropogenic pressure is also felt in relation to the decline in agricultural soil fertility, with soils that are "tired" or "no longer produce as before," and lower yields than those obtained with slash-and-burn shifting agriculture or in the first years of migrants' settlement. Another indicator is the proliferation of weeds in the fields. The disappearance of fallow land in the densest areas no longer enables control of weeds, which therefore require the use of expensive herbicides. Some informants cited changes in agricultural techniques, with the use of inputs (fertilizers, pesticides) and plowing. Only one observed an increase in agricultural production, which is nevertheless self-evident given the increase in cultivated areas mentioned above.

Political, social, and economic changes are less widely mentioned, but in the end, these cover a rather wide range. They include changes related to the expansion of villages and the arrival of new equipment (creation of a market or school or the arrival of electricity) and services (vaccination). With the increase in the population, inhabitants are observing the appearance of new products on the markets but also an increase in the cost of living. Changes in local activities are also mentioned, in particular the disappearance of tourists who no longer come to an Upper Benue that is empty of wild animals and changes in agricultural practices with the replacement of millet and sorghum crops – previously the staples of indigenous populations – with corn that is faster to grow and commands higher prices on the market. "People collect grass to sell *sekos* (woven mats)" refers to increased pressure on the resource but also to the appearance of a commercial use of gathered products transported to the city. Several isolated changes also highlight negative perceptions of the social changes observed: "There are many ethnic groups" or "immigrants are invading" reflect a feeling of being overwhelmed in the face of an uncontrolled situation and a power struggle that is reversed between indigenous people and migrants who have become the majority in village territories. We could have classified here also the quotations "the Fulani invade the bush" or "the Fulani came to the village" (coded above in "rising pressure on pastures"), because they evoke the deep antagonism between the indigenous Duupa and nomadic herders. "The bush is less safe" refers to the growing insecurity in the region, where the arrival of new stakeholders (*zargina*, armed nomadic pastoralists from the Sahel zone, gold prospectors) makes it difficult to control the area, particularly in protected areas where these activities are illegal (Seignobos, 2011; Raimond and Aboubakar 2018).

6.2.1.2 Specific Perceptions of Climate Change

Figure 6.2 gives the distribution of the recoded responses. We note three changes that predominate in the interviewees' perceptions: more than 60% of the informants mention change in the rains, change of the bush, and the disappearance of wildlife.

Rain-related changes are cited by the largest number of informants (nearly 75%), who often also cite other climate changes (length of the rainy season, increase in temperatures). The local populations have a detailed perception of climatic variations, which validates the results of climatologists:

- The unpredictability of rains is the best indicator they perceive: "It doesn't rain like before."
- The causes of the disturbances of the agricultural calendar are well-identified: "The beginning of the rainy season is delayed," "We sow almost one month later than 10 years before," "The dry periods during the rainy season are more frequent," and "The seasons have changed."

Other climate change indicators are also perceived by local people: 34% of informants mention high temperatures and 26% hydrological changes. To clarify the importance of the perception of climate change in relation to all other types of change, it appears here that variations in water resources in the broad sense (rainfall, rivers, and groundwater) clearly dominate the observations and concerns of informants in Upper Benue. If we group all the changes in the "rains/hydrology" type, this one appears very clearly above all of the others.

When looking at the distribution of change types (Fig. 6.3), "rains/hydrology" groups the most citations by informants, who cite several of them in their list. "Land cover and diversity of flora" comes next, with informants referring mainly to "the bush has changed" but also to other changes in vegetation, such as the disappearance of grass, trees, or useful species. "Demography/pressure on rural land" includes social changes that are cited by fewer people but which when combined in the same category is more cited than the observed changes on the "fauna." Indeed, many informants cite the disappearance of wild animals, but they specify few changes in other categories of animal biodiversity, while anthropogenic pressure is expressed by several indicators, in particular: "The population has grown," "Farms have extended," or "Rising of breeding pressure."

6.2.1.3 Anthropogenic Pressure Dominates Local Residents' Perception of Changes

An analysis of the order of citation of the three most frequently cited changes adds nuance to the predominance of climate change in informants' perceptions (see Fig. 6.4).

Informants first cited the retreat of the bush, the disappearance of animals, and, later on in the list, climate-related changes. "The bush has changed" is cited by 67% of informants (Fig. 6.2), slightly less than "Rains have changed," but it most often appears at the top of the list of those who cite it. Similarly, the disappearance of large wild mammals is cited by 60% of informants and also appears at the beginning of the citation list, between first and third place. "Rains have changed" comes later, between fourth and fifth place. Thus, while the perception of rainfall changes is obvious, it appears on average less important to informants than landscape changes and the loss of wildlife diversity. The latter two are essentially causes of changing practices and population growth.

The drying out of streams is mentioned by pastoralists and farmers. Their mobility and search for water for cattle explain why they are more sensitive than others to this fact.

> *The changes I have seen are numerous, if one must count them, one cannot. First, bush patches are destroyed by farmers and one does not know where to lead his cattle. Grass has diminished (...). People have created farms everywhere and they treat grasses with chemicals. And even where there are no farms, grass does not grow normally. The space has become very small for us (...). If we go back 10 years from now, you could not cross (the river) to come where we are. All the streams were full. Today, it is not raining normally and everything has become dry* (IB 43, nomadic herder, Boumba).

This citation is a good example of the local discourse about changes. This herder mentions the main changes in the bush, in relation with his activity and the needs of herds. He identifies the causes of these changes: the climate is cited last, after anthropogenic changes such as farm extension and use of chemicals.

It is interesting to compare this with the speech of an indigenous farmer, who identifies the same trend of dry rivers but does not attribute them to the same causes. For this person, the change in rainfall is partly responsible for the drying of rivers,

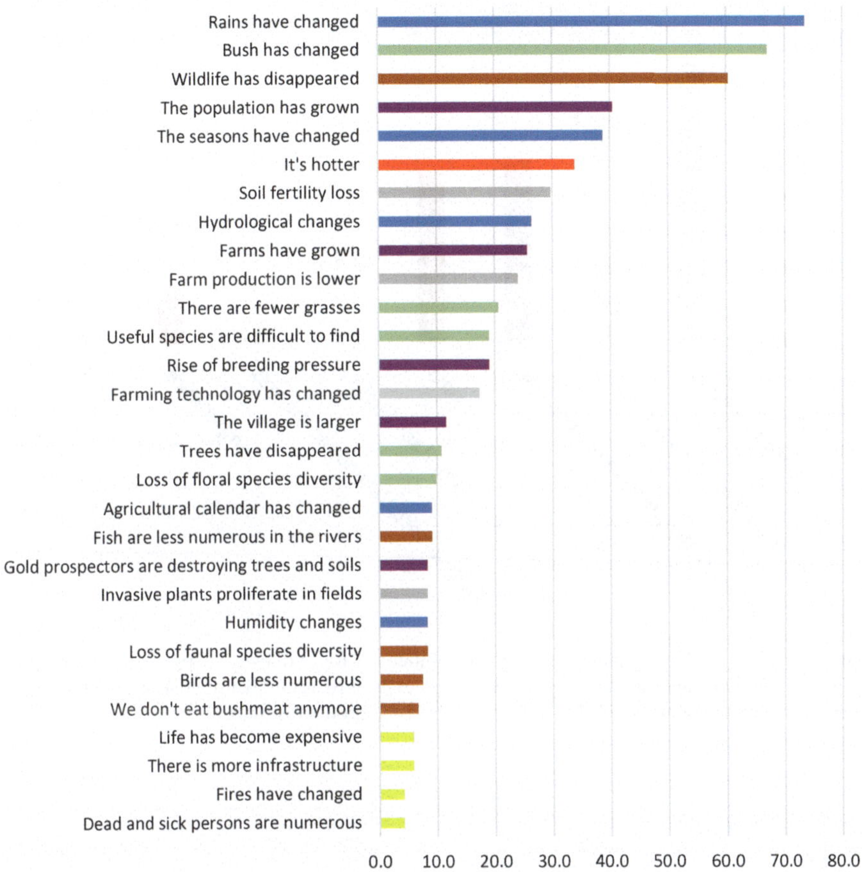

Fig. 6.2 Changes perceived by residents of four Upper Benue villages (verbatim responses recoded into 47 categories, several responses possible, as a percentage of 121 interviewees; changes given by fewer than five people are not included). Blue, rains/hydrology; green, land cover and diversity of flora; purple, demography/pressure on rural land; brown, fauna; red, temperature; gray, farmland fertility; yellow, political/social/economic changes

but larger livestock herds also contribute to higher water withdrawal for their watering.

If we look at all the responses, we observe many converging answers about climate but an even greater diversity of non-climate-driven changes. With "the population has grown," we can summarize the main changes perceived by local people.

> *I'm doing many different things [economically] nowadays hoping for a little gain. In farms, when there were too many weeds, we used to leave the place and slash another swidden. Today, space has become small, we just spray chemicals* (IB 20, native farmer, Boumba).

Climate varies, and people can observe its consequences locally, but the main cause of landscape, biodiversity, and life changes is population growth.

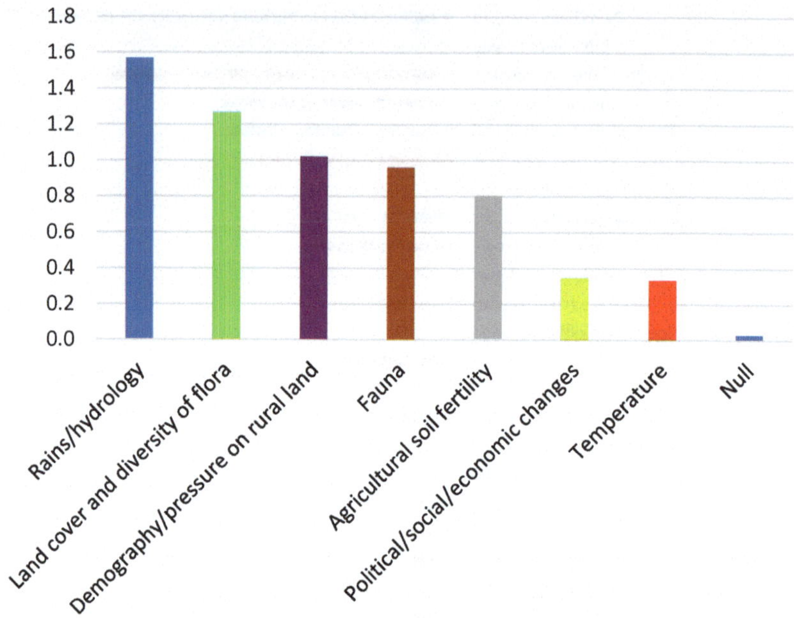

Fig. 6.3 Types of changes perceived by the inhabitants of four villages in Upper Benue (grouping of the 47 recoded responses into 8 types of change, expressed as number of citations of changes by person, 121 respondents; see Table 6.2 for details of the groupings made)

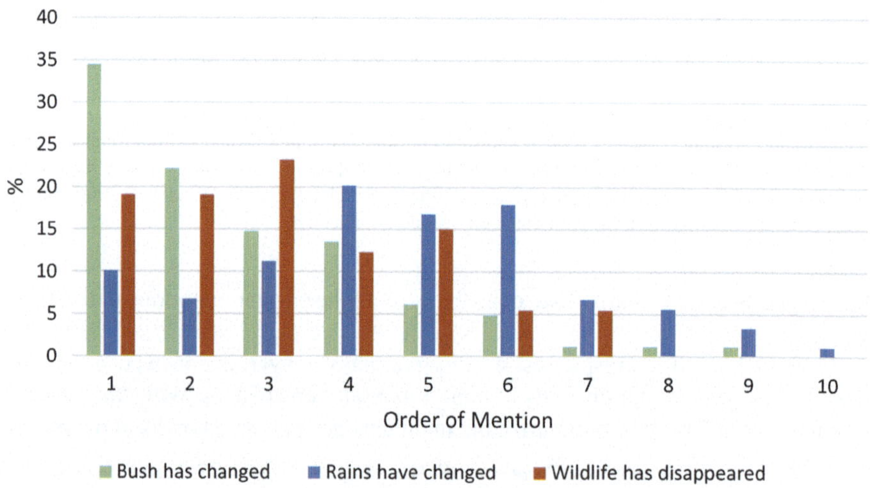

Fig. 6.4 Citation order of the three most frequently cited recoded verbatim (as a %, 81 respondents "The bush has changed," 89 respondents "Rains have changed," 73 respondents "Wildlife has disappeared")

6.2.2 *Environmental Changes in Upper Benue Analyzed by Geographers and Ecologists*

6.2.2.1 Land Use Changes and Demographic Growth

While the number of migrants has only been counted at the regional level for those who have been enrolled in development projects (between 90,000 and 130,000 people between 1975 and 1993), the continuous arrival of unenrolled migrants has driven a population increase of 4% per year between 1987 and 2005. Census figures and projections indicate that the population doubles every 20 years (Fig. 6.5).

Land-use changes are also measured by scientists. A satellite survey of the area shows the transformation of landscape in the village territory of Djaba and the edge of the park (Fig. 6.6). Farm extension is shown in white on the left of the image and red on the graph and indicates in-migration in the area between those two dates. Pastoral pressure and various activities by people in the bush (collecting firewood, grasses for buildings) transform dry forest to wooded savanna (Aoudou et al., forthcoming). All those landscape changes are caused by population growth, especially driven by internal migrations. In the village of Djaba, the population rose from 240 native individuals in 2000 to 6000 in 2014, with farmland of 80 hectares in 1992, 530 hectares in 2004, and 2200 hectares in 2012.

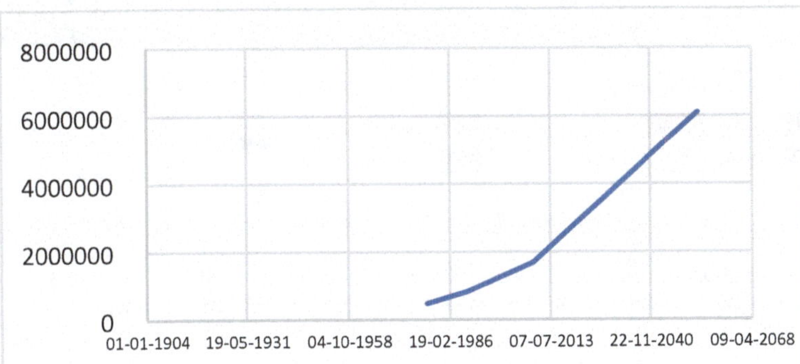

	1976	1987	2005	2050
Population	480,000	830,000	1,690,000	6,120,000 to 7,950,000
Density (inhabitants/km2)	7.5	12.3	25.5	90-120

Fig. 6.5 Population growth in the northern region of Cameroon. Two projections are proposed for 2050 using (a) the average intercensal rate for Cameroon as a whole (2.9% per year) and (b) the average rate for the period 1987–2005 in the North (3.5%), which is more likely to take account for internal migration. (Source: RGPH)

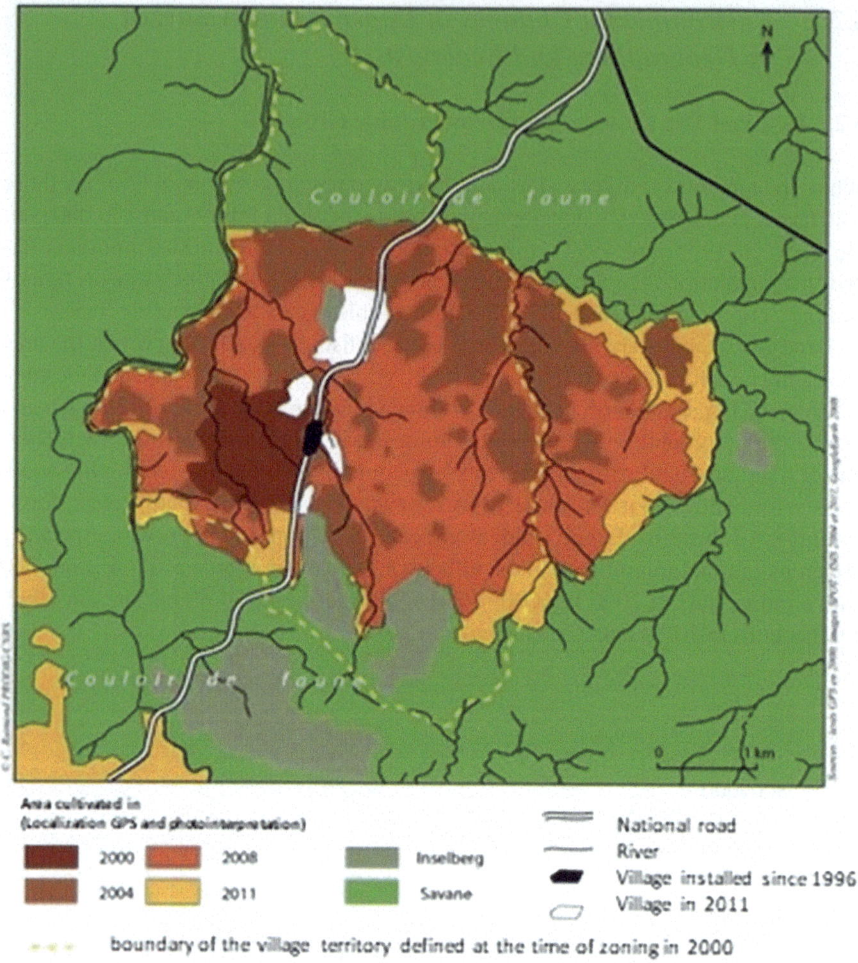

Fig. 6.6 Land-use changes in Djaba, 2000 and 2012 (images: SPOT XS, 087-333/4 dated December 3, 2000, and SPOT DEM dated December 2012; ISIS/CNES contracts)

These landscape changes are visible in all the countryside along the agricultural frontier and are spreading even faster because they are not covered by a development plan, unlike the area around the Benue National Park.

6.2.2.2 Climate Changes

Climate research in the Sudanese zone shows a high variability in rainfall during the twentieth century, marked by several major droughts (1973, 1984), a net deterioration in climate during the 1990s and 2000s, and an increase in rainfall in some Sahel zones that has major consequences on the environment and agricultural practices (Niel et al.

2005; Dardel et al. 2014; Sultan et al. 2013). To verify whether the perception expressed by stakeholders in Upper Benue corresponds to current developments, in particular on changes in rainfall seasonality and temperature increases, rainfall data from five observation stations (Garoua, Guider, Fignolé, Poli, and Tchollirè) were analyzed over the period 1960 to 2013. The southern movement of isohyets observed during the major droughts of the 1970s to 1980s was reversed, and they have generally returned to their original position, although some stations lost an average of 200 mm of rainfall between 1960–1975 and 1990–2013. As often mentioned in interviews with stakeholders in Benue, rainfall trends (calculation of moving averages) confirm the increase in rainfall variability, which makes it very difficult to predict when rainfall will start and how long it will last (Aoudoua Doua; unpublished data). In particular, farmers have observed a significant decline over the past several decades, as confirmed by station surveys. The rainy season start date is determined using the Sivakumar method (Sivakumar 1987). The criterion used is to collect 20 mm of rain in three consecutive days after May 1, with no dry period exceeding 7 days within the following 30 days. At the Fignolé station (Fig. 6.7), there is a delay of almost a month between the beginning of the rainy season measured in the 1960s and that of the 2010s. The rainy season now begins later than before the 1970s, which is a major change for agriculture: if they sow too early, farmers can lose their seeds. The farming schedule has changed, and they have to adapt landraces and species of crops.

Moreover, as in the Sahel area, we observe a rise in temperatures. It is proved that it is now hotter than 50 years ago (Fig. 6.8). This 1 °C increase, on average, over the past century confirms local populations' perceptions and also has major ecological effects. Upper Benue residents often relate this higher temperature to deforestation, which is also said to reduce local rainfall. While scientists have not confirmed this direct effect, the link with global climate change is not proven, either, and is not referred to as an explanation for changes in the local climate.

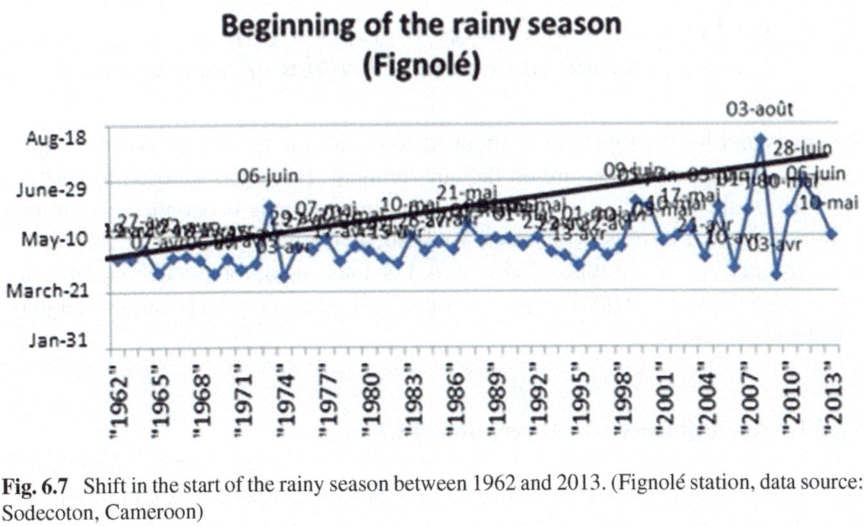

Fig. 6.7 Shift in the start of the rainy season between 1962 and 2013. (Fignolé station, data source: Sodecoton, Cameroon)

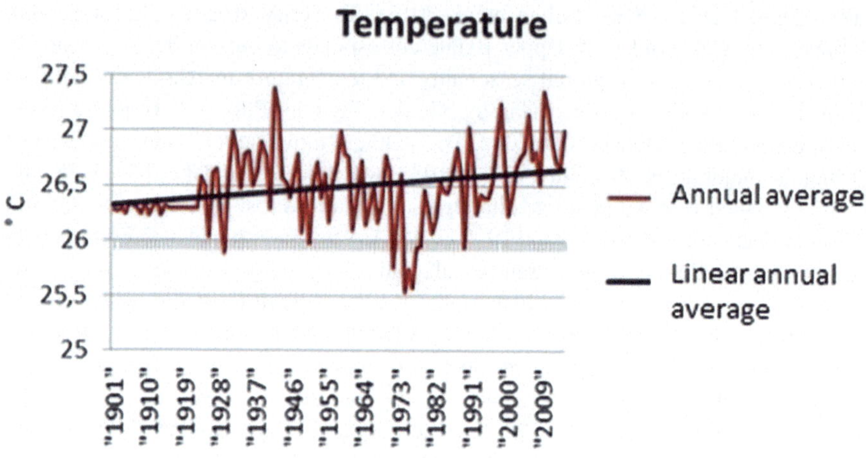

Fig. 6.8 Rise in temperatures between 1900 and 2009. (Data source: cruts_3.23)

Table 6.3 Number of recoded responses per person and type of changes (descriptive statistics)

Variable	Observations	Min	Max	Average	Stand. Div.
Rains/hydrology	121	0	5	1,57	1109
Land cover and diversity of flora	121	0	3	1,27	0,796
Demography/pressure on rural land	121	0	4	1,02	0,899
Fauna	121	0	3	0,97	0,806
Agricultural soil fertility	121	0	3	0,80	0,928
Political/social/economic changes	121	0	2	0,36	0,590
Temperature	121	0	1	0,34	0,475
Null	121	0	1	0,03	0,180

6.2.3 *Do Perceptions of Change Vary According to the Economic and Cultural Profiles of Stakeholders?*

Scientists and local people agree about features of climate change and the importance of demographic pressure on the environment. However, we tried to verify if the perception of changes is different according to one's economic and cultural profile. We tested various hypotheses. To do so, all the items cited by the informants were grouped into eight types (Table 6.3). For these data, the p-value significance test is good (less than 0.0001), except for the Null category, which is too small to be statistically reliable.

6.2.3.1 Age Influences the Perception of Change

We test first an age hypothesis, assuming that personal experience and reference to more or less remote periods of the past can influence the perception that one has of changes. We were expecting three contrasted answers:

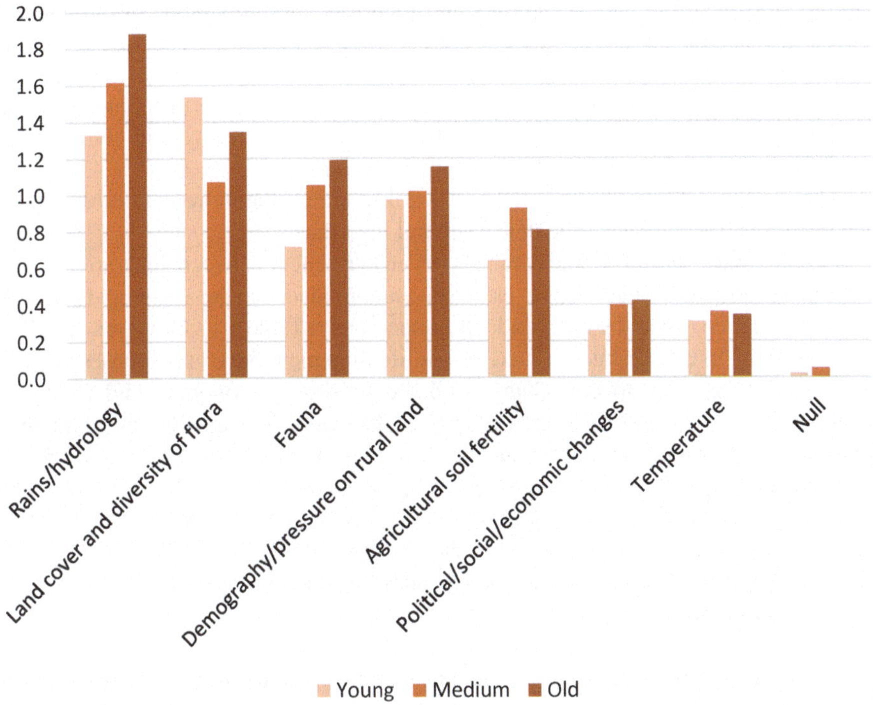

Fig. 6.9 Perception of changes by age (39 young people aged 20–35, 55 aged 35–50, 26 over 50 years old; expressed as number of citations of changes per person by age type)

- Elders know and observe biodiversity changes better (flora and fauna).
- Elders notice the rise in temperatures more than young people.
- Young people, because their activity depends directly on natural resources, suffer from the loss of such resources more intensely than elders.

Overall, the results are relatively consistent across the seven categories of change (Fig. 6.9). Logically, because they have less experience than older people, young people under 35 perceive less change than others. This is particularly true for changes related to climate (rainfall, hydrology, and temperature), wildlife loss, and agricultural soil fertility. On the other hand, young people report more changes in vegetation than their elders because they are particularly dependent on this resource. Thus, of the 47 responses "Useful species are difficult to find," 22 are provided by young people: natural resources collected in the bush (grass, wood, fish are the most cited) are indeed an important source of income, especially for young unmarried people who do not yet have plots to farm. The disappearance of these resources in the vicinity of villages and the increasingly remote location of supply sites are therefore particularly perceived by this category of actors.

The changes linked to the fertility of agricultural soils most often cited by the 35–50 age group can also be explained by their activity. Among the population interviewed, whose activity is mainly agriculture, this age group corresponds to people who have the largest families and therefore the most labor to farm large fields. They therefore tend to cite more the decline in soil fertility and agricultural production and the invasion of plots by weeds. This is particularly true, as we will see below, in villages on the periphery of the national park where cropping systems have evolved from slash-and-burn agriculture to a fallow-free system confined to the human areas delineated by the development plan.

The changes observed in rainfall and wildlife are all the more noticeable when the person is older, which validates our initial hypothesis.

Despite these variations, the result shows an agreement between generations. We obtain the same result between sexes, despite an imbalance in the sampling between women and men (26 and 94 respondents, respectively; Table 6.1). However, women are less likely to cite changes related to wildlife than men (54% of responses versus 77%) and less likely to cite changes related to social, economic, and political changes (15% versus 44%). Some of these results can be explained by the gendered distribution of work, in particular the fact that most of the hunting work is done by men.

6.2.3.2 Length of Residency in the Locality Affects Perceptions in Complex Ways

The second hypothesis tested here is that more time spent in one place gives a more acute perception of gradual changes (wildlife, demographic growth, soil fertility, and climate). The results are contrasted between people living in the region for a short or a long time.

Contrary to our hypothesis, it appears that the longer the period of residence in the place, the less people cite changes, on the one hand, and that the results are mixed according to the types of changes, on the other hand (Fig. 6.10). The perception of demographic pressure, climate variability, and temperature depends little on the duration of residency in the region, because it has the same impact on all the stakeholders in the territory. On the other hand, the decline in soil fertility is strongly felt by people who have been settled for less than 10 years, who are primarily migrants (Table 6.4).

In the migratory context of northern Cameroon, migrant farmers have been living in the villages studied for less than 15 years. They come either directly from northern areas where the shortage of available land pushed young people to leave to clear plots in other regions or from former migration areas where they flee the increased number of land conflicts with indigenous people. They arrive with already large families and a desire to clear large plots of land in the still uninhabited bush. Their perceptions of change are mainly focused on the result of their actions on the environment, i.e., the modification of vegetation cover, the development of agricultural land, and the decline of wildlife. They also strongly feel the vagaries of rainfall (less the faster drying of rivers due to a lack of perspective to compare), as well as the constraints of soil fertility in savanna areas, where the disappearance of fallow land in favor of an intensive input-based system makes it very difficult to control

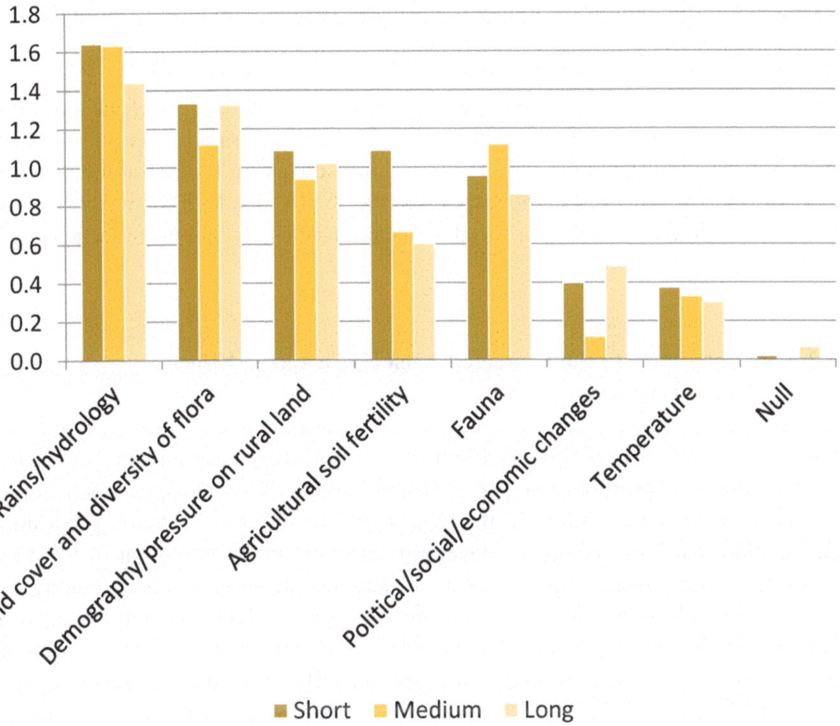

Fig. 6.10 Perception of changes according to the length of stay in the place (45 people who stayed less than 10 years ("short"), 33 people between 10 and 20 years ("medium"), 43 people over 30 years ("long"); expressed as number of citations of changes per person by length of stay)

Table 6.4 Number of changes cited by categories of people based on time living in the place (121 respondents)

	Short	Medium	Long	Total
Sample				
Native	9	16	35	60
Migrant	36	17	8	61
Total	45	33	43	121
No. of citations of changes				
Average	6.9	5.9	6.1	6.4
Std deviation	1.81	2.21	3.13	2.47
Min	3	1	1	1
Max	11	10	13	13

weed growth in plots and the renewal of organic matter. These changes are also felt by the natives (Table 6.4).

What is remarkable about this analysis is that newly settled people report on average more changes (7), with a smaller standard deviation, than those who have lived in the region for longer (6). Among people who have lived in the place for longer, the variability in citation is the greatest, with people who cite no change (an easy way to escape the researcher's tedious questions) or a lot. The comparison between the two categories of indigenous and migrant (regardless of the duration of the relationship, an elderly migrant can live in the place for a longer period of time than a young indigenous person) shows a consensus on the variables observed. The differences observed in the responses (Fig. 6.10) therefore show that it is duration in place that is the relevant variable to capture these variations in the perception of changes.

6.2.3.3 Proximity to the Protected Area Has a Large Effect on Perceptions

One assumption of participatory biodiversity protection programs is that environmental education promotes the perception of nature-related changes and thus interventions for their mitigation. In the region studied, the rehabilitation program for Benue National Park includes a development plan, the participation of local residents in the monitoring of protected areas, and also awareness-raising among local populations to limit tree cutting and prohibit poaching. Thus, we hypothesized that populations located near the protected area and benefiting from this awareness would perceive changes in vegetation and wildlife more than those living in an "ordinary" rural area (i.e., one with no biodiversity protections) who would perceive other types of changes better (or who have not been trained to respond with conservation statements). We tested this hypothesis on our dataset by comparing responses according to the proximity of informants to the protected area.

Overall, residents of the protected area perceive more changes than others, and the responses are particularly distinct for three types of changes: those related to wildlife, farm land, and temperatures (Fig. 6.11).

The increased perception of changes, on rains/hydrology, temperature, fauna, farmland fertility, and pressure on land, on the periphery of the Benue National Park is explained not by the effect of awareness of conservation issues but by the policies of protected area management which has concentrated and accelerated regional trends inside the zones of human use. Indeed, the Benue National Park development plan consists of zoning between biodiversity areas where no human activity is permitted, wildlife corridors to facilitate wildlife movement between the national park and other protected areas, and human use areas where all agricultural, pastoral, and collection activities must take place. By concentrating these activities within village areas without controlling the continuous arrival of migrants, the development plan has artificially increased population densities and human pressure in these areas. One of the consequences has been the disappearance of wild animals near these overcrowded areas, perceived not only as a loss in relation to the supply of bushmeat (which everyone consumes in the region, as well as in the large Cameroonian cities that represent the main outlet for this product) but also as a benefit with the reduction of crop predators. It is possible that the environmental education programs

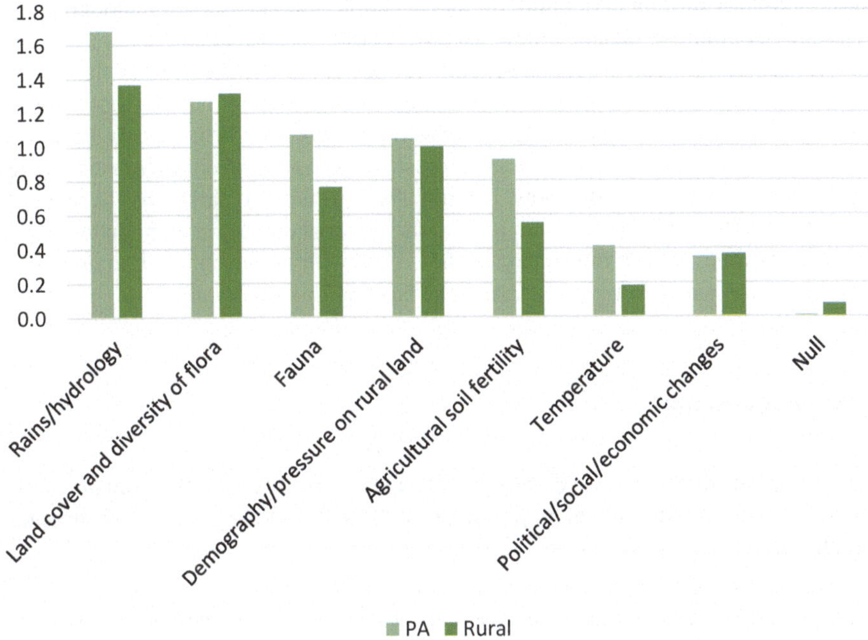

Fig. 6.11 Perception of changes according to proximity to the protected area (82 people in the villages of Djaba and Sakje near the national park ("PA"), 38 in the villages of Wante and Boumba remote from the park ("rural"); expressed as number of citation of changes per person by type of residence)

conducted by Benue National Park, particularly to ensure compliance with the delimitation of wildlife corridors, which is a major objective of the development plan, may be an explanation of the particular attention given to the disappearance of wildlife by riverside residents when asked about changes in their environment. However, the interviews show that personal experiences are at the root of this perception. In the two villages far from the protected area, the disappearance of wildlife is older and less cited by the interviewees, both by native people and migrants. Anthropogenic pressure is also very high, particularly in Boumba, where migrants began settling in 2010, which explains why the perceived changes in demography, vegetation cover, anthropogenic pressure, and social, economic, and political changes are not very contrasted.

The greater perception of soil fertility loss in villages bordering the PA is also explained by the development plan: by forcing the location of cultivated plots within authoritatively delineated human areas, it has also accelerated the disappearance of the fallow practice that allowed farmers to control grass cover and renew soil fertility. This change in farming system has been imposed on both migrants and native people, despite an attempt to set aside land to maintain their practices and set aside land for future generations (Raimond et al. 2017). In villages located outside the development plan, a system of more or less short fallow cultivation for migrants and slash-and-burn shifting agriculture for native people still persists, even though practices are beginning to change ("now we grow corn," "we use inputs").

The contrast in the perception of temperatures is more difficult to explain, as it combines a perception related to the practice of artisanal gold panning that extends near and inside the national park, in excavations where the heat quickly becomes stifling, with the passage from a shaded forest cover to an agricultural season where very few trees are kept in the plots. It could also be an effect marked by the level of education (47% of informants with a secondary level of education mention temperatures, compared to 26% in primary studies, 12% with no educational attainment), but this factor is difficult to interpret on the basis of this survey alone.

6.3 Conclusion

All classes of informants have observed changes in biodiversity, but climate variations are rarely cited as the main proximal cause of these changes in environmental conditions.

> *So many things have changed. First, the trees you see there, they are just the few that remain from slashing. Before, where we are now, it was the bush. Now, everything has been cut, there are no animals left. Today, it is difficult to see animals. Even monkeys are scarce, but there were plenty of them before. They were even coming home to eat, they were not afraid of us. Today, even ordinary birds, like fowls and guinea fowls, are scarce. We can hear them sometimes, but we don't see them anymore. They are much too afraid of humans. For sure, it is because they are hunted, and because the places they used to stay have deteriorated* (IB 13, native farmer, Boumba).

> *Changes are affecting me in everything I do or eat. Today, you cannot come back from the farm with a guinea fowl, and it is scary. For sure, it is better for us that there no more animals and birds close to the fields (less depredation), however one would like to see or have it* (IB 40, migrant farmer, Boumba).

It is difficult to draw conclusions, on the basis of this survey, about the effects of global discourses on perceptions of local change in Sub-Saharan Africa, but it is shown that these discourses focus on changes that are directly perceived by all respondents. Disappearance of wildlife, population pressure, and climate change are not only ideas conveyed by whistleblowers from various sectors who are seeking answers to problems coming from other places, such as the Benue National Park rehabilitation project for wildlife conservation (2000–2005), or successive development projects that have promoted tree conservation in cultivated fields to preserve soil and soil fertility and tree cover favorable to rainfall (North East Benue 1975–1993, Water Sol Arbres 2002–2011). They are also actual daily reality, which northern Cameroon residents perceive through a number of indicators. Changes are perceived by local people and their consequences are anticipated and managed. We have not underlined a clear signal of differing perceptions of change according to stakeholders' specific characteristics (age, ethnicity, education, etc.). Perceptions of change appear to be weakly influenced by global media narratives about global change and are more influenced by local experience and native strategies for mitigation and adaptation to change. Despite differences in ratings of the various changes and priorities, there is an overall consensus about ongoing changes. Biodiversity indicators are recognized according to their value for the subsistence of various stakeholders (fodder, firewood, pasture, bushmeat, etc.). However, whether for domesticated plant crops and animals or wild fauna and flora, the climate is neither the sole nor the most important perceived driver of ecological change.

The explanations for these changes differ completely from the global to local scale. For the inhabitants of the rural area of northern Cameroon, the changes are not due to increased greenhouse gas emissions that disrupt atmospheric circulation and warm the planet but to population increases in local villages that lead to more people clearing the bush, killing wild animals for food, increasing their consumption of grass and wood, and removing trees that promote rainfall and dew. Different causes and explanations, but identical effects: global and local discourses on change should be able to come together to define appropriate public policies.

While the relationship between deforestation and dewfall has long been demonstrated (Ellison et al. 2017), the correlation with the decline in local rainfall is not, although it is an idea that is well-rooted in the minds of an African population that has been aware of the issue of deforestation for several generations. It is perhaps in this explanatory trait that we must see, today, the effects of an awareness discourse on the perception of a very diverse population.

Acknowledgements This material is based upon work supported by the ANR Young Researcher Program ♯ANR-13-JSH1-0005-01.

Bibliography

Chevrillon-Guibert R, Magrin G (2018) Ruées vers l'or au Soudan, au Tchad et au Sahel: logiques étatiques, mobilités et contrôle territorial. Bulletin de l'association des géographes français (BAGF), septembre-novembre 2018, pp 272–289

Dardel C, Kergoat L, Hiernaux P, Mougin E, Grippa M, Tucker CJ (2014) Re-greening Sahel: 30 years of remote sensing data and field observations (Mali, Niger). Remote Sens Environ 140:350–364

Ellison D, Morris CE, Locatelli B et al (2017) Trees, forests and water: cool insight for hot world. Glob Environ Chang 43:51–61

Garine E, Langlois O, Raimond C (2005) «Le territoire est-il bien un patrimoine? Approche comparative de deux sociétés de la Haute Bénoué (Dìì, Duupa, Nord- Cameroun)». In M.C. Cormier-Salem, D. Juhé-Beaulaton, J. Boutrais, B. Roussel (éd.): Patrimoines naturels aux Suds: territoires, identités et stratégies locales. Paris, IRD, Colloques et séminaire, pp. 178–235.

IPCC (2014) Climate change 2014: synthesis report. Contribution of Working Groups I, II and III to the Fifth Assessment Report of the Intergovernmental Panel on Climate Change [Core Writing Team, R.K. Pachauri and L.A. Meyer (eds.)]. IPCC, Geneva, Switzerland, 151 pp

IPCC (2018) Summary for Policymakers. In: Global Warming of 1.5°C. An IPCC Special Report on the impacts of global warming of 1.5°C above preindustrial levels and related global greenhouse gas emission pathways, in the context of strengthening the global response to the threat of climate change, sustainable development, and efforts to eradicate poverty [Masson-Delmotte, V.P. Zhai, H.-O. Pörtner, D. Roberts, J. Skea, P.R. Shukla, A. Pirani, W. Moufouma-Okia, C.

Péan, R. Pidcock, S. Connors, J.B.R. Matthews, Y. Chen, X. Zhou, M.I. Gomis, E. Lonnoy, T. Maycock, M. Tignor, and T. Waterfield (eds.)]. In Press

Magrin G, Pérouse de Montclos MA (éds) (2018) Crise et développement. La région du lac Tchad à l'épreuve de Boko Haram. Paris, AFD, 294 p

Niel H, Leduc C, Dieulin C (2005) Caractérisation de la Variabilité Spatiale et Temporelle des Précipitations Annuelles sur le Bassin du Lac Tchad au Cours du 20ème siècle. Hydrol Sci J 50:2–243

Niang I, Ruppel OC, Abdrabo MA, Essel A, Lennard C, Padgham J, Urquhart P, 2014: Africa. In: Climate Change 2014: Impacts, Adaptation, and Vulnerability. Part B: Regional Aspects. Contribution of Working Group II to the Fifth Assessment Report of the Intergovernmental Panel on Climate Change [Barros, V.R., C.B. Field, D.J. Dokken, M.D. Mastrandrea, K.J. Mach, T.E. Bilir, M. Chatterjee, K.L. Ebi, Y.O. Estrada, R.C. Genova, B. Girma, E.S. Kissel, A.N. Levy, S. MacCracken, P.R. Mastrandrea, and L.L.White (eds.)]. Cambridge University Press, Cambridge, United Kingdom and New York, NY, USA, pp. 1199–1265.

Raimond C, Aboubakar M (2018) Politiques environnementales en Afrique subsaharienne : arbitrer les interactions entre projets paysagers contrastés d'une diversité croissante d'acteurs sur le territoire. In Grégoire E, Kobiane JF, Lange MF (eds). L'Etat réhabilité: les nouvelles politiques publiques en Afrique, Karthala

Raimond C, Garine E, Langlois O (2017) Territoire et biodiversité : mises en patrimoine et anticipations des acteurs. Une relecture à partir de la Haute Bénoué (Cameroun). *Bulletin de l'Association de Géographes Français*. 2017-2 CNFG « Paysanneries et patrimonialisation dans les Suds. Ressources, conflits, arrangements », pp 211–227

Seignobos C (2011) Le phénomène zargina dans le nord du Cameroun. Coupeurs de route et prises d'otages, la crise des sociétés pastorals mbororo. Afrique contemporaine, vol. 239, pp. 35–59.

Sivakumar MVK (1987) Predicting rainy season potential from the onset of rains in Southern Sahelian and Sudanian climatic zones of West Africa. Agric For Meteorol 42:295–305

Sultan B, Roudier P, Quirion P, Alhassane A, Muller B, Dingkuhn M, Ciais P, Guimberteau M, Traore S, Baron C (2013) Assessing climate change impacts on sorghum and millet yields in the Sudanian and Sahelian savannas of West Africa. Environ Res Lett 8(1):014040

Sultan B, Lalou R, Amadou Sanni M, Oumarou A, Soumaré MA (éds) (2015) Les sociétés rurales face aux changements climatiques et environnementaux en Afrique de l'Ouest. Marseille: IRD, coll. Synthèse, 464 p

Chapter 7
Climate Change in a Floodplain of the Brazilian Amazon: Scientific Observation and Local Knowledge

Esther Katz, Annamária Lammel, and Marie-Paule Bonnet

Abstract In the Lago Grande de Curuaí, a floodplain of the Amazon River, located close to the city of Santarem (Brazil), the inhabitants have always adapted their activities (centered on agriculture, fishing, cattle raising) to the seasonal fluctuations of the water level. Historically, strong floods occurred about every 20 years, but recently, they have occurred more frequently, every 2 or 3 years, reaching water levels previously unrecorded. In recent years, dry seasons have also been drier. Scientists mainly attribute these changes to global climate change, which can be observed and measured at the scale of the Amazon basin. This chapter compares the scientific data with the observations, knowledge, and interpretations of the inhabitants of the Lago Grande. Local people do not connect directly floods, drought, and climate change, because at the scale of the floodplain the water level is little impacted by local rainfall. Environmental change cannot be dissociated from social, economic, and political changes. It is generating uncertainty, but floodplain dwellers are more able to adapt to it than other societies.

Keywords Brazilian Amazon · Floodplain · Climate change · Local knowledge · Interdisciplinarity

---

E. Katz (✉)
IRD (Institut de Recherche pour le Développement) – UMR 208 PALOC IRD/MNHN, Paris, France
e-mail: Esther.Katz@ird.fr

A. Lammel
Université Paris 8-Vincennes-Saint-Denis – Laboratoire Paragraphe, Saint-Denis, France

M.-P. Bonnet
IRD (Institut de Recherche pour le Développement) – UMR 228 ESPACE-DEV, Montpellier, France

M. Welch-Devine et al. (eds.), *Changing Climate, Changing Worlds*, Ethnobiology, https://doi.org/10.1007/978-3-030-37312-2_7

7.1 Introduction

In the Lago Grande de Curuaí, a floodplain of the Amazon River, located close to the city of Santarem (Brazil), the inhabitants have always adapted their activities (presently centered on agriculture, fishing, and cattle raising) to the seasonal fluctuations of the water level. Historically, strong floods occurred about every 20 years, but recently, they have occurred more frequently, every 2 or 3 years, reaching water levels previously unrecorded. In recent years, dry seasons have also been drier. Scientists who study climate change and its impact on the Amazon basin have indicated that there has been an increase in rainfall over the past several years, strongly concentrated in the rainy season and leading to extreme events. However, the inhabitants of the Lago Grande do not refer to climate change when discussing extreme floods and droughts. Examining data on local perception of environmental change alongside scientific observations, the authors seek to understand how climate change fits into environmental changes in this area, how people interpret these changes and adapt to them, and how much local and scientific knowledges coincide.

7.2 Location of the Study

The Lago Grande do Curuaí (the Great Lake of Curuaí) is a floodplain (*várzea*) of the Amazon River, in the shape of a triangle, of about 200 km (approximately 120 miles) long. The floodplain, together with the local upland catchment, represents a surface of 4000 km^2. It is located between 2.3° and 1.9° South and between 56°10 and 55° West, between the cities of Santarem to the east, Óbidos to the north, and Juruti to the west. Its territory is divided between these three administrative units. It is delimited to the north by the Amazon River and to the south by the *terra firme* ("firm land," lands that do not get flooded). It is composed of several interconnected lakes seasonally or permanently connected to the Amazon by small channels (Bonnet et al. 2008). It is separated from Santarem by the Tapajos River, which is about 15 km (9 miles) wide where it flows into the Amazon (Figs. 7.1 and 7.2).

7.3 Methodology

The authors conducted field research in the Lago Grande as part of the interdisciplinary project *CLIM-Fabiam – Climate change and floodplain lake biodiversity in the Amazon Basin*, coordinated by the hydrologist Marie-Paule Bonnet. The anthropologists Esther Katz and Annamária Lammel participated in the work package *Local population and climatic and environmental changes: perceptions and adaptations*. The aim of the project was to understand, through different scientific approaches, the interaction between nature and society in that floodplain and to

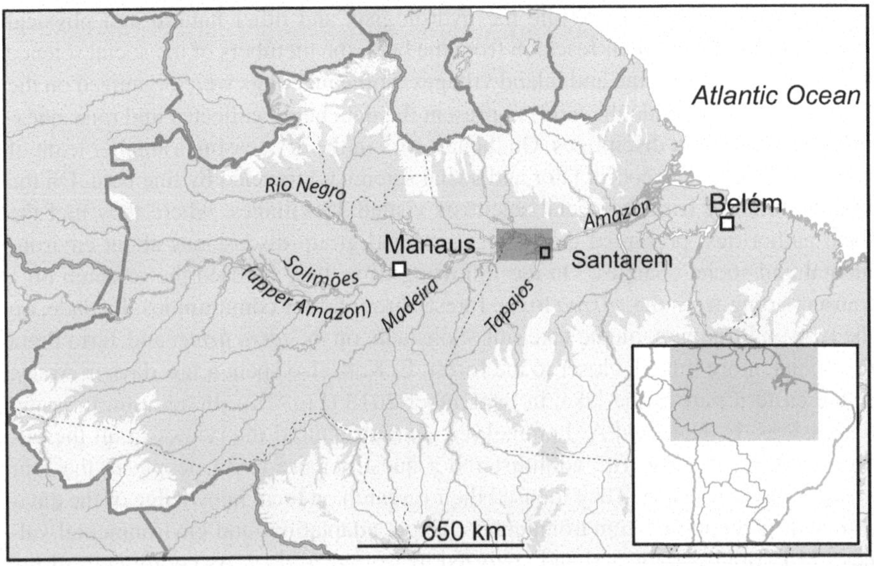

Fig. 7.1 Location of the study in the Amazon basin and in Brazil. (Figure courtesy of Elisabeth Habert (IRD))

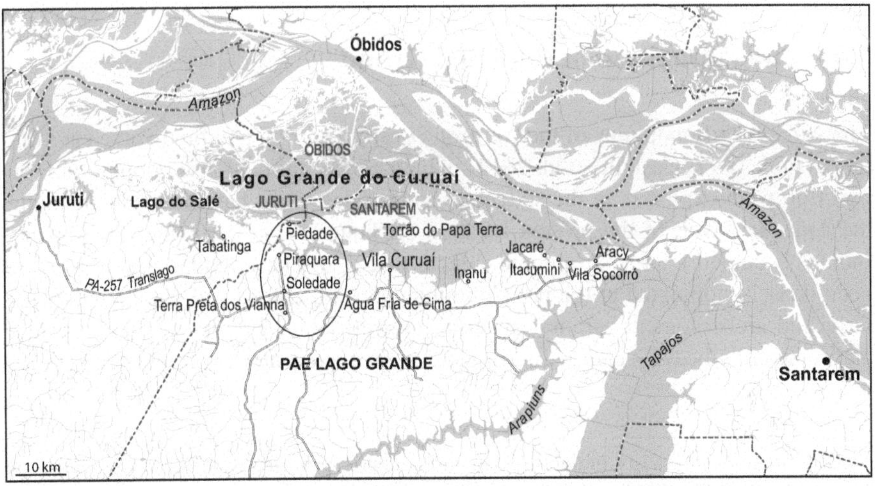

Fig. 7.2 Location of the study. (source: IBGE. Figure courtesy of Elisabeth Habert (IRD))

assess local biodiversity in a context of environmental and climate change, involving local stakeholders and proceeding to a restitution of the research to local communities. The data were collected over four periods of about 2 weeks each, two during flooding and two during flood recession: in March and September of 2013, March and November–December of 2014. During the first three field trips, a multidisciplinary team of about 20 people traveled to the lake on a boat outfitted to serve as a scientific laboratory. While the hydrologists and other natural and physical scientists[1] conducted their research from the boat, the members of the social science teams visited the riverine and inland villages. Some meetings were organized on the boat (as well as outside the field) to present the data being collected and raise questions from different disciplines. On the last trip, in the dry season, a smaller team of about 10 people (the coordinator and social scientists) traveled by line boat. On the first exploratory trip, the social scientists visited 18 villages, where they met the local authorities, presented the project, and had group discussions about environmental and social changes. On the following trips, they focused the research on a transect going from the *várzea* to the forest, through four communities, Piedade, on the *várzea*; Piraquara, on the lake side; Soledade, on the *terra firme*; and Terra Preta dos Viana, more inland, closer to the forest. E. Katz also spent a few days in Aracy, in the eastern part of the lake, in September 2013 (Fig. 7.2). In addition to group discussions, E. Katz and A. Lammel led semi-structured interviews in all the villages and, on the last trip, administered a questionnaire to 34 people in the four villages of the transect. They focused their research on local knowledge of the environment, perception of environmental change, adaptation, and environmental values.[2] M.P. Bonnet collected and analyzed hydrological data. As coordinator of the project, she performed a synthesis of scientific data collected in Curuaí floodplain over several years, which will be presented here.

7.4 The Population of the Lago Grande

The *várzeas* or floodplains of the Amazon constitute a rich and biodiverse environment (Junk 1997; Castello et al. 2013), which can sustain a dense population. The first European explorers in the sixteenth century described large settlements along the Amazon River. In the present location of Santarem, in particular, the Tapajos Indians occupied a large stronghold (Stenborg 2016). The Lago Grande must have been densely populated at the pre-Columbian period, since local inhabitants regularly find archeological ceramic material, as well as *terras pretas* (black earths), anthropic soils heavily charged in organic matter, presently exploited for agricul-

[1] Geochemists, geophysicists, biologists, ecologists, limnologists, geographers, and specialists of remote sensing participated in the project. They collected samples of water and soil, measured fluxes, led the first analysis of the material in the boat, and then established a landscape map and an analysis of cover change over the last 20 years.

[2] Both of them have researched these themes for over 30 years, E. Katz with an approach in ethnobiology and A. Lammel in cognitive anthropology and psychology. They started working on local knowledge on weather and climate in 1992, wrote articles, and coordinated three books on the subject (Goloubinoff et al. 1997; Katz et al. 2002; Lammel et al. 2008). Although it existed previously under other formulations, the field of indigenous and local knowledge of the environment was really defined as such after the 1992 Convention on Biological Diversity, when it started being taken into account for sustainable development and biodiversity conservation. It contributed to the promotion of participatory research (cf. Gadgil et al. 1993; Sillitoe 1998; Nakashima and Roué 2002).

ture. At the time of the first contact, the indigenous population around Santarem was struck by epidemics and then, at the end of the seventeenth century, was gathered in Jesuit missions, while the Portuguese Crown granted lands to colonists who developed cocoa plantations and raised cattle, with indigenous and African slave labor. At the end of the eighteenth century, after the expulsion of the Jesuits, Mission Indians were dispersed. According to Harris (2011), a "heterogeneous peasantry" composed of Indians, mestizos, poor Portuguese settlers, and freed slaves occupied the small lands in the interstices and the periphery of the large plantations; some of them settled permanently in the *várzea* in order to secure their access to its resources and adapted to that way of life, in spite of its hardships. Present-day inhabitants of the Lago Grande descend from this heterogeneous population. As other mestizo people who live on Amazonian riversides, they call themselves *ribeirinhos* ("riverine people"). Since the 1988 reform of the Brazilian Constitution, the *ribeirinhos* have gained the status of "traditional population." Responding to several years of social claims and to the threat of illegal mining and agribusiness exploitation, the Land Reform Institute (INCRA)[3] secured the land of traditionally established riverine and inland communities of the Lago Grande, by creating in 2005 the *PAE Lago Grande* (*Projeto de Assentamento Agroextrativista do Lago Grande*, Lago Grande Project of Agroextractivist Settlement), an area of approximately 290,000 hectares, included in the municipality of Santarem, following the limits of Santarem territory to the west, the Lago Grande to the north, the Amazon River to the east, and the Arapiuns River to the south (Fig. 7.2). It results today in 133 communities with a total of 30,000 inhabitants (Folhes 2010:17).[4]

7.5 The *Várzea*: An Ecosystem Marked by Seasonal Changes

"Floodplains are wetlands which oscillate between terrestrial and aquatic phases" (Junk 1997: 3). "The landscape of the floodplain is in a constant process of redefinition and becoming" (Harris 1998: 70). The flood covers riversides and islands for several months, leaving rich nutrients in the earth, and then it recedes for the second half of the year (Fig. 7.3). When uncovered, the *várzea* lands, a real "gift of the Amazon River," are particularly fertile (Bahri et al. 1991). In the Amazonian

[3] INCRA: *Instituto Nacional de Colonização e Reforma Agrária*, National Institute for Colonization and Agrarian Reform.

[4] The PAE allows the traditional population to secure its land through a concession led by a communal regime. The inhabitants are supposed to exploit the local resources through viable, socially fair, and ecologically sustainable economic activities (Brazil, Portaria/INCRA/P/N° 268 of October of 1996, quoted by Folhes, ibid.). But some *fazendas* (large agricultural properties) remain in between the communal lands. In 2007, the FEAGLE (*Federação das Associações das Comunidades do Assentamento Agroextrativista da Gleba do Lago Grande*, Federation of the Association of the Communities of the Lago Grande Agroextractivist Settlement) was created. It is a civil organization responsible for representing the settlement in front of government institutions and civil society (Folhes ibid.).

Fig. 7.3 Hut and trees on an island of the floodplain in the flood season (March 2013). (Photograph by E. Katz)

floodplains, the water level can fluctuate about 10 meters between the flood stage and the dry period (WinklerPrins 2002). When the flood is high, it drags fish from the Amazon River to the floodplains, and when it recedes, the fish are trapped in smaller lakes or ponds and are easy to capture. Rainy and dry seasons punctuate the year, but for the inhabitants of the floodplains, the alternation between flood and drought is more important.

7.5.1 Climate, Weather, and Seasons

According to the *ribeirinhos* of the Lago Grande, the rainy season (*inverno*, "winter") starts in December and ends in May–June, with a maximum of rain in January and February. The dry season (*verão*, "summer") starts in June–July and lasts until November. Actually, this season is rarely totally dry. According to measures taken by scientists since 1971 (Fig. 7.4), March is the wettest month of the year, but maximum rainfall can occur in January (Bonnet 2017).

The *ribeirinhos* pay attention to nature signs indicating the coming rain, especially the direction of the winds. They distinguish the winds coming from four directions: the *várzeiro,* from the *várzea* (from the north); the *terral*, from *terra firme* (from the south); the *vento de cima*, "upriver wind" (from above), from the west; and the *vento de baixo*, "downriver wind" (from below), from the east (Fig. 7.5). They observe that usually the clouds come from "below" (downriver), "from Santarem,"

Fig. 7.4 Mean mensal rainfall measured in Curuaí over the 1971 to 2016 period with minimum and maximum (Bonnet)

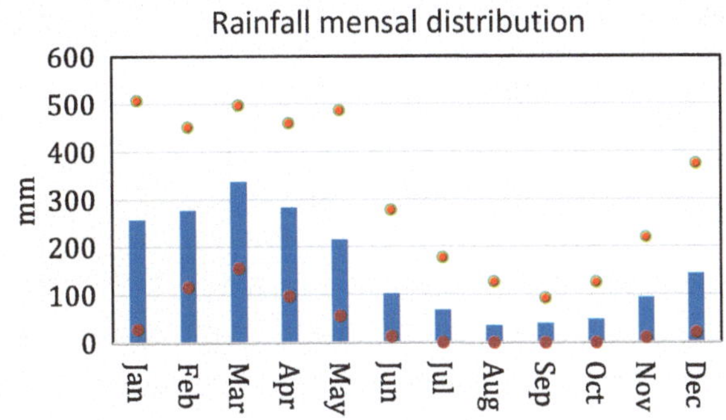

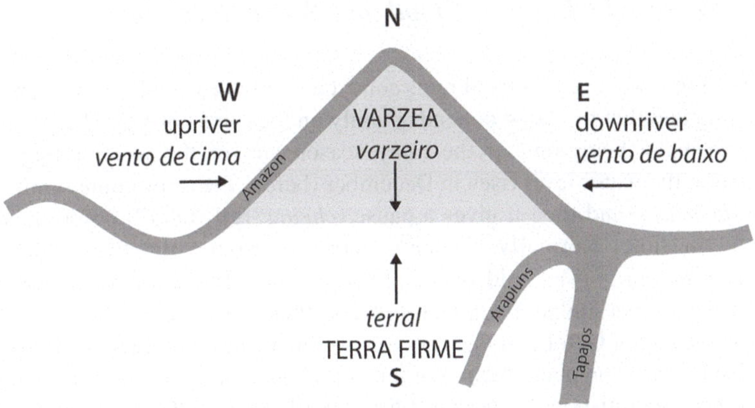

Fig. 7.5 Direction of the winds in the Lago Grande de Curuaí. (Source: field data)

but sometimes the *terral* brings clouds too. The upriver wind is the strongest and the coldest; it produces waves on the lake. It passes rapidly, while the downriver wind is not strong. One person told us that rain comes from the evaporation of the river or the lake and another one from the evaporation of the rain fallen on the earth. No one mentioned that the clouds could come from the Atlantic Ocean (located to the east, downriver). Some older people predict the coming of the rain or drought and the rainy or dry season by observing animals (birds, monkeys, toads, ants, termites, fish, river dolphins) and celestial bodies (sun, moon, Orion's Belt), but they only perform short-term weather forecasting.[5] For instance, the song of the parrots, the howling of the howler monkeys, or the flight of the termites announce the rain.

[5] In other societies where E. Katz and A. Lammel did fieldwork, in Mexico in particular, where the dry season is very dry and most traditional agricultural systems depend on rainfall, Amerindian farmers forecast the weather both for short- and long-term periods, through observation of nature as well as symbolical systems (Lammel et al. 2008).

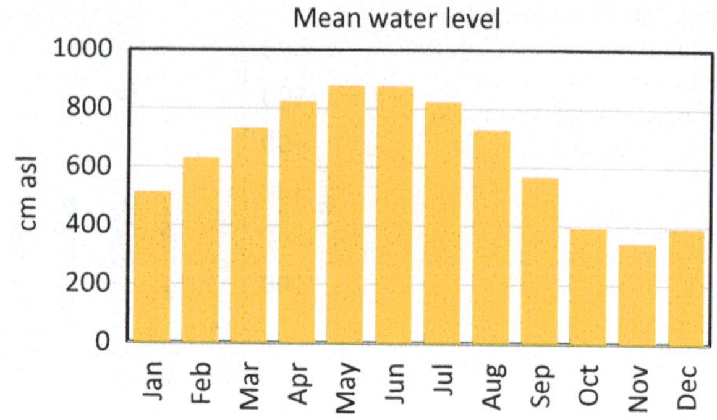

Fig. 7.6 Mean water level measured at Óbidos based on the 1981–2016 period (Bonnet)

7.5.2 Seasonal Changes: Flood and Water Recession

The *ribeirinhos* believe that local rains contribute only in a small proportion to the flood rising and that the flood depends mainly on the rains upriver. There is a time lag between the beginning of the rainy season and of the water rising. They observe that the water level rises in December (before the rains come) until about April (*enchente*), and then it gives a pause (*cheia*, "full water") before receding (*vazante*) in June. Normally, "it starts receding around the festival of Saint Anthony (*São Antonio*)," held on the 13th of June. The flood recession slows down in September and stops in October, and then the water of the lakes stands still from the end of October to the end of November (they call it *seca*, "drought").[6] They also observe the phenomenon of the "tide" (*maré*), ocean water flowing into the Amazon. Scientists indicate that it flows until Óbidos, 800 km from the estuary of the Amazon.

In a hydrological study led between 1997 and 2003 in the Lago Grande, based on in situ and satellite data, Bonnet et al. (2008) calculated that the Amazon River is the main source of water inputs to the flooded area year-round, accounting for about 77% of the annual total inputs; rainfall accounted for about 9% and runoff for 10%, while seepage from the groundwater system accounted for 4%. The hydrological residence time of the lake is about 3 months, and the floodplain makes a net contribution of water to the river. They observed that each year, the storage stage of the floodplain starts between November and January and lasts until May–June. The draining phase starts in July and lasts until November; the largest exported volume occurs from August to October. Bonnet (2017) recalculated recently the mean water level at Óbidos for the 1981–2016 period (Fig. 7.6).

Therefore, the *ribeirinhos*' observations coincide roughly with those of the hydrologists, except that the local inhabitants perceive moments of pause in the water rising and recession while the hydrologists calculate a slowing down of these fluxes. The inhabitants, like the hydrologists, believe that local rains contribute only

[6] Inhabitants of Manaus region and of the Tapajos also divide the year in periods of flooding, "full water," recession, and "drought" (Pinho et al. 2014).

in a small proportion to the flood rising and that the flood depends mainly on the rains upriver.

The *ribeirinhos* observe nature carefully in order to predict the magnitude of the coming flood and recession; they are particularly sensitive to the way the water rises (especially in January–February) and recedes, at which speed, which moment, and which levels it reaches compared to the previous years. People also seek information about the flood rising upriver, in Manaus or on the Madeira River, through radio, television, or phone calls from relatives living in Manaus, as this phenomenon occurs there at an earlier period. This way they get an idea of the importance of the coming flood. Some older people also forecast the floods and droughts from symbolical interpretations of natural signs, but few people now hold this knowledge. For instance, in Piraquara, none of the interviewed persons held it. The predictions usually relate to the next season, not to a longer period. The indicators are water, plants, animals (toads, birds), and celestial bodies (moon and constellations such as the Pleiades and the Southern Cross). As one example, an informant told us, "when the water recedes a lot, the next flood will be low; when it does not recede much, the next flood will be strong." Another offered, "if the 'tide' appears before All Saints' Day, the water will rise low; if it appears after that date, the flood will be strong." The most common prediction is based on the position in which, in October, the husks of the *jauari* (*Astrocaryum jauari*), a riverside palm tree, fall from the trunk. The husk is called *barca* or *canoa* (canoe), referring to its shape. "If the 'canoe' falls with its 'mouth' (the hollow part) up, the water will rise high, and with the 'mouth' down, it will rise low."

7.5.3 Seasonal Changes: Activities

In all the floodplains of the Amazon basin, the *ribeirinhos* organize their activities according to the rhythm of the seasons (Harris 1998) (Fig. 7.7). Their main activities have always been agriculture and fishing, but some people also raise cattle; they also practice some hunting and gathering, as well as extractivism in some areas. Agriculture is practiced on *várzea* lands in the drought period and on *terra firme* in the flood period. Short-cycle annual plants such as corn are typical of *várzea* agriculture. Corn, beans, squash, melon, watermelon, tobacco, and even short-cycle cassava (6 months cycle) are cultivated in the *várzeas* of the Lago Grande, as well as in other floodplains of the Amazon (Bahri et al. 1991; Harris 1998; Lima 2005; WinklerPrins 2002). In *terra firme*, cassava (including long-cycle varieties) is the main crop. Cattle have been raised in the Lago Grande since the colonial period, first by Portuguese colonists and currently by small farmers and large landowners. During the dry season, large pieces of land (*restinga*) appear, as water recesses. The inhabitants rebuild their wood huts with palm roofs (cf. Fig. 7.3) and settle there for a few months with their cattle that graze on the natural grasslands. During the flood season, the cattle remain on floating corrals (*marombas*) or are taken to *terra firme*. As fish is very abundant in the floodplain lakes, fishing is an important

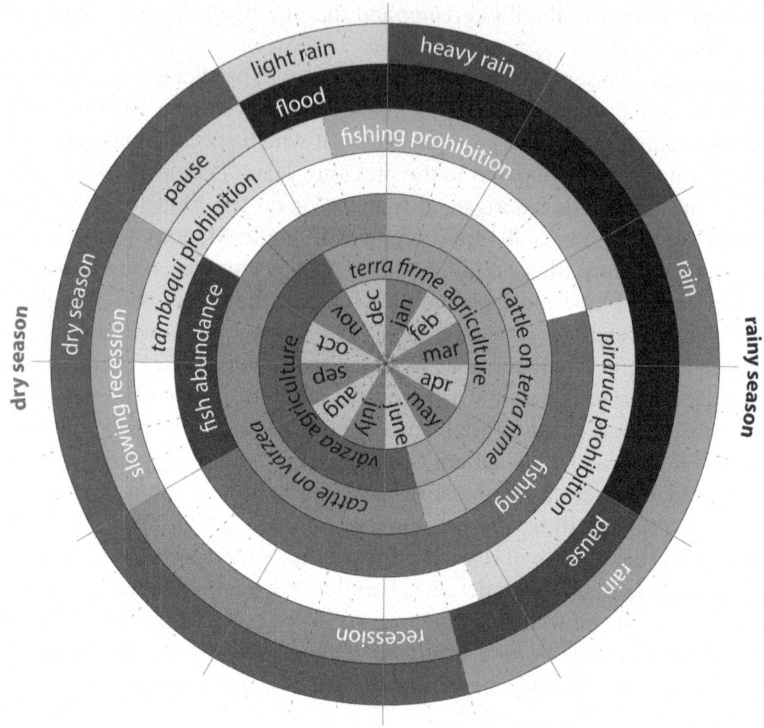

Fig. 7.7 Calendar: seasons, water fluxes, and activities (source: field data). (Figure courtesy of Laurence Bilhaut (IRD) and Esther Katz)

activity, but techniques have evolved over time and also differ from drought to flood period. Now fishing with nets is the more common technique, but some fishermen still use harpoons on big fish in the dry season.

7.6 Climatic Change in the Amazon According to Scientists

Over the last two decades, scientists have realized the likely impact of global warming on the Amazon basin and its moist forests as well as its consequences for the whole planet, since it plays a major part in the global water and carbon cycles (Nobre et al. 2012; IPCC 2014). This region experiences inter-annual variability in the amplitude, height, and timing (onset and end) of the rainy seasons, which is influenced by the El Niño-Southern Oscillation (ENSO) and the tropical Atlantic Ocean (Pinho et al. 2014). Some climate modeling analyses have predicted a drying trend. Nevertheless, hydrologists have reported that the catchment of the Amazon has been experiencing a substantial wetting since approximately 1990. This intensification of the hydrological cycle is concentrated overwhelmingly in the wet season, driving

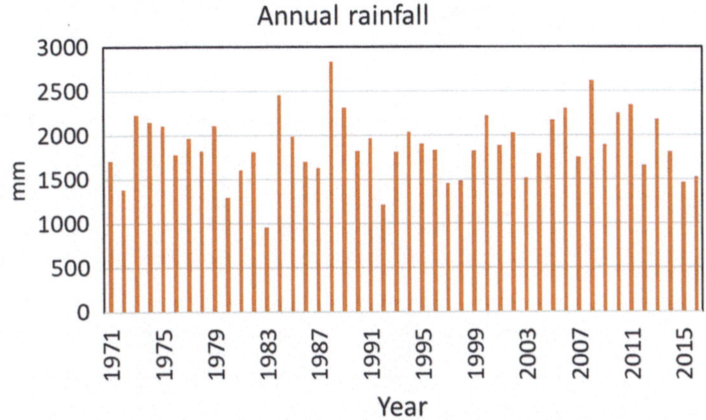

Fig. 7.8 Annual rainfall amount from 1971 to 2016 measured at Curuaí (Bonnet)

progressively greater differences in the Amazon peak and minimum flows, with an increase in the severity of flood and drought events (Gloor et al. 2013).[7] The increase of extreme events is effectively the main concern presently, for scientists as for riverine people. Scientists have measured water levels in the harbor of Manaus since 1903, in Fazenda Vista Alegre on the Madeira River since 1967, and in Óbidos from 1928 to 1947, then again since 1970 (Marengo & Espinoza, 2016). Recent analyses of these data show that major floods occurred about every 20 years until about 1990. Then they became more frequent, and this frequency has even increased since 2006. Until 2009, the highest water level recorded was that of the flood of 1953. In 2009, the water level went well above that of 1953, and in 2012, it reached still a higher level, at least in Manaus. Other major floods were recorded in 1976, 1989, 1999, and 2014. In the meantime, severe drought events took place in 1979–1981, 1982–1983, 1995, and 1997–1998 and extremely severe ones in 2005 and 2010 (ibid.). In Curuaí, the annual rainfall amount measured over the period from 1971 to 2016 ranges between 958 mm (in 1983) and 2836 mm (in 1988), with a mean value of 1883 mm (Fig. 7.8) (Bonnet 2017). There is no systematic unidirectional long-term trend toward drier or wetter conditions (Marengo 2009). At the basin scale, Gloor et al. (2013) reported an intensification of annual rainfall amount for the 2001–2010 period of about 10% when compared to the 1981–1990 period. In Curuaí, an increase of 4% between these two periods was registered (Bonnet ibid.).

At the Curuaí gauge, the water level has been measured since 1982. It reflects the measures taken at the Óbidos gauge (Fig. 7.9) (Bonnet 2017). It indicates, as the measures cited above for the Amazon basin, important floods in 1989, 2006, 2009, 2012, and 2014 but with a higher flood in 2009 than in 2012 and a very severe drought in 2016.

[7] These authors propose that the Amazon precipitation changes since 1990 are related to increasing atmospheric water vapor import from the warming tropical Atlantic.

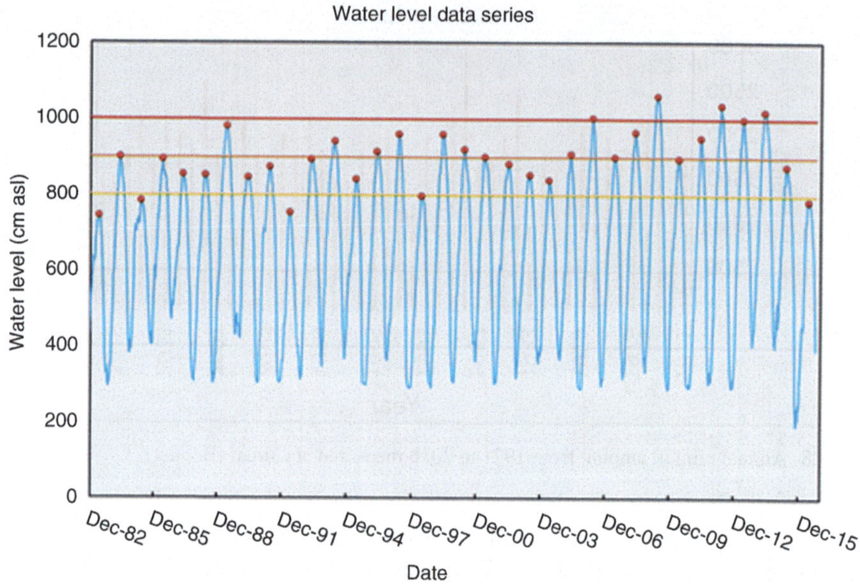

Fig. 7.9 Water level measured at the Curuaí gauge (Bonnet)

7.7 Perception of Climate Change in the Lago Grande

Very few interviewed people in the Lago Grande had heard about global climate change. This is probably due to the fact that they have had a limited access to the *media*. Electricity came in only around 2010, as well as television; at least until 2014, internet was available only in Vila Curuaí, and in most places, there was no connection for cell phones. A man in Tabatinga (Lago do Salé) had heard about "global warming" (*aquecimento global*) and ice melting (*derretimento das geleiras*). A young man in Piraquara talked about the ozone layer (*camada de ozono*) and the melting of the polar cap (*derretimento da calota polar*). A lady in Piedade said that strong floods were more numerous because "ice stones" were melting, due to a warmer climate (*pedras de gelo deretem porque esquenta*).

Sixty-two percent of people who answered the questionnaire noticed changes in the rainfall pattern, more irregular rainfalls, or changes in the moment of arrival of the rainy season[8]; a stronger intensity of the rains in the rainy season was the most cited feature (*o temporal é mais forte*), but some people said it rained less, especially in the dry season. They attributed the decrease of rain and the disappearance of morning mist or dew (*orvalho*) to deforestation. Their observations on the rainfall pattern fit with those of the scientists: according to the measures taken in Curuaí, the months of December, July, August, and October have been nearly 20% drier over the 2001–2010 period than over the 1981–1990 period, while the months of March

[8]We thank S. Nasuti who helped us synthesize the results of the questionnaires and calculate the percentages in the answers.

and April have been roughly 30% wetter. This means that the total annual rainfall is higher than before, but the rains are stronger in the rainy season, and the dry season is dryer (Bonnet 2017).

However, while the scientists cited above consider the changes in rainfall pattern an important factor in explaining the increase in extreme flood and drought events at the level of the Amazon basin, the inhabitants of the Lago Grande do not correlate flood and rainfall. This is because they rely only on their local observations, and at the local level, the floods do not depend on the local rainfall.

As far as temperature is concerned, 82% of the people who answered the questionnaire mentioned it had risen over the last years.[9] As a consequence, they found it difficult to work in the fields after 10–11 AM, while they used to work the whole day. Several people related it to deforestation, especially in Piraquara, where the forest used to be closer to the village.

Many informants mentioned that lightning, formerly a rare phenomenon, is occurring more often now, associated with the strong rains of the rainy season. They referred to persons being struck by lightning in several places of the Lago Grande and on the Arapiuns River. According to a resident of Agua Fría de Cima, thunder and lightning come along with strong winds from the *várzea*. Dwellers of several villages mentioned the emergence of occasional whirling winds (*remoinhos*) over the last 10 years. A man in Tabatinga said it used to happen only on the lakeside, but not inland. Inhabitants of Vila Socorro indicated an exceptional hail event (*granizo*) that struck violently for 10 minutes, on an early morning, during the 2012 dry season. In Aracy, it occurred in May 2013. Articles on climate change in the Amazon do not refer to these phenomena observed at a micro level, but a Brazilian research group on atmospheric electricity, ELAT-INPE, indicates that lightning tends to increase with global warming, affecting the tropics more.[10]

7.8 Observation of the Major Floods and Droughts by the Local Inhabitants

In the Lago Grande, when we asked about environmental changes, the first answer was the change in the flooding regime. Every informant (except in Terra Preta dos Viana, located inland, at some distance from the lake) mentioned immediately the increase of extreme events, especially the floods of 2009 and 2012 that greatly surpassed the previous water levels. They also mentioned the increase of severe

[9] People we interviewed in French Guiana in 2010 in the framework of the ANR-ACOCLI Project directed by A. Lammel (*Cognitive Adaptation to Climate Change*, 2008–2011) also complained of a higher temperature felt on land and sea.

[10] http://www.inpe.br/webelat/homepage/menu/infor/relampagos.e.efeitos/aquecimento.global.php, information from the website of ELAT-INPE (*Grupo de Eletricidade Atmosférica – Instituto Nacional de Pesquisa Espaciais*/Group of Atmospheric Electricity – National Institute of Space Research), accessed 11 jan 2019. We thank Laure Dentel for indicating research being led on lightning. However, we were not able to find more precise information.

Fig. 7.10 Stilt houses in Piedade in the flooding season (March 2013). The nearest house in the photo was rebuilt one meter higher after the 2012 flood. (Photograph by E. Katz)

droughts, but they were more concerned by the floods, especially these two major ones. Like the scientists, they observed that the 2009 flood was more severe than the one in 2012, unlike what happened in Manaus. In different villages, older people mentioned the 1953 flood as the most severe one in the twentieth century. Its memory was very clear for all of them, whether they had lived through it or had it reported by their parents.[11] Several people also indicated major floods in 1949 (there are no records for Óbidos at that time), 1989, 1996, 1999, and 2006 that did not reach the level of the floods of 2009 and 2012 but are also reported by scientists (Fig. 7.6).[12] In 2014, there was another major flood; its recession was slow (it lasted until October) and not complete. After our 2013–2014 fieldwork, a terrible drought hit the whole Amazon basin, from November of 2015 to April of 2016.

In Piedade, located on the *várzea*, in 2009 and 2012, the water rose about 1 meter above the floors of the houses. Some had to be rebuilt (Fig. 7.10). One couple reported spending 4 months in their boat. Other families left for other homes or relatives' houses in Curuaí or in Óbidos. Another couple even completely moved out in 2009 with their cattle and went to live in Piraquara, where they now own a butcher's shop; however, in the dry season, they still take their cattle to the *várzea*. In 2009, the water started receding only after the 30th of June. In 2014, the flood severely damaged the church of the

[11] For instance, according to a lady who was born in July of that year, her mother always told her that she had not grown up well because she was born during a major flood.

[12] Several people also had an approximate memory of the flood dates and gave erroneous dates, such as 1956 instead of 1953.

community, which had to be rebuilt. As a consequence of repeated major floods, several *várzea* communities disappeared. The inhabitants moved to the firm land. In Tabatinga, on the Lago do Salé, a man reported that his family moved from a nearby community called São Miguel after the 1953 flood and that all the inhabitants moved out after the last three major floods (2006, 2009, 2012); even the cemetery was flooded. In communities in the eastern part of the lake, Aracy and Itacumini, most of the inhabitants used to live on the lakeside, but they moved in the early 1990s because their houses and crops were getting flooded too often. In Aracy, only seven families out of 85 now remain on the lakeside, but most of their fruit trees died. In Itacumini, the church that used to be on the lakeside did not withstand the floods. M.P. Bonnet wonders whether these lakesides have been abandoned because of floods or lakeside land erosion. Lima et al. (2007: 48-49) mention such a phenomenon of "fallen lands" (*terras caídas*) in Manaus region, as does Harris (1998: 70) for the Parú floodplain near Óbidos.

WinklerPrins (2002: 416) points out that "The annual flood is an anticipated yet anxiety-provoking event because residents do not know how high the water will rise, how long the flood will last, or the speed with which it will build and recede." However, we described above that some people still perform predictions by interpreting signs from nature and receive information about the state of the flood upriver and, above all, that they observe the evolution of the water rising or receding in comparison with the previous years, a method that has also been reported by Pinho et al. (2014) about communities on the Tapajos and in the State of Amazonas. For instance, in March of 2014, a man in Piraquara told us that the flood would reach a high level, but not as much as in 2009, because in 2009 the water was already higher at the same period. He turned out to be right. Nevertheless, in spite of their observations and predictions, all the interviewed people expressed that they had not anticipated such dramatic floods in 2009 and 2012. Pinho et al. (2014) also show that, in the communities they studied, people cope with normal floods, but these recent major floods and droughts reached unexpected levels, making it harder to cope with.[13] Nakashima et al. (2012) argue that all over the world, traditional knowledge is challenged by rapid and unexpected environmental change.

7.9 Social and Environmental Changes

For the *ribeirinhos* of the Lago Grande, extreme floods and droughts are linked to other environmental and social changes affecting their activities. Several people indicated deforestation upriver as being the cause of extreme floods and droughts. M.P. Bonnet thinks that they are likely to be right. For many inhabitants, local deforestation, due to population growth, agriculture, and cattle raising, also causes small rivers to dry out and temperature to increase. In its turn, many activities are disrupted by extreme floods and drought and are likely to evolve.

[13] On the tributaries of the Tapajos, when they decided to move out during the drought, it was too late; the level of the river was already too low to navigate.

Substantial deforestation in the *várzeas* of the Lago Grande, as in other floodplains of the Amazon, was caused by jute (*Corchorus capsularis*), intensively cultivated as a cash crop from the 1940s to the 1990s (WinklerPrins 2006; Reno et al. 2011). Several decades ago, the forest used to be close to the lake, and jaguars used to attack the cattle in the riverside villages,[14] but nobody has seen them for a long time, especially after the Translago road was built, around 1970. Agriculture was a major activity of the Lago Grande until the late 1990s. In *terra firme*, land dedicated to slash-and-burn agriculture increased by 26% between 1985 and 1997 but decreased by 18% between 1997 and 2014 (Peres 2016). The villages of the Lago Grande produced cassava flour for their own consumption and also commercialized it in Santarem. But, according to local people, when the government started providing social assistance, especially family allowance and cultivators' retirement fees, in 1997, cassava production decreased tremendously, as confirmed by the satellite images. Currently, inhabitants produce cassava only for their own subsistence, but are not fully self-sufficient, so they import cassava flour from Santarem. Agriculture decreased even more in the *várzea* than in *terra firme*. *Várzea* crops described above can still be observed at the entrance of the lake (*boca do lago*), in the eastern part, but already in the villages of Aracy and Itacumini, people do not cultivate on the lakeside anymore, as it floods. In Piedade, in the western part of the lake, dwellers also used to cultivate *várzea* short-cycle plants and fruit trees, but they lost all their fruit trees and ceased that form of production. Now they only plant a few vegetables, herbs, and spices in raised pots and dedicate their activities to fishing and cattle raising.

Fishing has experienced a tremendous boom. Currently, fishing is the main economic activity in all the floodplains of the lower Amazon. It is estimated to be practiced by 84% of the households of these areas, who fish on average three times a week during 4–5 hour trips, catching approximately almost 1200 kg over a year, of which half is consumed by the family and the other half is sold (Almeida et al. 2011: 109). In the Lago Grande, the *ribeirinhos* used to fish only for their own subsistence (Fig. 7.11). They all recall how abundant (*farto*) it used to be. They occasionally sold a little bit of fish to Santarem, preserved in salt. When ice became available in the 1970s, they were able to sell bigger quantities.[15] It became more lucrative to commercialize fish than to sell cassava flour. While fishing intensified, agricultural activities decreased even more due to government social help. But in the 1990s, commercial fishery boats coming from other parts of the lower Amazon basin (Belém, Macapa, Parintins, Monte Alegre, Alenquer) started invading the Lago Grande, competing with local fishers. With the intensification of big commercial fishery threatening the local fish resources, fishers in *várzea* communities around Santarem initiated a process of closing off lakes and defining fishing laws (Almeida et al. 2011). In the PAE Lago Grande, a period of fishing prohibition (*defeso*) was established during the period of fish reproduction, from the 15th of November to the 15th of March, starting on the 1st of October for the *tambaqui* (*Colossoma macropomum*) and extended until the 15th of June for the *pirarucu*

[14]A middle-aged lady reported this information that she had learned from her father who had worked in a cattle farm on the lakeside when he was young.

[15]WinklerPrins (2002: 219) made the same observation on the *várzea* of Ituqui, east of Santarem.

Fig. 7.11 Local fishermen on the lake (March 2013). (Photograph by E. Katz)

(*Arapaima gigas*) (see Fig. 7.7).[16] Commercial boats were prohibited but are still seen every year on the lake. Guards from the IBAMA[17] office in Santarem often arrest them, seize their nets, and tax them, but they continue to return. Scientists estimate that only the commercial species are overexploited, while the impact on most other fish species (locally eaten) is only moderate (Almeida et al. 2011: 112). Fishermen of the Lago Grande observe that commercial species like *pirarucu*, *tambaqui*, *sorubim* (*Pseudoplatystoma fasciatum*), and *tucunaré* (*Cichla* spp.) have disappeared from the eastern part of the lake and are getting scarcer in the western part. This is also the case of *mapara* (*Hypophthalmus* spp.), a fish commonly eaten in the region. Manatees seem to have totally disappeared from the lake. Local people also fish in the small rivers (*igarapés*) that flow into the lake. Deforestation of the riparian vegetation has lead several rivers to dry, which was exacerbated during the extreme droughts. Extreme droughts also have had an effect on the fish in the lake: ponds where the fish get trapped during water recession dried out, leading the fish to die.

Cattle raising also increased when agriculture declined. Presently the "white cattle" (*gado blanco*), the Indian zebu, predominates, but water buffaloes were also introduced in the region in the 1970s, expanding around 2000. Then people realized that the water buffaloes were damaging the *várzea* environment by trampling grasses eaten

[16] *Pirarucu* is a huge fish that can reach 3 meters. It is very appreciated all over Amazonia. The scientific names for fish were taken from Furtado (1993: 153, Table 9: "Amazonian ichthyofauna found in Óbidos").

[17] IBAMA: Brazilian Institute of the Environment and Renewable Natural Resources (*Instituto Brasileiro do Meio Ambiente e dos Recursos Naturais Renováveis*).

Fig. 7.12 Cattle on the *várzea* before the flood (March 2013). (Photograph by E. Katz)

by the cattle and vegetation where the fish hide. Since 2005, the FEAGLE has advised inhabitants to get rid of the buffaloes or not let them in the *várzea*. With the increase of floods, the time spent by the cattle on the *várzea* is being reduced (Fig. 7.12). Inhabitants of *várzea* villages, like Piedade, used to leave their cattle on floating corrals (*marombas*) when the area was flooded. Now that the floods are higher, they have to drive their cattle to rented lands in communities in *terra firme*. Members of the Work Package *Modelling local population practices* identified that this process was inducing deforestation in the inland communities, such as Soledade and Terra Preta dos Viana (Bommel et al. 2016). By analyzing satellite images and going to the field, another member of the team was able to determine that 27% of the forest surface disappeared in favor of grazing land from 1997 to 2012 (Peres 2016). Deforestation has also had an impact on wild fauna that became scarcer. Dwellers of Terra Preta even observed that species such as white-lipped peccary have disappeared.

Ribeirinhos, used to an unstable environment, are very mobile and able to adapt to changes. Dwellers have moved out of the *várzeas* that now get flooded for too long. When the flood rose too high, they spent time on their boats or moved to another house. If their house got flooded, they built it again at a higher level; if it was damaged, they totally rebuilt it. When jute brought cash, *ribeirinhos* cultivated it, but when its price was too low, they dropped it. When floods on *várzea* lands lasted longer than usual, they gave up agriculture on these lands. When selling cassava flour became less lucrative than receiving family allowances, agriculture declined in favor of fishing and cattle raising, and when cattle could not remain on the *várzeas* all year long anymore, the raisers rented land in *terra firme*. But now we wonder whether *ribeirinhos* can adapt to changes that occur at a much faster pace.

7.10 Conclusion

The *ribeirinhos* of the Lago Grande observe their environment in-depth and have a strong working knowledge of it. They have developed and refined their skills to achieve maximum benefit from this specific habitat, and their livelihood strategies are highly adapted to this "pulsing" environment. Their knowledge is pragmatic, and they are clear on what they observe at a local level. They do not, however, extrapolate on what they have not seen, even in neighboring regions of the Amazon basin. When they forecast the weather or the water level, *ribeirinhos* base it more on tangible observations than on symbolical systems, and they perform the forecasting for a short-term period, the next day or the next season, not for a long period. It differs from other societies[18] and also from the scientists who observe and gather information at a wider scale. In this case, scientists can assert that climate change is taking place at the scale of the Amazon basin. On the other hand, without a global vision of the whole region, the *ribeirinhos* do not connect extreme floods and droughts to climate change, because the water level depends more on the flux from the Amazon River than on local rainfall. They observe changes in rainfall, in temperature, and in the presence of lightning or whirling winds. They perform fine observations at the local scale, even for phenomena like dew or lightning, not fully taken into account by scientists. Rather than climatic phenomena, they think that upriver and local deforestation is more likely to cause extreme floods and droughts, an opinion shared by scientists. Through their holistic way of thinking, they perceive environmental change in general as totally intertwined with social, economic, and political changes. Deforestation has been caused by population growth, cattle raising, and, until recently, agriculture. Fishing, especially when performed by outsiders, has led to an overexploitation of some aquatic resources. In the meantime, extreme floods and droughts have affected floodplain agriculture, fish stocks, and dwellings. Some people also mentioned that floods were a divine punishment as a response to moral sins and loss of respect for nature. Rapid environmental change is generating uncertainty (cf. Nakashima et al. 2012), but *ribeirinhos*, who have lived for generations in such a fluctuating environment, are more able to adapt to changes than other societies. Their knowledge of the local environment in the past and the present, at a micro level, can offer precious data for scientists and trigger new directions in their research.[19] In the *Clim-FABIAM* project, bringing together social and environmental scientists and local stakeholders has been very helpful to fully under-

[18] For instance, it differs from the Mexican Amerindians mentioned in note 5, who deal with symbolical systems for year-long weather forecasting, and the Kali'na Amerindians we interviewed in French Guiana (ANR-ACOCLI project). The Kali'na attributed local climate change to global environmental change, pollution caused by industrial countries, and disturbance in the atmosphere caused by rocket launching, and they were sensitive to the fate of polar bears on melting pack ice.

[19] Local and indigenous knowledge is increasingly taken in consideration in environmental assessments. See, for instance, the UNESCO program Local and Indigenous Knowledge Systems (LINKS) and the recent Intergovernmental Panel on Climate Change (IPCC) and Intergovernmental Panel on Biodiversity and Ecosystem Services (IPBES) assessments.

stand the diverse interactions between local production practices and the environment. In addition to confirming and interpreting scientific environmental observations, it has been possible, thanks to the social sciences team, to gain the trust of the local population, to share scientific results with them, and to collectively identify scenarios for a better management of the region.

Acknowledgements Project *CLIM-Fabiam – Climate change and floodplain lake biodiversity in the Amazon Basin* (2012–2015), funded by the French Foundation for Research on Biodiversity (FRB), with the participation of French and Brazilian teams. The analysis of the data is also integrated in a present project funded by the European Union, *Odyssea – Observatory of the Dynamics of Interactions between Societies and their Environments in the Amazon* (2016–2019) also coordinated by M.P. Bonnet (Horizon 2020 Research and Innovation Programme under the Marie Skłodowska-Curie grant agreement N° 691053).

In the field, the research project was led in partnership with the FEAGLE, the main authority for the Lago Grande (see note 4). Its president, Antonio Oliveira, facilitated it by introducing us to the local authorities of several villages during our first trip. We thank him as well as all the authorities and inhabitants who received us in the villages.

A. Lammel, P. Faulhaber (anthropologist, MAST), and the students A. Blaser (anthropologist, CDS-UnB) and A. Trégourès (MNHN) participated in Work Package 3 *Local population and climatic and environmental changes: perceptions and adaptations* coordinated by E. Katz. E. Coudel (agronomist, CIRAD), S. Nasuti (geographer, CDS-UnB), G. Melo (sociopsychologist, postdoc CIRAD), and M.P. Bonnet (also involved in other WP) participated to the Work Package 4 *Modelling local population practices and impacts on biodiversity: present and prospective analyses with stakeholders* coordinated by P. Bommel (modeller, CIRAD); they organized participatory workshops on scenarios of environmental change and were particularly active in the restitution of the research to the communities (about the results of this work package, see Bommel et al. 2016). P. Faulhaber, A. Blaser, P. Bommel, E. Coudel, S. Nasuti, and G. Melo and the three authors participated in the first exploratory trip. S. Nasuti synthesized the results of the questionnaires (see note 8).

References

Almeida O, Lorenzen K, McGrath DG, Rivero S (2011) Impacts of the co-management of subsistence and commercial fishing on Amazon fisheries. In: Pinedo-Vasquez M, Ruffino ML, Padoch C, Brondizio ES (eds) The Amazon Várzea: the decade past and the decade ahead. Springer / New York Botanical Garden Press, Dordrecht/Heidelberg/London/New York, pp 107–116

Bahri S, Grenand F, Grenand P, Guillaumet JL, Lourd M (1991) La *várzea* est-elle un don de l'Amazone? Les techniques traditionnelles de mise en valeur de la *várzea* face à la modernité. In: Sahel, Nordeste, Amazonie: Politiques d'aménagements en milieux fragiles. UNESCO/ l'Harmattan, Paris, pp 105–143

Bommel P, Bonnet MP, Coudel E, Haentjens E, Nunes Kraus C, Melo G, Nasuti S, Le Page C (2016) Livelihoods of local communities in an Amazonian floodplain coping with global changes: from role-playing games to hybrid simulations to involve local stakeholders in participatory foresight study at territorial level. International Congress on Environmental Modelling and Software. Paper 72, 8 p. http://scholarsarchive.byu.edu/iemssconference/2016/Stream-D/72. Accessed 27 Nov 2017

Bonnet MP, Barroux G, Martinez JM, Seyler F, Moreira-Turcq P, Cochonneau G, Melack JM, Boaventura G, Maurice-Bourgoin L, León JG, Roux E, Calmant S, Kosuth P, Guyot JL, Seyler P (2008) Floodplain hydrology in an Amazon floodplain lake (Lago Grande de Curuaí). J Hydrol 349:18–30. https://doi.org/10.1016/j.jhydrol.2007.10.055

Bonnet MP (2017) Dynamiques socio-environnementales des plaines inondables amazoniennes: la modélisation comme outil d'intégration des connaissances. HDR en Sciences de l'Univers, de l'Environnement et de l'Espace. Toulouse: Université Toulouse III-Paul Sabatier

Castello L, McGrath DG, Hess LL, Coe MT, Lefebvre PA, Petry P, Macedo MN, Reno VF, Arantes CC (2013) The vulnerability of Amazon freshwater ecosystems. Conserv Lett 6(4):217–229. https://doi.org/10.1111/conl.12008

ELAT-INPE. Informações: Relâmpagos e efeitos. Aquecimento global. São Paulo: Grupo de Eletricidade Atmosférica – Instituto Nacional de Pesquisas Espaciais.http://www.inpe.br/webelat/homepage/menu/infor/relampagos.e.efeitos/aquecimento.global.php, Accessed 11 Jan 2019

Folhes RT (2010) Cenários de Mudanças no Uso da Terra na Amazônia: explorando uma abordagem participativa e multi-escala no PAE Lago Grande. Master Thesis in Environmental Sciences. Belém: Universidade Federal do Pará, Instituto de Geociências

Furtado LG (1993) Pescadores do rio Amazonas. Um estudo antropológico da pesca ribeirinha numa área amazônica. Museu Paraense Emilio Goeldi, Belém

Gadgil M, Berkes F, Folke C (1993) Indigenous knowledge for biodiversity conservation. Ambio 22(2–3):151–156

Gloor M, Brienen RJW, Galbraith D, Feldpausch TR, Schöngart J, Guyot J-L, Espinoza JC, Lloyd J, Phillips OL (2013) Intensification of the Amazon hydrological cycle over the last two decades. Geophys Res Lett 40:1–5. https://doi.org/10.1002/grl.50377

Goloubinoff M, Katz E, Lammel A (eds) (1997) Antropología del clima en el mundo hispanoamericano. Abya-Yala, Quito

Harris M (1998) The rhythm of life on the Amazon floodplain: seasonality and riverine village. J R Anthropol Inst 4:65–82

Harris M (2011) The floodplain of the lower Amazon as a historical place. In: Pinedo-Vasquez M, Ruffino ML, Padoch C, Brondizio ES (eds) The Amazon Várzea: the decade past and the decade ahead. Springer/New York Botanical Garden Press, Dordrecht/Heidelberg/London/New York, pp 37–51

Junk WJ (1997) General aspects of floodplain ecology with special reference to Amazonian floodplains. In: Junk W (ed) The Central Amazon floodplain: ecology of a pulsing system. Springer, Heidelberg, pp 3–20

IPCC (2014) In: Core Writing Team, Pachauri RK, Meyer LA (eds) Climate change 2014: synthesis report. Contribution of Working Groups I, II and III to the fifth assessment report of the intergovernmental panel on climate change. IPCC, Geneva

Katz E, Lammel A, Goloubinoff M (eds) (2002) Entre ciel et terre: Climat et sociétés. Ibis Press/IRD, Paris

Lammel A, Goloubinoff M, Katz E (eds) (2008) Aires y lluvias. Antropología del clima en México. CIESAS/CEMCA/IRD, Mexico City

Lima D (2005) Diversidade socioambiental nas várzeas dos rios Amazonas e Solimões: perspectivas para o desenvolvimento da sustentabilidade. ProVárzeas/Ibama, Manaus

Lima HN, Wenceslau GT, Souza KW (de) (2007) Os solos da paisagem da várzea com ênfase no trecho entre Coari e Manaus. In: Fraxe TJP, Pereira HS, Witkoski AC (eds) Comunidades ribeirinhas amazônicas: modos de vida e uso dos recursos naturais. EDUA, Manaus, pp 35–52

Marengo JA (2009) Long-term trends and cycles in the hydrometeorology of the Amazon basin since the late 1920s. Hydrol Process 23:3236–3244. https://doi.org/10.1002/hyp.7396

Marengo JA, Espinoza JC (2016) Extreme seasonal droughts and floods in Amazonia: causes, trends and impacts. Int J Climatol 36:1033–1050. https://doi.org/10.1002/joc.4420

Nakashima D, Roué M (2002) Indigenous knowledge, peoples and sustainable practice. In: Munn T, Timmerman P (eds) Encyclopedia of global environmental change, Social and economic dimensions of global environmental change, vol 5. Wiley/Unesco, Chichester, pp 314–324

Nakashima DJ, Galloway McLean K, Thulstrup HD, Ramos Castillo A, Rubis JT (2012) Weathering uncertainty: traditional knowledge for climate change assessment and adaptation. UNESCO/ UNU, Paris/Darwin

Nobre CA, Reid J, Veiga APS (2012) Fundamentos Científicos das Mudanças Climáticas. Rede Clima / INPE, São José dos Campos

Peres LGM (2016) Análise temporal do uso e cobertura da terra na bacia hidrográfica do Lago Grande do Curuai, Pará. Master Thesis in Geography. Brasília: Universidade de Brasília

Pinho PF, Marengo JA, Stafford-Smith M (2014) Complex socio-economical dynamics driven by extreme events in the Amazon. Reg Environ Chang 15:643–655. https://doi.org/10.1007/s10113-014-0659-z

Renó VF, Novo EMLM, Suemitsu C et al (2011) Assessment of deforestation in the lower Amazon floodplain using historical Landsat MSS/TM imagery. Remote Sens Environ 115(12):3446–3456

Sillitoe P (1998) The development of indigenous knowledge: a new applied anthropology. Curr Anthropol 39(2):223–252. https://doi.org/10.1086/204722

Stenborg P (2016) Towards a regional history of pre-Columbian settlements in the Santarém and Belterra regions. In: Stenborg P (ed) Beyond waters. Archaeology and environmental history of the Amazonian inland. University of Gothenburg - Department of Historical Studies, Gothenburg, pp 9–22

WinklerPrins AMGA (2002) Seasonal floodplain-upland migration along the lower Amazon river. Geogr Rev 92(3):415–431. https://doi.org/10.1111/j.1931-0846.2002.tb00151.x.

WinklerPrins AMGA (2006) Jute cultivation in the lower Amazon, 1940-1990: an ethnographic account from Santarém, Pará, Brazil. J Hist Geogr 32:818–838

Chapter 8
The Year People Helped Zebras to Stand Up: Climatic Variability and Extreme Weather Observed and Portrayed by Kenyan Maasai Pastoralists

Joana Roque de Pinho

Abstract In 2009, an extreme drought devastated herds of livestock and wildlife across southern Kenya and impacted local livelihoods. For local Maasai pastoralists, this event was both unprecedented and perceived as part of a trend in increasing climatic instability. In this chapter, I contextualize Maasai pastoralists' experiences of this extreme drought by examining their climatological and meteorological knowledge, and how it informed their interpretations of the drought. Based on mixed-methods research, including visual ethnography, I first examine Maasai perceptions of recent and current climatic conditions in Kajiado and Narok counties, and their weather forecasting expertise (which includes observation of fauna and flora). Second, I explore Maasai observations of biodiversity changes associated with climatic changes. I do so by analyzing images of unusual animal behavior photographed during the 2009 drought by Maasai agropastoralists who acted as collaborative researchers and photographers. Across both counties and over a decade, perceptions of evolving climatic variability feature longer dry seasons, more intense rains and droughts, altered spatial and temporal rainfall patterns, and the notion that traditional forecasting knowledge is increasingly irrelevant as weather events and indicators are no longer synchronized. The 2009 drought was not predicted and, as my collaborators documented in photography, caused animals to behave strangely and in unprecedented ways, e.g., letting themselves be touched by people, entering settlements, and dying of starvation. This participatory visual research project resulted in unexpectedly rich, fine-grained information on local drought impacts and on human and nonhuman coping, which was later shared with the community and local policy-makers. Importantly, the participatory process fostered critical debate on future livelihood options under climate change among the project participants and their local audience.

J. Roque de Pinho (✉)
Instituto Universitário de Lisboa (ISCTE-IUL), Centro de Estudos Internacionais, Lisboa, Portugal

Natural Resource Ecology Laboratory, Colorado State University, Fort Collins, Colorado, USA
e-mail: roquedepinho.joana@gmail.com

M. Welch-Devine et al. (eds.), *Changing Climate, Changing Worlds*, Ethnobiology, https://doi.org/10.1007/978-3-030-37312-2_8

Keywords Pastoralism · Climate change · Drought · Indigenous knowledge · Weather forecasting · Animal behavior · Collaborative research · Participatory photography · Photovoice · Kenya · Maasai

8.1 Introduction

In 2009, as an extreme drought assailed Kenya (Zwaagstra et al. 2010; Mwangi 2016) and northern Tanzania (Goldman and Riosmena 2013), the relationship that exists between Maasai pastoralists, their livestock, and wild animals around Amboseli National Park (ANP) intensified[1]: not only did human-wildlife conflict increase, with starving wild herbivores raiding cultivated fields and predators killing debilitated livestock (Hazzah et al. 2013), but some wild animals, seemingly unafraid, let themselves be touched by people (Fig. 8.1), while others tried to enter homesteads.[2] Astonished, people further witnessed wild animals roaming beyond their habitats and mingling with species they do not usually encounter, and livestock and wildlife eating foods out of their dietary ranges. Most shockingly, individuals of wildlife species known to survive droughts *died*.

Maasai are intent observers of wildlife behavior for both practical reasons (Goldman 2007) and pleasure (Roque de Pinho et al. 2014). For them, these unprecedented observations spoke to the exceptional character of this drought, which led to the *en masse* death of livestock and wild animals (Kenya Wildlife Service et al. 2009; Wangai et al. 2013; Western 2010) and to novel human (Goldman and Riosmena 2013; Rutten 2016) and nonhuman responses – a drought so severe that older pastoralists, holders of long-term memory of climatic variability, characterize it as the worst they can remember.

Increasingly frequent and intense extreme weather events like the 2009 drought are likely outcomes of global warming processes (IPCC 2012). Pastoralists throughout the East African drylands are reporting longer droughts in a context of ever more erratic rainfall patterns (Gichangi et al. 2015; Pas 2018; Roque de Pinho et al. 2009), which are rendering their livestock-reliant, rainfall-dependent livelihoods riskier. Like other pastoralists in East Africa (Luseno et al. 2002) and elsewhere (e.g., Marin 2009), Maasai have developed forecasting expertise based on long-term observation of weather patterns, natural phenomena, and biodiversity. However, like others, they are nowadays finding this body of knowledge less reliable. In 2009, Amboseli Maasai acknowledged that their traditional forecasting system did not help in predicting the drought's beginning; neither did rainmaking rituals make it stop nor could "confused" elders predict whether/when it would rain again based on signs Maasai have historically interpreted.

[1] For Amboseli Maasai pastoralists, independently of how they feel about specific wildlife species, people, their livestock, and wild animals have always "stayed together" on the same land, all being part of God's creation (*Enkai naitayuo*) (Roque de Pinho 2009).

[2] *Enkang*, pl. *inkangitie*: homesteads composed of several households that may (or not) be family related.

Fig. 8.1 Photo by Jackson ole Korduni

At the same time, as the drought unfolded, people observed, interpreted, and dealt with the behaviors of domestic and wild animals as they, too, reacted to the extreme biophysical conditions in unprecedented ways. Some of these behaviors were photographed, narrated, and interpreted by a group of Maasai women and men who were collaborative researchers in a participatory visual anthropology project. Within the project's participatory photography framework (also known as photo-voice: Gubrium and Harper 2013; Wang et al. 1996), these men and women trained their cameras on the devastation and people's and animals' attempts at survival; collectively discussed and interpreted each other's pictures' meanings; and later shared their perspectives with their community, local policy-makers, and civil society representatives, all critically debating drought and pastoralism issues.

In this chapter, I contextualize Maasai experiences of the 2009 drought by examining Kenyan Maasai climatological and meteorological knowledge and how it informed their interpretations and reactions to that disaster. I first examine Maasai perceptions of recent past and current climatic conditions in Kajiado and Narok counties and Maasai traditional forecasting expertise, which includes monitoring abiotic and biotic indicators. Second, I focus on how Maasai agropastoralists in the Amboseli ecosystem made sense of the 2009 drought, and I analyze observations, photographs, and associated narratives of animal behavior created by my research collaborators at its height. Finally, I discuss how participatory photography, innovatively employed in the study of local perceptions of environmental changes *during* an extreme weather event, can inform public debate and contribute to communicating local knowledge of environmental changes to policy-makers.

8.1.1 Variability in the Kenyan Rangelands and the "Worst Drought Ever"

Maasai are one of the pastoralist groups inhabiting Kenya's arid and semiarid lands, an area covering 80% of the national territory and supporting about seven million people (Herrero et al. 2010). These rangelands are characterized by nonequilibrium ecological dynamics, i.e., by highly variable temporal and spatial distributions of precipitation and grazing resources (Ellis and Swift 1988). Subsisting on livestock in these drylands requires responding flexibly to strong seasonality in rainfall and availability of resources, as well as to sporadic extreme events like droughts and floods that affect rangeland production, livestock condition and productivity, and human welfare (Scoones 1995). Nomadic and seminomadic pastoralists do this by strategically moving their livestock to harness temporally and spatially transient nutritious forage (Krätli et al. 2012; Krätli and Schareika 2010), together with reserving dry season pastures and using their social capital to access resources at specific times (Galvin 2009). Awareness of rainfall patterns (past, present, and future) and pasture status is thus crucial to inform communal grazing management, and livestock breeding and trading decisions.

Droughts are currently the main hazard affecting the food security of Kenyan pastoralists (Zwaagstra et al. 2010; Mwangi 2016). While simulation models for East Africa are unclear about future changes in drought severity (reviewed in Herrero et al. 2010), increasingly frequent extreme weather events are expected to profoundly impact dryland livestock systems, which are among the most vulnerable to climate change in Africa (Thornton et al. 2009). Across the region, pastoralists and farmers are reporting increasingly shorter rainy seasons and more severe, sometimes clustered, droughts (Mwangi 2016; Orlove et al. 2010; Speranza et al. 2010).

In 2009, Kenya as a whole suffered a record-breaking drought. In the drylands, historic monthly precipitation was below average, rangeland production was the lowest ever, and livestock mortality was even higher than in the 1999–2000 and 2005–2006 severe droughts (Mwangi 2016; Mworia and Kinyamario 2008; Nkedianye et al. 2011; Zwaagstra et al. 2010). In the Amboseli ecosystem, people responded with time-tested coping strategies, such as long-distance migration of cattle, and experimented with new ones. Still, most of the cattle and small stock died. By December 2009, over 3000 families had lost *all* their livestock to starvation. Wild herbivore populations collapsed between September and November 2009 (Wangai et al. 2013; Western 2010). Amboseli National Park, a key tourism site, suffered high revenue losses (Kenya Wildlife Service et al. 2009). Endowed with adequate pastures, neighboring Narok County initially hosted herds from Kajiado County and other areas. When grazing ran out, both resident and guest herds emigrated, many animals dying in the process (Interview data, 2011).

Maasai distinguish droughts according to the degree of their impact: *olameyu* is used generically for any dry spell, including "dry season"; in *olameyu sapuk* ("big drought"), a long period without rain leads to extensive forage deficit and long-distance migrations (as in 2000 and 2005), while *emboot* is an extreme drought featuring massive livestock and wildlife mortality, e.g., the 2009 drought (Field notes, 2002–2004, 2009; personal communication by project photographer Isaac ole Mutunkei, 2018).[3]

In Maasailand, *emboot* droughts causing hunger, extensive migrations, and livestock mortality are recurrent, occurring every ten to twenty years. Why did Amboseli Maasai consider the 2009 drought as the worst ever (Interview data, 2009; Western 2010)? For Western (2010), the failure of consecutive rainy seasons in 2008–2009, coming after the 2000 and 2005 droughts, pushed Amboseli – a social-ecological system already suffering from forage deficit caused by changes in livestock and wildlife population dynamics and in land use – to a "tipping point." In recent decades, factors like land privatization, cultivation, sedentarization, fencing, land use restrictions for conservation, and urbanization have interacted to fragment East African pastoral ecosystems, thus curtailing livestock and wildlife mobility and people's capacity to respond to environmental fluctuations, and worsening their vulnerability to climatic shocks (Galvin 2009; Goldman and Riosmena 2013; Thornton et al. 2009; Watson and Kochore 2012; Western and Manzolillo Nightingale 2005). These combined effects were strongly felt in 2009 in Kajiado County (Rutten 2016; Mwangi 2016), particularly so in the Amboseli ecosystem (Kenya Wildlife Service et al. 2009; Western 2010), causing the extraordinary devastation and the human and nonhuman responses that the Maasai photographers poignantly portrayed (this chapter).

In defining what droughts are, Maasai acknowledge linkages between the ecological, climatic, and social dimensions of their livelihoods. Excluding the notion of precipitation amount, Maasai actually consider droughts to be conditions of forage depletion caused by altered spatial and temporal patterns of rainfall in interaction with socioeconomic forces, which worsen livestock condition and productivity, and human welfare. For instance, in-migration of cattle to an area with nutritious pastures may "bring the drought" to that area (Goldman and Riosmena 2013). For Mwangi (2016), a "pastoralist drought" is a multivariate and multiscalar event that simultaneously manifests meteorologically (rainfall delays/failures), ecologically (forage deficit), and socioeconomically (household food shortage) and is differentially perceived and experienced across microscales and households.[4] Speranza et al. (2010) also show that non-meteorological factors shape how agropastoralists determine *onset*, *duration*, and *end* of droughts. Going further, Goldman et al. (2015) argue that beyond investigating how Maasai pastoralists *know* droughts, we should examine how they *enact* them through practice, experiencing changed rain-

[3] Rutten (2016) mentions *emperi* as another drought category, corresponding to a long-lasting drought (more than a year) featuring drying of water holes, long-distance migration of animals, and massive deaths of domestic and wild animals. This would correspond to the 2009 drought, but for my informants, the 2009 drought was *emboot*. Rutten (2016) suggests that the term *emperi* is less known among younger pastoralists, partly because food relief nowadays prevents human casualties.

[4] Drought somewhere may be "lack of drought" elsewhere: in 2002–2004, the fact that my informants living in the northern part of the study area continuously complained of drought was gently mocked by my friends in the south, who saw them as actually spoiled by an abundance of good pastures that did not exist further south.

fall timings, long-distance mobility, numbers of cattle in pastures, lack of forage and water for animals, livestock health, crop yields, and cattle market prices. Throughout this chapter, this is the cultural framework my informants and collaborators articulate when they explain their experiences of climatic variability, and visually and narratively represent the 2009 drought's extraordinary character.

8.2 Methods

For Crate and Nuttall (2009), anthropologists increasingly encounter effects of climate change in their study areas even when not focusing on climate. Their observation echoes my experience working on non-climatic issues yet hearing a lot about changing weather patterns from my collaborators in African rural communities. This chapter thus relies on partly serendipitous ethnographic data on local knowledge of climate collected through mixed methods between 2002 and 2011 in Kajiado and Narok counties in Kenya Maasailand. My main sources of information are interviews and conversations in which people spontaneously explained transformations in rainfall dynamics and how they made sense of them. Insights on climate-related observations of environmental processes and biodiversity also originated in conversations I had with friends while observing night skies, herding cattle, and strolling through cultivated fields. A research partnership with Amboseli pastoralists (described below) allowed for the collection of visual and narrative data on the 2009 drought.

8.2.1 Study Areas and Data Collection

The Amboseli study area is a 8,500 km^2 semiarid savanna ecosystem in southeastern Kajiado County (BurnSilver and Worden 2008), whose boundaries are the range historically covered by rainy season dispersal of wildlife out of the Amboseli basin and swamps – now Amboseli National Park (ANP)[5] – onto Maasai-owned land (Western 1973). Precipitation is temporally and spatially patchy, ranging from 500–600 mm/year in the north, and on the Kilimanjaro slopes and Chyulu Hills, to 250–300 mm/year further south. Its pattern is bimodal, featuring a "long rains" season (approximately March–May) and a "short rains" season (approximately November–January) (Altmann et al. 2002). For twenty months between 2002 and 2004,[6,7] I conducted 192 semi-structured interviews with randomly selected pasto-

[5]Amboseli NP is a 392 km^2 unfenced protected area created in 1974.

[6]That is, Imbirikani, Olgulului-Lolarrash, and Osilalei group ranches. Group ranches are ranches collectively owned by a group of registered Maasai household heads.

[7]Maasai sections (*olosho*, pl. *iloshon)* are historically politically autonomous territorial units with their own celebration of ceremonies, architecture, dress, and linguistic specificities.

ralist and agropastoralist household heads and one dependent each in three Maasai-owned group ranches spaning the Kisonko and Matapato sections.[8] When asked to free-list challenges to their livelihoods, my informants frequently cited and then explained altered rainfall and ecological patterns. During this period, I collected additional data on these topics in focus group interviews with elders, doing so as well as in July–August 2009, during the drought, and in its immediate aftermath (December 2009).

Data collected two years *after* the 2009 drought in an ecologically and culturally different area, the Maasai Mara in Narok County, eastern Kenya, offer a broader perspective on Maasai climatological and meteorological knowledge. Part of the Purko Maasai section and located in the north of the 30,000 km^2 greater Serengeti-Mara ecosystem, the Maasai Mara is a more mesic area that encompasses Maasai-owned, now privatized, savanna rangelands around the Maasai Mara National Reserve (MMNR). Most of the land surrounding the Reserve is now in conservancies. Rainfall is bimodal, with long rains (March to June) and short rains (November to December). Land uses include pastoralism, rain-fed cultivation, conservation, and tourism. In September 2011, on the outskirts of the MMNR, near Talek town, I interviewed 26 Maasai pastoralists, with a range of diverse occupations, about changes in rainfall patterns and livelihoods (see Galvin et al. 2013).[9] In the process, I again learned about climate-related observations of the environment and people's experiences of the 2009 drought.[10]

8.2.2 Participatory Photography in the Amboseli Ecosystem

This chapter's pictures and narratives portraying the 2009 drought are outputs of a participatory visual ethnographic project I implemented in Kalesirua, an agropastoralist area near Amboseli NP.[11] By engaging residents as collaborative researchers and photographers, my initial goal was for perspectives on people-wildlife interac-

[8] *Olmarei*, pl. *ilmareta* ("family"; Mol 1996: 245) is conventionally used as meaning "household" although there is no single word in Maa (the Maasai language) that corresponds to "household." Household are considered to include a husband (or a widow), his wife/wives, children, and other economically dependent persons.

[9] I interviewed community organizers, research assistants, political leaders, elders, church leaders, conservation practitioners, tourism business owners, a nurse who is also a balloon pilot, and performers at cultural villages. Some of these men and women participated in the participatory filmmaking project that resulted in an award-winning participatory documentary "Maasai Voices on Climate Change (and other changes, too)" (2013) (available here: https://vimeo.com/73980798).

[10] See chapter by Galvin et al. in this book for additional information on the *Pastoralist Transformations to Resilient Futures* research project.

[11] More information on the participatory research process is available in the documentary "Through our Eyes: a Maasai Photographic Journey" (2010), by Lindsay Simpson and Joana Roque de Pinho: https://vimeo.com/29026437. This project resulted in two photo exhibits, collaboratively curated by the Maasai photographers, in the USA.

tions to emerge from their photography and storytelling in ways that would be less constrained than by interviewing.[12] However, conducted at the drought's paroxysm, this project became a medium through which participants represented, discussed, and interpreted this disaster's many facets.

My methodological choice to collaborate with pastoralists as visual researchers was influenced by my previous interactions in Amboseli. In this research- and tourism-intensive context, my interlocutors frequently displayed "interview fatigue" and frustration with outsiders photographing people and cows and – as they perceive it – "selling" their image. Participatory photography, or photovoice (Wang and Burris 1997), in which members of marginalized groups and/or people conventionally characterized as "respondents," "subjects," or "informants" become co-researchers in their own right, is one way to address such power imbalances while giving them tools for self-representation (Gubrium and Harper 2013). The theoretical roots of photovoice lie in the Participatory Action Research (PAR) tradition that combines Paulo Freire's approach to community problem-solving through critical consciousness (Freire 1970) with feminist theory, and in documentary photography (Wang 1999). PAR's goal is to break down barriers between analysis and praxis, and between the researcher and the "researched" (Gubrium and Harper 2013). This participatory photography project, whose participants decided what to shoot and narrate – i.e., what information to collect and share – furthermore addressed the image exploitation issue and provided them with a way to share their perspectives within and beyond their community.

Located in southern Imbirikani GR, in southeastern Amboseli ecosystem, Kalesirua is a key dry season grazing area for wildlife and livestock including swamps that have been settled for horticulture since the 1970s by non-Maasai farmers and Maasai agropastoralists (the latter taking up cultivation after a major drought). It includes permanent Maasai homesteads,[13] a trading center, and a large area of privately owned irrigated horticultural plots (*shambas*, Swahili).[14] While drought impacts in 2009 were severe for domestic and wild animals, Kalesirua residents nutritionally coped well thanks to cultivated foods and relief efforts.

Twenty-six volunteers, i.e., three women and 23 men, all of them agropastoralists (some were also schoolteachers and businesspersons), volunteered to join the project after hearing about it at an explanatory meeting with members of the community and through word of mouth. When I realized that most volunteers were men in their early 30s, I then recruited three women and three teenagers, as well as one older senior elder (c. 60 years old), all residents of nearby homesteads and interested in participating.

Most participants had neither formal education nor previous experience operating cameras. However, all were familiar with cell phone and mobile money transfer

[12] Evolving human-wildlife relationships was the focus of my previous research in the area (Roque de Pinho 2009).

[13] Permanent homestead: *Emparnat*, pl. *imparnati*.

[14] The remainder of Imbirikani GR is communally managed.

technologies.[15] After receiving their cameras (analog or digital point-and-shoot), the participants received training in photography and visual research ethics (see Marion and Crowder 2013; Roque de Pinho 2013).[16,17]

Despite the dismal circumstances, my collaborators freely took pictures during their daily activities. At regular intervals, I would meet with small groups (two to six participants) in which each photographer "told the story" of a few selected pictures. Group participants collectively debated the images' meanings and commented on each other's pictures in what constituted preliminary collaborative data analysis (Gubrium and Harper 2013). Most pictures were associated to complex narratives.[18] At the project's conclusion, the photographers held a community-wide meeting where each explained one screened picture to an audience of relatives, friends, neighbors, invited religious and political leaders, and NGO representatives.

This approach resulted in multiple layers of qualitative data (Gubrium and Harper 2013), which include visual content, as well as textual data from transcribed audio and video records of photographers' individual narratives and group discussions, community screening audience commentary, and interviews with elders.[19] I then analyzed these data by following a strategy of first coding the images according to the main themes depicted in them and then into increasingly finer categories. I followed the same procedure to code the narratives associated with the photographs and the transcripts of interviews and audio recordings of the small group discussions. I then merged my 2009 data codes with my earlier (2002–2004) and later (2011, Maasai Mara) data, using previously existing codes and creating new ones. This allowed for complementarity of information and comparison.[20] Respecting my collaborators' request to be acknowledged as the photographs' authors and contributors to the research – i.e., as producers of knowledge in their own right – I identify them by their real names.

[15] This was pre-smartphone time. In contrast, in 2011 in the Mara, some Maasai collaborators were using smartphones.

[16] We debated the topics of authorization to shoot and publish images, informed consent, reciprocity, and danger of taking pictures for both photographers and their subjects. All the photographers signed consent forms granting me the right to display their photographs in photo exhibits for noncommercial purposes and to publish them in scientific and nonscientific publications.

[17] Further key ethical considerations included distributing printed pictures to all their authors and them giving a second copy to each person appearing in their pictures. At the end of the project, the photographers became the owners of the cameras they had used, as well as of other equipment they had used collectively (i.e., tripods, extra lenses, battery chargers etc.).

[18] This was usually independent of the degree of photographic skill involved. Some of the conventionally "worst" pictures had very elaborated narratives.

[19] I thank Lindsay Simpson (and the Center for Collaborative Conservation at Colorado State University that funded her trip) for filming the meetings and transcribing the videos' soundbites.

[20] I am grateful to Nicolas Tapia and Lindsay Simpson for transcribing interviews from the Maasai Mara (2011).

8.3 "God's Politics": Perceptions of Changing Rainfall Dynamics in Maasailand

> Rains have decreased, the droughts are prolonging. We don't know why because we haven't gone up there [informant looks at the sky]. I don't remember [when rains started to decrease] but now very rarely do years provide enough rain (Senior elder, Amboseli 2004).

> Really some changes have occurred, such that at those months that we expect rain, rain doesn't happen. Now the months without rain are more than the months with rain (Elderly woman, Maasai Mara, 2011).

So central is rain to pastoralism that when Maasai comment on imminent rainfall, they interchangeably say "God is coming" (*eeou Enkai,* Maa*)* and "rain is coming" (*eeou encan*) (Roque de Pinho et al. 2009).[21] Or as I overheard during a heavy downpour in the Maasai Mara, "Bring the calves home! Don't let God beat the calves!"[22] (Field notes, 2011). In Maasai religion, Enkai is the divine entity who "created everything on top of the land" (Roque de Pinho 2009). Linguistically female-gendered (*en* – prefix; Mol 1996), Enkai is associated with the notions of fertility and birth (Hodgson 2005). Rain is thus a physical expression of the nurturing qualities of Enkai, who is understood to "take care of the land" (*eramatata enkop*) by "blessing it with rain".[23] Enkai's moods further manifest as drought/displeasure and lightning/anger (Roque de Pinho 2009).

Puzzled by changes in rainfall patterns he had experienced, an Amboseli elder commented in 2003 that there were "some politics going on with Enkai nowadays, like the politics we're are having on the land [...] because rain has really decreased".[24] This elder's quote reflects a notion, widespread in Amboseli and the Maasai Mara, in the early 2000s and about 10 years later, that precipitation is declining, i.e., rainy seasons are shorter, sometimes failing for consecutive years. In Amboseli (2002–2004), 42% of my informants (n = 189)[25] listed droughts as their families' main challenge. For 58% of those asked about rainfall patterns (n = 114), rain had decreased since their childhood. Prolonged dry seasons are seen as causing "lack of pastures" and livestock mortality, preventing rain-fed crops from maturing and forcing people to move their cattle frequently.[26] In response, people buy water and maize

[21] A conflation between weather and God also exists among Mongolian pastoralists (Marin 2009).

[22] See the video "Of God, Rain and Motorbikes" (2013): https://vimeo.com/65117460

[23] Reflecting the symbolic association between rain and fertility, a form of blessing consists in gentle spitting on someone's face or hands.

[24] This analogy between changing rainfall and people's politics is particularly poignant. In this elder's home area, i.e., the subdivided and privatized Osilalei GR, drastic changes have impacted people's lives, with land privatization blamed for increasing selfishness, isolation, enmity among neighbors, and alcohol consumption. Elsewhere in Amboseli, people have complained about increasingly aggressive clan-based politics, land grabbing, and corruption (Roque de Pinho 2009).

[25] The sample size varies according to interview question because not all 192 informants answered all the questions.

[26] In more mesic Osilalei GR, in the north of the Amboseli ecosystem, cultivation is rain fed.

stalks for weakened livestock and food for people when the herds are away, also paying for traveling and managing dispersed family members.[27] A senior elder notes that to face those extra expenses, "Now we sell cows even when it's raining. [Before], it rained a lot, there was grass and milk so people didn't sell cows to buy food."

In 2011, Maasai Mara pastoralists also mentioned suffering from shorter rainy seasons, which sometimes fail for two years in a row: "Rains when I was growing up [were enough for] our cows to stay in or near the *bomas.*[28] Cows would only move out because of the [accumulated] dung they made [from eating grass]," explains a senior elder. "However, nowadays, the rains we get don't [produce] enough grass for the cows." People also complained of a lack of "continued rain": "It rains today, then after three days it rains a little bit," clarifies another senior elder. Past rainfall, instead, "rained everywhere",[29] more continuously, and produced enough forage, in quality and quantity, to see the livestock through the dry seasons without their having to move extensively.

Another prevalent sentiment in both areas is that past rain was "organized." Now, rainy season onset is seen as less regular and predictable. In Narok, informants of all ages conveyed a strong shift in seasonality, as in this representative quote:

> April, May and beginning of June used to be rainy. But it has changed so that it's highly raining now [September 2011] and it shouldn't be. Now is not a time for rain, but, you see, it's raining. And when you're expecting rain, then it doesn't rain (Junior elder).

As a striking illustration of this observation, unexpected, short and intense bursts of rain periodically flooded the Talek plains during my 2011 fieldwork (Roque de Pinho and Galvin 2015; Roque de Pinho 2013[30]). Some characterized these heavy rains, coming during what should be the dry season, after the previous long rains had failed, as a "little help from Enkai." To others, these rains were harmful, intensifying a trend noticed in recent years of rainfall that causes "cracks on the ground, [...] coming with storms and [destroying] infrastructure and trees" (Maasai community organizer). A Maasai tour operator highlights how this shift affects the ecological dynamics that sustain his industry:

> The usual September rains weren't flash rains like now. It was just short showers that moved the wildebeest migration around the Mara. Wildlife would get enough to drink for that day. Beautiful green grass would come up for the little babies, for their health. But now we're experiencing storms and floods. And it's not the entire Mara getting a flood, a storm. Just one particular area gets a huge storm that washes everything out.

Others blamed the altered seasonality for changes in pasture quality:

---

[27] In the words of a junior elder, "There is a Maasai saying: drought is happy when people are separated in many places" (Amboseli, 2003).

[28] *Boma*: fence (Swahili); understood as homestead when used in English speech.

[29] When speaking in English, my Maasai interlocutors frequently used "rain" as both subject and verb in the same sentence.

[30] See also the participatory documentary by Maasai pastoralists in Talek "Maasai Voices on Climate Change (and other changes, too)" (2013; 10 minutes): https://vimeo.com/73980798

> Because of these dry spells in the past ten years, we have bare land [with no] grass growing anymore. We used to have tall grass throughout the Mara, but now frequently they're not green. [Before], when it rained, short rains, [grasses] were green and everything was ok, but now, however much rain falls, we don't experience the tall grasses we used to have (Maasai conservancy employee).

A senior elder reinforced that raining does not end the drought: "Before, after a drought, we'd have enough grass. But, now, even after the rains, there isn't enough grass for the cows, which means the drought continues day-by-day." Additionally, people have observed ecological changes such as "new species appearing in open grasslands, vegetation blocking and changing the course of rivers and streams, and erosion closing dams" (Senior elder, 2011).[31]

A godsend to some, the September 2011 rains confounded others' decision-making. To cope with the prolonged dry season, some families had moved their cattle out. This caused nutritional hardship, extra expenses, lost livestock trading opportunities, and migration-related livestock mortality. Then, it unexpectedly rained in Talek, benefitting the families whose animals had stayed put and coming too late for those who migrated and paid the price for it. Mara rainfall patterns have changed so much (as also reported in Amboseli in 2002-2004) that "cows are migrating monthly," even during the long rains, which is unprecedented. This change has also disrupted the synchronicity between seasons and crop planting and harvesting, livestock breeding and trading,[32] and even house construction (from lack of dung).

In both counties, most informants ascribed these changes to "just God" (*Enkai ake*). Some formally educated informants invoked concepts from climate change science: "In scientific knowledge, the change of climate, or global warming, is bringing confusion in rain, so rain is raining at unpredictable times" (Senior elder, 2011). Finally, others linked the effects of altered rainfall dynamics to the changing political economic context:

> Land privatization is now done. Rain is little now. If the rain stops for a few days, grass just stops growing and cows lack food. In the past, even in the dry season, some places always remained green because the land wasn't so much settled (Senior elder, 2011).

8.4 Ostriches like Lions: Maasai Observations of the Environment and Weather Forecasting

Enkai being endowed with the power to provide and withhold rain, Maasai strive to content Her by maintaining societal harmony and conducting sacrifices (Roque de Pinho 2009).[33] At the same time, pastoralists also try to predict

[31] This is also illustrated in the participatory documentary "Maasai Voices on Climate Change (and other changes, too)" (2013).

[32] In the Mara, cattle births were synchronized with the short rains. Recently, calves are being born during prolonged dry seasons, which hurts the mothers.

[33] This includes ritually slaughtering sheep and cattle of holy colors (*asajaki Enkai*). Other rituals

Table 8.1 Maasai rainfall schedules in the Amboseli and Maasai Mara ecosystems

Amboseli		Maasai Mara	
Month	Maasai season	Month	Maasai season
January	*Oladalu*: Short dry season. Hottest sun marks the end of the short rains	January to mid-January	*Oladalu*
		Mid-Jan. to February	*Erat*: Occasional showers. Grass in swamps.
February	*Oladalu*	February	*Erat*
March	*Inkokua* (Pleiades): Long rains (the Pleiades are visible)	March	*Erat*
		Mid-march to April	*Eni Oikok*: Bulls become fat thanks to the beginning of the long rains
April	*Inkokua*	April	*Ilookokua* (Pleiades): Long rains
May	*Inkokua*	May	*Ilookokua*
June	*Oloirujuruj:* Small showers	June	*Kujerok/Kuju orok* ("black grass"): Little rain, end of the long rains. Best grass left from long rains
July	*Enkijape* ("cold"): Dry, cold month	July	*Ilekijape*: Coldest month with little rain
August	*Olameyu*: Long dry season	August	*Ilekijape*
September	*Olameyu*	September	*Iladorukejek:* "Need to walk long distances to get water and grass"
October	*Olapa loorkirisat:* Month of "small rains"	October	*Olameyu/Kara obo*: Single month when nothing is available. Totally dry month
Late October to November	*Oltumuren:* Short rains	November	*Olookisirat*: Some rain in some parts of the land
December	*Oloibor are* ("white water"): Heavy rains	December	*Oloitushul idapan:* Little but heavy rain everywhere

onset and duration of seasons through traditional methods that involve long-standing knowledge of rainfall timings (see also Goldman 2006) and monitoring natural phenomena and fauna and flora (Hollis 1905; Roque de Pinho et al. 2009; Rutten 2016). Explaining how they know when rains will come and for how long, informants explained they follow a well-known "pattern in how the rain rains" (Senior elder). In Narok and Amboseli, these "rain schedules" reveal geographically specific understandings of rainfall patterns that are finer and more precise than the bimodal classification (Table 8.1) (see Goldman (2006) for northern Tanzania).

performed when rains fail include naked pregnant women, young girls and men wearing sheep skins and carrying milk and beer (as fertility symbols) walking over pastures to bless the land, and women's fertility songs and dances (holding green grass) (Roque de Pinho et al. 2009). Certain elders would also lie naked in dams to invite rainfall. "Now, it is shameful to do these things, now you just go to church" explained an elder in Amboseli (2002–2004); "these things mostly stopped with the Iseuri age-set" (i.e., men now in their 70s).

By "counting the months," pastoralists analyze departures from expected timings and anticipate seasonal events: "No rain in June shows that a serious drought is coming with death of cows" (Elderly woman, 2011). The unseasonal September 2011 rains in Talek had some predict the next long rains' failure: "September is always dry. Because we're getting rain now, changes will happen in the expected rainy time next April or May, it might be a dry season" (Junior elder). In an opposite interpretation, it rained then because the *previous* long rains had failed: "In April and May when we expected the long rains, it didn't rain, so we're getting this rain now because of that" (Senior elder).[34]

These "schedules of rain" associate astronomical indicators with weather events: "Before, we used stars (*olakerai*, pl. *ilakir*, Maa) to know when it's going to rain. If a certain star isn't there, we know there won't be rain" (Senior elder, 2011). Celestial bodies' physical appearance and positions also structure local understandings of rainfall dynamics: in Amboseli and the Mara, position and brightness of the *Inkokua,* the Pleiades constellation, signal the onset of the long rainy season, giving it its name (Table 8.1). In Amboseli, relative positions of Venus (*Kileken*) and the moon (*Olapa)* indicate onset of rainy and dry seasons. People may also elatedly notice a "moon bended as if water is flowing from it" or how the moon's "horns" are positioned relatively to Mount Kilimanjaro, promising "flowing rivers" (Roque de Pinho et al. 2009; Field notes 2002–2004).

Another set of physical indicators that signal future rain include cloud formation and behavior (i.e., low, heavy black clouds), wind direction, air temperature, and sunset color. Distant lightning indicates impending rain (Field notes, 2002–2004; Elderly woman, 2011). In Amboseli, thunderstorm and lightning are interpreted as *Enkai narok* ("the god-which-is-black") "beating the drought." Observation or knowledge of rainfall in other areas informs predictions for people's home area s: "Another way to know whether there'll be rain here is [that] there's a kind of rain that rains in the west called *enkoroine*; when it happens people expect rain here" (Senior elder, 2011). Assessments can combine multiple elements, as articulated by a senior elder in Amboseli, January 2003:

> We realized [we were in a drought] after December 1st [2002] because rain hadn't happened yet even in places that usually get lots of rain, so we knew [this drought] is dangerous. These rains are the short rains coming late. But they might also be the long rains coming early because rain is coming from all directions and the December sun was very hot and red, which shows the end of the short rains. […] We're still hoping for rain in March or April [long rains]. If that doesn't happen, this [drought] will be terrible because most [of our animals] have given birth between November and now.

Both nonhumans and humans are sensorially attuned to very fine changes in environmental parameters: in Amboseli, I have witnessed people "smelling rain" thirty km away from where we stood and cattle running toward pastures

[34]Also, as explained by a senior elder, "[This year, 2011], we didn't get the long rains, the *loonkokua*. The *Onkokua* (the Pleiades) are supposed to rise at the beginning of the rains. And then they descend. This year, they went down and it was still dry. There was no really long rains. Only short rain for a few days and then again a few days and then it stopped."

where, according to Maasai onlookers, they "smelled rain." Friends, on certain nights, would point at their cattle's restlessness, "excited from smelling the coming rain." Observations of biotic indicators are key in the Maasai forecasting system. Some wild animals' behaviors auspiciously signal the end of dry spells. Wildebeests are known as the first animals to reach pastures where rain will fall (see also Goldman 2006, 2007). Ostriches "roaring like lions" signal the end of droughts: "[They] are like *iloibonok*[35,36] When they roar, we know there's something good ahead, God will rain (*esha Enkai*)" (Senior elder, 2004). Specific sounds of the *oltilo* – a woodpecker, considered a bad omen –, interpreted by forecasting experts, signify forthcoming drought or rain. Other birds, such as a rare crow species, are only seen when it is raining (Junior elder, 2009). Maasai also monitor specific plant species' phenology. Greening of *Osilalei* (*Commiphora* spp.) and some acacias (*Acacia* spp.) during a drought indicates incoming rain. Soil humidity is assessed by uprooting grasses. Additionally, some experts forecast rainfall by reading intestines of sacrificed cattle and small stock (Field notes, 2002–2004).[37]

The domain of specialists, sign-based forecasting is currently perceived as disappearing. The "elders who knew about the rain" are dying or dead.[38] Modernization is also blamed for "people [not] using the methods like the stars and that certain rain. People think these things aren't true because of modernity" (Elderly woman, 2011). Finally, climatic instability itself hampers forecasting as environmental indicators and weather events are no longer synchronized: "There was that kind of rain. When it came closer [we knew] it would rain here [...]. Nowadays this isn't working because that rain is not existing" (Senior elder, 2011). As summarized by a Talek elderly woman, "Now people don't know how to predict rain or drought because at the expected time, no rain comes, and at the non-expected time, it rains."

These widely shared perceptions of multifaceted changes in seasonality and knowledge of rainfall dynamics form the backdrop against which I, next, examine Maasai understandings of the 2009 drought and how people made sense of an event that defied their expectations, predictions, and even prayers.

---

[35] Ostriches (*esidai*, pl. *isidan*, "the beautiful one") have a special significance in Maasai culture (Roque de Pinho 2009).

[36] *Oloibon*, pl. *iloibonok*: ritual experts with divinatory powers.

[37] Described as "regular people with that natural power," they are the same experts who predict "to the day" when women will give birth (Field notes, 2002–2004) and translate the speech of hyenas, coming mostly from the Ilkisikon subclan of the Ilatayiok clan (Field notes, 2009). One of them, in Kalesirua, was famous: "*Menye Isaiah* [Father of Isaiah] used to read goat intestines. He'd say 'this is dangerous, no rain this time' or 'people, get ready, lots of rain coming' or 'rain is coming and also a worm problem.' He was very specific!" (Project photographer, Richard ole Supeet, 2009).

[38] In the Mara, the Ilnyankusi age-set members (i.e., men now in their 80s) are reputed to be the last ones to know some signs.

8.5 "Closing the Gate": Impacts of the 2009 Drought

A little before starting fieldwork in 2009, a Maasai friend told me that his herd had left Amboseli in search of forage after several rainy seasons had failed. Despite his warning, on my way to Amboseli in July 2009, I was shocked to see Maasai cattle grazing in Nairobi's city parks and sidewalks and camels from northern Kenya transiting in the peripheral neighborhoods – both unusual sights. Arriving in Kalesirua, I could barely recognize this familiar horticultural area. The air was so dusty that breathing was challenging.[39] The contrast offered by patches of remaining swamp vegetation and irrigated *shambas*[40] visually accentuated the overall aridity. By then, *ilmurran* herders had already taken herds to the coast, Nairobi or Tanzania – for many of them, the farthest they had ever traveled (see also Carabine 2014).[41] Carcasses of large wild herbivores like zebras and wildebeest (and in the nearby swamps, hippos, and buffaloes) lied rotting close to people's homes, sometimes inside water points used by people and livestock. The few cows around could barely stand up, and people kept lifting them up. Kalesirua residents had never experienced such devastation and repeatedly quoted the elders as saying "this is the worst drought ever."

Understandingly, this catastrophe and people's survival strategies became the main subject of my collaborators' photography and storytelling (see pictures below; the captions are verbatim transcriptions of the narratives their authors shared in small groups and at the community meeting). They photographed countless corpses of cattle and wild animals (and "healthy, fat dogs")[42]; fields raided by starving zebras and neighbors' bulls; children, waiting and hungry, while their mothers engaged in exceptional farming casual labor (*kibaruwa*, Swahili); women carrying enormous loads of crop by-products on their heads and backs to feed animals too weak to move[43]; and beloved, previously protected trees turned into charcoal for sale. Their images illustrate the extent to which people went to save their remaining animals: selling some at throw-away prices to save others; feeding and lifting them up through the nights; offering human food to calves and small stock; and, for the richest elders, renting trucks to transport animals or purchased forage (facilitated by mobile money transfers) (Fig. 8.2).

Despite their best efforts, most families lost *all* their cattle, suggesting a drought of unprecedented scale across Maasailand:

> Some people had more than five hundred cows and now have nothing. This is called "you've closed the gate" (*eikelo kishomi*), it means you have no more cows. You're left with nothing [...] Members of your family come and contribute cows to you and that's called *ewolore*, a fundraiser. Now it's done in money. This year is tough because *ewolore* must be done for

---

[39] "Amboseli" derives from *empusel*: dry, parched land.

[40] Cultivated plots (Swahili).

[41] *Olmurrani*, pl. *ilmurran*: "the circumcised ones," known as "warriors."

[42] In the Mara, hyenas are also said to have benefitted from the 2009 drought: "That time was a feast for the hyenas: they really fattened (Maasai female community organizer).

[43] Crop by-products fed to livestock included maize stalks and cobs (powdered and made into porridge), onion stalks, and bean leaves.

Fig. 8.2 This is a very angry *mzee*. He's looking at a plastic bag that was in the stomach of his favorite cow. It died. It was hungry so it ate a plastic bag with wires inside. The cows are desperate. I asked him if he'd seen a drought like this before, and he said "never." He's very famous, the chief of my father's age-set, the Iseuri. He's 80 years old. He moved his cow to many places to make sure it ate good grass. It was too weak, so he rented a truck to bring the cow back home. He bought a lot of maize stalks to feed the cow, but it had already eaten the bag. So, he used a lot of money trying to save it, in vain. He told me "people never give the drought credit," which means that whenever cows die, people say "it's the plastic bag's fault" or "it's disease." But the drought says "No, I'm the one killing your animals!" People do this because if they keep mentioning the drought, it might prolong. This *mzee* is not very comfortable about keeping livestock anymore because this cow died. He feels very sad and was giving me the story of his cow all the time. (Photo and narrative: Richard ole Supeet)

> everyone because everyone has closed their gates. So there's no need to go into it because everyone needs to be helped equally (Project photographer Daniel Koisinget[44]).

Stock associateship is another Maasai institution affected by the "terrible drought.... You're even ashamed to give a goat to someone to make a friend because you know it's going to die" (Daniel Koisinget). A young woman implicitly suggests that Maasai identity is also a casualty of the cattle losses: "Soon, Maasai won't know about *engorno* (butter) anymore." Already challenged by land privatization, Maasai reciprocity norms further suffered: "the drought is making people not accept others'

[44] This chapter contains many quotes by this photographer, Daniel Koisinget, as he was a prolific, detail-oriented informant. He is the oldest in the photographers' group (in his 50s). A respected leader of several community organizations, he became the dedicated chairman of the *Maasai Photographers for Conservation,* the association the photographers created at the end of the project. During the project, he took almost no pictures but extensively facilitated debate among the photographers and contributed in-depth information. The participatory photography approach was interesting in that it brought to the fore my collaborators' contrasting skills.

Fig. 8.3 I found this *mama* who had gone to the swamp very early in the morning at 6 a.m. This is the bag she brought back. She reached home carrying lots of maize stalks and found that the cow that she had carried these stalks for had died and people had already removed its hide. The *mama* is very tired. Maybe she's sick in her back and elsewhere, and the cow has died. So she's very sad, and she's returning the stalks to the bag. This was her last cow. This woman had so many cows! She lost everything. Her son is a security guard in Nairobi. She really tried to save this cow. She really struggled. Even the son became crazy. He's refusing to come home. He isn't working anymore. This has really affected him. (Photo and narrative: Joshuah ole Katapa)

Fig. 8.4 This cow was affected by the drought. It's already dead. People put thorns on it because they're going to burn it. We always burn dead cows because when a cow dies, other cows might pass by and smell it. If a cow smells a dead cow, it gets shocked and might also die. If you don't burn a dead cow, then you must at least avoid the other cows coming near it. (Photo and narrative: Kilowa ole Korduni)

cows" (Senior elder, 2011). In Amboseli (Western 2010; Carabine 2014) and the Mara, depression and suicide weighed heavily among people's concerns, as this junior elder explains:

> In 2009 I moved my cows from Talek to Trans-Mara, about 100 km away. They were very skinny. I was waking them up because they couldn't walk, and I saw lots of carcasses. I was so depressed to see my cows dying. I saw people lose hundreds and hundreds of cows and commit suicide. […] I remember two old men after all their cows died. They said "this is the end of us." They put a rope around the neck and hung themselves from trees. They died because they said "How can I go back without my cows? This is the food my family depends on. What will I give my children?" […] In Maasai culture, [suicide] is not accepted. It's very, very rare. Very unusual, but it happens because of the love for the cows. Especially the older generation, if they lose their old cows, they see that as the end of their life. That's it, there's nothing more […]. They went to raid other tribes; you know they fought very hard to get their cows. So you know they'll be very sad and may commit suicide.

Some photographers portrayed the drought's toll on both human and nonhuman mental health (Figs. 8.3 and 8.4).

8.6 Making Sense of the "Worst Drought"

Confronted with untold devastation, the photographers, most of whom were not alive during the 1960s *emboot*, spontaneously, and as part of our collaborative research process, engaged in making sense of this climatic shock, reflecting on how it compared with previous droughts and what it said about future rainfall and grass, pondering whether they should rebuild their herds or further diversify their livelihoods, and questioning their future as pastoralists. My collaborators surmised that the drought had been years in the making:

> Bad times started in 2006. We missed the short rains, mostly in the north, in Kaputei. Not so much here. All the Kaputei and some Kangere and Matapato Maasai brought their cows here because we had received rain and had grass […].[45] A lot of people moved here. We bought their cows at low prices, improved breeds for Ksh 1,000 [Kenyan Shillings], and took them to our *shambas*. They fattened because we had good long rains in March-April 2006. So 2006 was fine here. Then 2007 started with very nice short rains that kept us going. But 2007 long rains were not successful. So our problems started there! 2008: we got no short rains, no long rains. Now, we're having short rains but they're not very good. (photographers' group discussion).

To some observers, specific rainfall indicators forewarned the drought: "That small rains month, *Olapa loorkisirat* (October; Table 8.1), in 2008, this sign [of the forthcoming short rainy season] didn't appear. The month when we receive heavy rains and cows create lots of dung, *Oloibor are* (November–December), it didn't happen, it never rained [in 2008]" (Daniel Koisinget). Others contended that "this drought

[45] Kaputei, Kangere, and Matapato are neighboring Maasai sections (*iloshon*).

wasn't predicted because those specialists who can predict the rain [through] intestines, the stars and the birds [...] have died" (Richard ole Supeet).

Equally worrisome was the fact that elders could not predict the drought's conclusion: "*Wazee*[46] are saying they've never seen a drought like this. They're confused. The other droughts weren't like this. Now, they don't know if it's going to rain more." Eventually, people resorted to rainmaking prayers and sacrifices, "but we did all that and nothing happened" (Daniel Koisinget). In December 2009, after some timid rains and the land now greener, people mentioned that these recent rains were not ending the drought – i.e., cows were still not having grass: "Now, the signs are not satisfying, they're not clear. A star hasn't risen yet. When that star isn't up, even if it rains, grasses won't grow, like right now: it's raining but the grass is not growing and the cows are not satisfied, and they should be" (photographers' group discussion).

These developments were puzzling to Kalesirua residents, who explained the gravity of the 2009 drought by comparing it to previous droughts. In Maa, years featuring extreme events are named after their impacts, where cows migrate to, animal behaviors, or foods people survived on. For instance, 1971–1972 and 1996 are both called *Ilare loolkurto* ("seasons of worms") because of worms devastating pastures[47] (Field notes, 2002–2004). In the early 2000s and in 2009, the previous worst drought that informants consistently cited was *emboot enkuruma sikitoi* ("the drought of the yellow maize"), which caused widespread famine and deaths in Maasailand around 1961 and is named after the government-provided relief food. My informants used it as the reference to assess the 2009 ordeal: "I asked my father, he's very old, he really knows, if there was any drought like this before and he said 'yes, *Olari loonkamirika,* the year of powder milk'" (the other airdropped relief food), explains photographer Daudi Oloomoni. That drought is also evocatively named "the drought of the bones" because "people boiled old bones and drank that water. Already this drought smells the same!" (Daniel Koisinget).

These comparisons highlight the 2009 *emboot*'s distinctive features, such as the rainfall failure all over Kenya: "In normal droughts, there's always a place with some rain. But in this one, there's been rain nowhere" (Daniel Koisinget). No matter where people moved their cows, they found no grass, indicating a truly bad drought: "A drought is only bad if it rains nowhere. If it rains somewhere, we go there" (photographers' group discussion). The fact that rainmaking was not successful was also intriguing.[48]

The 2009 drought is seen as the worst drought within a worsening trend noticed even by younger pastoralists:

> We've been experiencing drought after drought. [...] Every school holiday I looked after cows. I grew up herding. I can say it's getting worse and worse. I remember a drought in 1984 when I was very young; we lost lots of cows. I remember a very bad drought after high-school in

---

[46] *Mzee*, pl. *wazee* (Swahili): elder, old man.

[47] Other examples, among others: 1980 is "the season of the rats" (*Olari loondero*) after rodents that ate grass and people's stored food; 1997–1998 is *Olari lenkare*: "the season of the water," named after El Niño rains.

[48] In contrast, a drought situation in 2003 in Amboseli was allegedly resolved by a sheep sacrifice (the preparation of which I witnessed). It rained shortly thereafter.

> 1993; we lost a lot of cows, but not like in the droughts of 2006 and 2009. Year after year it's getting worse and worse and more challenging and more problematic (Junior elder, 2011).

People recall the 1960s drought for its high human mortality, which did not happen in 2009 in Maasailand. Still, people considered the 2009 drought worse and blamed specialized pastoralism for the impacts of the 1960s drought – clearly articulating their socioeconomic conceptualization of droughts:

> Why this 2009 drought didn't kill people is because we tried other ways of living, mostly farming. This drought is worse than the "Drought of the bones." Then, people didn't have knowledge: the swamps were here but people didn't bother to farm. The Chyulu hills had lots of grass but people weren't using vehicles to take water there for their livestock. Otherwise, they would have made it very well. Finally, it rained a lot after that. [...] Although there were so many resources, people weren't wise in terms of helping themselves through other ways; they were not diversifying, only depending on cows. This is a bad drought but people are wiser trying to save themselves in many ways. But still the cows died and up to now the rain hasn't been satisfying. And we're not sure if it's going to rain (Daniel Koisinget).

Maasai Mara informants also credited their economic diversification, in particular market-based activities like livestock trading and renting out of urban plots, with better surviving recent droughts.

8.6.1 Giraffes Like Goats and Cows Like Gazelles: Wildlife Behavior During the 2009 Drought

Domestic and wild animals behaving in ways never seen before caught the eye of the photographers, who used stories of "mad animals" to illustrate and assess this drought's exceptional character. Here, I examine a selection of their most representative photographs and narratives.[49]

One set of frequently portrayed behaviors includes wild animals being so hungry that they get dangerously close to humans, let themselves be handled by them, and even try to access food reserved for livestock. As photographer Daniel Koisinget explains:

> This is a really bad drought because wild animals have become mad. Warthogs come inside the *shamba* when you're there. They're eating and just looking at you. Even zebras are mad: they come inside the *shamba* and if you chase them, [they don't run away], you can beat them inside the *shamba*! I learned that there's a lot of drought because if you offer maize stalks [livestock drought fodder], the wild animals will come to eat them.

A remarkable instance of this behavior was captured by Lois Lanet (Fig. 8.5):[50]

---

[49] The resolution of some images is low because of some cameras' low quality. No image was improved. Some pictures by different authors represented the same "story," so I was able to choose the most representative one.

[50] In another instance, elephants were photographed inside a settlement at night. However, the photographer was (fortunately) too distant for the camera's flash to reach it and allow for a clear picture.

Fig. 8.5 I took this picture because of the drought. In the evening, giraffes were coming home, standing near our boma, with nothing to eat. This giraffe was trying to get inside our boma to eat the maize stalks [reserved] for the goats. It saw me and started to go away slowly. I took this picture to remember that everything is hungry. You see, even the wild animals are trying to be fed at home. (Photo and narrative: Lois Lanet)

Jackson ole Korduni, an *olmurrani*, the author of the iconic image introducing this chapter (Fig. 8.1), in which two young men are pulling a zebra up by its tail (as people do with drought-distressed cattle), explains:

> This is really a unique thing happening because since time immemorial we've never seen zebras being helped to stand up; being so weak that they cannot get up by themselves. So this really explains this drought is bad to the extent that people help zebras to get up. Also, this picture is good for records. I might say "Do you remember the year when people were helping zebras up?"

This animal "madness" translates into exacerbated human-wildlife conflict, with people caught in between their crops and starving large herbivores. The story shared by Solomon ole Mutunkei at the community meeting sparked a heated exchange between pastoralists, farmers, political leaders, and conservation NGO representatives in the audience, illustrating how some images were politically deployed by their authors (Fig. 8.6).[51]

[51] For instance, photographer Naomi ene Jackson used her picture of skinny "desperate cows" waiting for women to feed them maize stalks to argue that, contrary to Maasai men's usual downplaying of women's livestock rearing competence, "in this drought, it is the *mamas* who saved the cows, not the men."

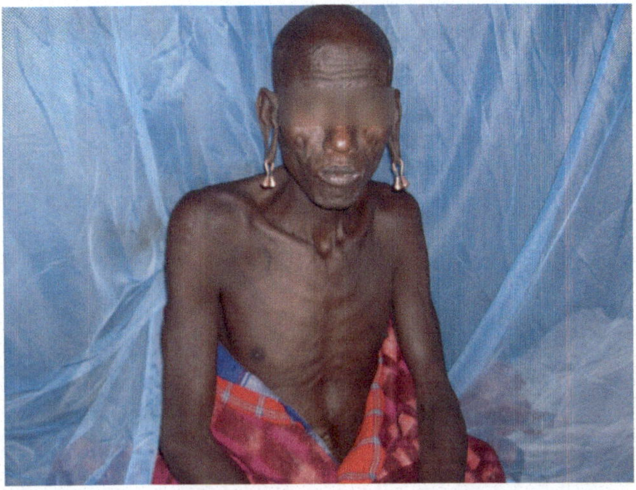

Fig. 8.6 This elder is a farmer. He was guarding his farm night and day. But one day, unfortunately, buffaloes came to his *shamba* at 7 a.m. He tried to chase them, but instead of running away, a buffalo decided to run into the man. He tried to escape but was unsuccessful and was attacked. I took this picture to show how these animals are attacking people, but no step is taken by the government. I took the picture so maybe government officials see this picture; they can be sympathetic with these people and do something about them. It's a serious problem. So many people are being attacked by hippos! In this drought, more than five people were attacked by hippos. Now, he is in hospital and needs help with medical bills. This drought is making buffaloes come out of the bush in search for food. His family lost their harvest, and there's no one to guard the *shamba* now because the man is injured, at home. The government should compensate. These people are really suffering because of the wild animals. (Photo and narrative: Solomon ole Mutunkei)

Maasai being keen amateur ornithologists (Galaty 1998; Roque de Pinho 2009), the photographers also depicted birds driven to eat unusual foods (Figs. 8.7 and 8.8).

Livestock feeding behavior was another frequent subject, with sheep and goats depicted not only eating habitual drought foods, like *osagararami* (*Acacia tortilis* pods), but also eating *unusual* foods "at home" (Fig. 8.9). Another series of images shows cows *browsing* on unpalatable vegetation, namely thorn bushes, *oltukai* palms (*Phoenix reclinata;* "We've never seen cows eating *oltukai*. This is really a bad drought"), and *oltulelai* (*Solanum incanum*). The most iconic representation is authored by Isaac ole Mutunkei (Fig. 8.10).

A most remarkable fact was that animals people had seen surviving previous droughts were dying (Figs. 8.11 and 8.12). Other wildlife species "killed by the drought" include giraffes, hippos (groups of them dead inside water), wildebeests, and buffaloes. Photographers additionally commented on how the extreme dryness changed some species' physical appearance (Fig. 8.13).

They also documented unusual interspecific dynamics, shooting multitudes of species brought together by hunger, and oddly mingling in unusual habitats (Fig. 8.14). Finally, even trees that survived major droughts were dying in this one (Fig. 8.15).

Fig. 8.7 This ostrich is eating *olngosua* (*Balanites* spp.) fruits. Ostriches usually eat grass, but now there's no grass, so it's eating these fruits because of the drought. (Photo and narrative: Peter ole Nkoyo)

Fig. 8.8 These crowned birds (African crowned cranes *Balearica regulorum*) are near the river. They're feeding on leeches because of the drought. They usually eat grass and little insects, when it rains. They also follow the cows to eat little things in the dung. (Photo and narrative: Peter ole Nkoyo)

Fig. 8.9 I took this picture because of the drought. This calf is eating maize porridge in a cooking pot because its mother is dead, so the owner is feeding it so it doesn't die. The calf is now dead. I took this picture because this is something to remember. The owner had lots of cows, and this is his last one. He lost all his cows. I gave him this picture. He said "I must show people I had lots of cows and that this was the last one". (Photo and narrative: Noah ole Kilelo)

Fig. 8.10 Cows normally graze but now they're competing with gazelles! It's unique to see cows raise their necks to take leaves from a tree. (Photo and narrative: Isaac ole Mutunkei)

Fig. 8.11 This is from when I was in the field doing this research. I found that not only cows died because of the drought: this drought is taking almost everything! In past years, we've never seen a drought of this kind killing cows and also wild animals. This drought is very serious. The 2000 drought was not like this one! (Photo and narrative: Joseph ole Shenaai)

Fig. 8.12 The drought has affected this warthog so that there's nothing else it can eat. I never heard of drought killing warthogs before because they usually eat leaves, but leaves are now dry, so it seems that the drought is extremely bad. It's not easy for warthogs to die because of drought. Usually, warthogs don't die in droughts. This warthog decided to come and eat small grass near the water, but it's so weak that it had no strength to walk away from the water. (Photo and narrative: Shadrack ole Metui)

Fig. 8.13 These elephants were trying to enter a *shamba* by force. They don't look healthy. They're reddish in color because it's dry and they're exposed to the wind. When it's not dry, there's grass so they're not exposed to the wind. Now it's dry to the extent that their color changed because this drought is very severe. (Photo and narrative: Simeon ole Kasaine)

Fig. 8.14 I took this picture to show different species of animals not previously seen in the swamps, like warthogs, zebras, gazelles, and birds. It shows that this is a serious drought because these animals do not usually come to the swamps. Zebras are known to come for water to drink and go back. Warthogs are known for staying in dry areas. But because of the drought, they're all coming to this area because it's a bit green and, the main reason, to look for food. It means they can't get their food where they used to. (Photo and narrative: Solomon ole Mutunkei)

Fig. 8.15 Trees are being cut for firewood. This is one effect of the drought. If there was rain, this tree could survive. This is a very bad drought to the extent that this *oltepesi* (*Acacia tortilis*) died. (Photo and narrative: Peter ole Nkoyo)

8.7 Discussion

8.7.1 Knowing the Weather and Adapting to a Changing Climate in Maasailand

Information on climate change in East Africa at both the national and local scales is still scant (Herrero et al. 2010). This study complements research findings on Maasai knowledge of rainfall patterns and meteorological system in Kenya and Tanzania (Goldman 2006, Mwangi 2016, Rutten 2016). Over a decade, in Kajiado and Narok counties, Maasai perceptions are broadly consistent, featuring longer dry seasons, more intense rains and droughts, and altered temporal and spatial distribution of rainfall, which all hurt livelihoods and mental well-being. Maasai observations are aligned with observations of increased climatic variability by other East African farmers and pastoralists (Speranza et al. 2010; Galvin et al. this book; Luseno et al. 2002; Orlove et al. 2010) and natural resource-dependent populations around the world (Savo et al. 2016).

Maasai perceptions of a drying trend are, however, neither in line with a projected wetter East Africa (Herrero et al. 2010) nor confirmed by instrument data collected in Maasailand (Roque de Pinho et al. 2009; Speranza et al. 2010; but see Mwangi 2016). Some authors (e.g., Savo et al. 2016) argue that the value of investigating indigenous knowledge of weather and climate lies precisely in this discrepancy: by assessing different parameters at different scales, indigenous and scientific knowledge of climate are, instead, complementary to each other. As we saw, Maasai analyses of the weather are fine-grained, reflecting subtle alterations and detecting quantitative and qualitative changes in rainfall, including in its spatial and day-by-day distribution, as well as the effects thereof on ecological dynamics and people's livelihoods. Such microscale information can thus enrich lower-resolution meteorological data (Speranza et al. 2010) that feature regional averaging of rainfall amounts (Mwangi 2016), and illuminate dynamics that are not captured by meteorological instruments and models (Marin 2009; Savo et al. 2016). The Maasai concept of drought itself emphasizes dimensions that are critical for coping with droughts and that may not adequately feature in policies based on scientific assessments of droughts (Goldman et al. 2015). Accordingly, for Maasai, alleviating the effects of a drought chiefly means "rescuing" their cattle. From their perspective, focusing relief efforts on livestock survival and condition is therefore crucial (but see Lesorogol 2008).

8.7.1.1 An Obsolete Forecasting System?

With Enkai's unfathomable "politics" at play today, Maasai are finding their forecasting system increasingly tested by climatic volatility and lack of inter-generational knowledge transmission. Other natural resource-dependent populations also report that their meteorological knowledge has become less effective (Savo et al. 2016; Speranza et al. 2010). This presents southern Kenya Maasai with a conundrum: their increasingly spatially constrained livelihoods mean that forecasting temporal and spatial distribution of rainfall *correctly* is ever more crucial for guiding decision-making across fragmented landscapes that offer obstacles to livestock movements. Yet, current climatic instability is hampering their very capacity to anticipate weather events and thus cope with them by moving their animals. In addition, while positively perceived by many of my younger informants, the economic diversification strategies they are trying and envisioning as solutions to survive ongoing climatic changes may further undermine their adaptive capacity by promoting more sedentary land uses.[52]

8.7.1.2 Maasai Understandings of the 2009 Drought and "Mad" Creatures

Through their photography and insightful narratives, my Amboseli research collaborators put to images and words the simultaneously climatic, ecological, and socioeconomic phenomenon of drought (Goldman et al. 2015; Mwangi 2016) in its extreme 2009 manifestation. In Africa, studies of changes in extreme events and on droughts' spatiotemporal distribution at fine scales are still lacking (Herrero et al. 2010; Mwangi

[52] See the *Maasai Voices on Climate Changes (and other changes, too)* (2013) participatory documentary for youth perspectives on those strategies.

2016). By sharing personal perspectives on how humans and nonhumans experienced this extreme drought at the microscale of a wetland environment located within a dryland, this participatory study contributes to the study of climatic extremes in Africa and the local scale contextualization that is key to devising adaptation policies.

With its impacts magnified by ecological and socioeconomic factors, the 2009 Kenyan drought signified a major "turning point" (Western 2010) for pastoralists. To Amboseli residents, this disaster represents a high number of "firsts," namely the first time herders traveled that far, the first time certain coping strategies were used, the first time animals ate certain foods, and the first time some wildlife species were seen dying. In this drought, even wealthy elders who hired vehicles to transport their cattle to better pastures lost them. Everyone, rich and poor, "closed their gates." Throughout the land, the drought this time was "the same" for everyone. That people did not predict this drought also highlighted to my collaborators both how outstanding it was and the decline of Maasai meteorological expertise. Yet, their photographic records suggest that their skills in observing and analyzing ecological dynamics were central to how they made sense of the drought. In particular, they used their pictures of unprecedented animal behavior to collaboratively analyze and critically reflect on the exceptional scale of this drought, simultaneously revealing their sophisticated knowledge about wildlife ecology and behavior (Goldman 2006, 2007; Roque de Pinho 2009). Some of the depicted animal behaviors reflect unusual coping strategies implemented by individual animals, e.g., eating out of their dietary range and habitats. If reoccurring in the future, such behavioral shifts may signal that local ecological conditions are becoming untenable for some mammals and birds. As species, too, are reacting to changing conditions, my collaborators' photographic records may be a harbinger of future conditions.

To the best of my knowledge, this study is the first that describes local observations of strongly altered animal behavior during a disastrous drought in Africa. In studies of local experiences with climate change in East Africa, animals usually feature as drought/rainfall indicators (e.g., Luseno et al. 2002; Speranza et al. 2010) and as protagonists of human-wildlife conflicts (Hazzah et al. 2013; Mariki and Svarstad 2015). Beyond these utilitarian aspects, the photographers, however, were interested in animal behavior *per se*, e.g., how hunger drove animals to madness and despair – even in species known to survive droughts.[53] Such observations confirm Maasai interest in watching wildlife independently of their instrumental value (Roque de Pinho et al. 2014). These data on animal drought responses emerged from the participatory photography process. It is information that I did not expect and deserves further investigation in Maasailand and elsewhere. For instance, it would have been interesting to ask elders who had lived through previous droughts which animal behaviors *they*, too, saw for the first time during the 2009 drought. Next, I discuss the value of participatory visual research approaches for researching local observations of changing environmental processes and related policy-making.

[53] Kenya Wildlife Service et al. (2009) noted the movements of buffalos, zebras, and wildebeests into the Amboseli swamps as signs of an unprecedented drought.

8.7.2 Collaborative Visual Research for Climate Change Research and Policy-Making

Mwangi (2016) stresses the need to engage multiple sources of data in assessing Maasai drought adaptive capacity and that "external analysts (scientists and policy-makers) should learn about drought from the affected" (Mwangi 2016: 13). Raygorodetsky and Green (2010) and Button and Peterson (2009) also argue that collaboration between researchers and community members is crucial for the integration of local knowledge of environmental processes with scientific knowledge with a view to develop responses to climate challenges. Finally, Reid et al. (2014) emphasize the value of the ecological knowledge held by the pastoral peoples that inhabit rangelands undergoing rapid socioecological changes globally.

During the 2009 drought, engaging Maasai agropastoralists as co-researchers of their own realities resulted in such complementary, personal, subjective information on local social and ecological processes, in the form of voluntarily contributed visual and narrative data. This participatory approach also allowed for vividly documenting unusual aspects of this drought's impacts. My collaborators explained that holding cameras made them focus their attention on what was unfolding around them and stimulated them to shoot aspects of their experiences they might not have thought of mentioning in interviews. For instance, some photographers actively sought wild species they had never seen before. Others brought to my attention situations whose significance I would have missed (e.g., a calf eating human food in a cooking pot). I contend that this method elicited better information on local observations of biodiversity during a drought than interviewing alone. In contrast, even though my Maasai Mara informants might also have witnessed unusual animal behaviors during the 2009 drought, they never mentioned those in interviews.

The 2009 drought was enacted differently by Maasai pastoralists across the international border (see Goldman et al. 2015) and across households (Mwangi 2016). In the Amboseli swamps, people did not starve, and water points did not dry, as in other parts of Kenya, but this drought was still the worst because everyone "closed their gates" – everyone lost all their cattle. My collaborators' photography and storytelling clearly highlighted unexpected dimensions of the complex socioecological dynamics (Western 2010) underlying the worst drought ever (for people *and* wildlife) in an area paradoxically endowed with permanent water sources. In so doing, they also showed that the Maasai concept of drought is dynamic, acquiring new meanings as people experience new impacts and observe evolving ecological dynamics, in this case expanding it to include unprecedented wildlife behaviors.

The participatory process proved suitable for collecting data during a climatic disaster as it avoided the time constraints of individual interviews, with photographers shooting while they conducted their daily activities and discussing their images in groups. It was also appropriate under the circumstances of families grieving the loss of their beloved animals. Previous studies of Maasailand droughts use quantitative parameters (i.e., precipitation data; mortality rates) to assess impact on human and nonhuman populations. However, Maasai are reluctant to "count children and cows," i.e., to reveal one's wealth in livestock and children,[54] and extremely reticent to dwell on dead or injured family members and livestock (Roque de Pinho 2009). This cultural characteristic affects the reliability of survey-based approaches, which will only very partially account for how Maasai experience climatic (and other) disasters. Instead, through photographing and being photographed, unexpectedly (given the local negative relationship with photography), people (even elders) felt comfortable and were even eager sometimes to create and share visual records of their woes "for [their] children and grandchildren," as many acknowledged.[55]

Importantly, besides contributing ethnographic data on how biophysical manifestations of global warming are locally understood, experienced, and dealt with, participatory visual research also helps marginalized voices and knowledge reach external and policy-making audiences and fosters local engagement in critical debate (see Gubrium and Harper 2013; Wang and Burris 1997; Wang et al. 1996). The photographers used their images and stories to educate the community meeting audience and share grievances with the conservation NGOs and leaders. Beyond the visual impact of images, though, the collaborative process itself may promote social change. My collaborators acknowledged that they acquired new knowledge through several processes that occurred during this project: by interviewing elders, by collectively reflecting on their representations of the drought, and by debating their concerns over the continued, even if partial, reliance on livestock. While producing relevant information for policy-makers, the project thus contributed to critical dialog and awareness among participants, and fostered a debate with and among their audience.

Also, the project was economically beneficial for some photographers who started photography businesses. A demand for records of the drought's impacts developed in the community, and photographers got paid for taking those pictures. Pride and heightened self-esteem of photographers and their community ensued ("Now, it's not only foreigners who can take photographs"). Finally, amidst the devastation, photographers also shared uplifting, well-appreciated stories, featuring "smart cows" climbing hills to wait for their owners bringing them food, Maasai elders (innovatively) eating cabbages,[56] and schoolchildren planting trees.

---

[54] This is believed to elicit jealousy in onlookers and thus bad luck through "having eyes" (evil eye).

[55] This also came as a surprise because, given the tough conditions and people's high levels of stress, I had expected that people would not want nor would have the time to participate and thus that the project would fail. Instead, the project elicited widespread enthusiasm, which suggests the appeal of visual and participatory approaches.

[56] Traditionally, Maasai pastoralists distinguish *endaa enkishu* (food from the cow) and *endaa enkop* (food from the land), the former being much more appreciated than the latter.

8.7.3 Limitations and Future Research

Except for the 2011 Maasai Mara study, I did not collect *systematic* data on local perceptions of climatic challenges nor on observed changes in biotic indicators of climate. It is unlikely, though, that people have not noticed the latter. Maasai knowledge of climate change through their observation of biodiversity should be further investigated. As indicator species are also adapting to changing conditions, are people identifying new indicators or new behaviors and phenologies, as suggested by Speranza et al. (2010)? Which abiotic and biotic indicators are people finding to be reliable and unreliable now a days ? Also, are people noticing emerging trends in rainfall seasonality in what they initially perceived as erratic? Finally, we might ask to what extent Maasai pastoralists are integrating climate information from other sources and whether the younger generations are leading the way in creating new knowledge about climate (see Sousa and Luz 2018).

8.8 Conclusion

This chapter examined two types of climate- and weather-related observations of the biophysical environment by Kenyan Maasai pastoralists covering different temporal scales: first, long-term, inter-generationally transmitted observations of weather patterns and associated environmental indicators, which inform their climatological knowledge and meteorological expertise; second, short-term, individual observations of biodiversity in the form of "snapshots" of animal behavior captured through the media of photography and storytelling *during* a recent climatic shock, the extreme 2009 drought. The latter partly underlie my Maasai collaborators' assessment of this extreme weather episode as a drought without precedent, part of a trend of more frequent and intense droughts. This drought might well be a forerunner of future extreme climatic events (disastrous floods hit Kenya in 2018) induced by global warming and whose local impacts are exacerbated by socioeconomic and cultural transformations that hinder people's coping. Policy-making targeting climate change must thus pay attention to local knowledge and local experiences of increasing climatic instability. Pro-pastoralism policies, such as the African Union's Policy Framework for Pastoralism in Africa (2010), which acknowledge the importance of pastoralist knowledge, are a step in the right direction.

Acknowledgments A *Fundação para a Ciência e a Tecnologia* (FCT) postdoctoral fellowship (SFRH/BPD/108780/2015) enabled my affiliation to the *Centro de Estudos Internacionais (Instituto Universitário de Lisboa* - ISCTE-IUL) (UID/CPO/03122/2013) and funded data analysis and the writing of this chapter. I am grateful to the following institutions for supporting my research in Maasailand over the years: the *Fundação para a Ciência e a Tecnologia* (PRAXIS XXI/BD/19669/99; SFRH/BPD/43242/2008); USAID's Global Livestock Collaborative Research Support Program; the Calouste Gulbenkian Foundation, and the *Fundação Luso-Americana para o Desenvolvimento*. The participatory photography project was funded by a Center for Collaborative Conservation Fellowship (2009–2010) at Colorado State University. The Ice Care Project

(Portugal) funded a post-drought visit to Amboseli in December 2009. Critical insights came from my dedicated Kalesirua research collaborators (and the Kalesirua community who encouraged their efforts). Their names are Naomi ene Jackson, Panian ole Kakami, Susan Lenet, Ismael ole Karori, Joshua ole Kasaine, Simeon ole Kasaine, Joshua ole Katapa, Noah ole Kilelo, Rafael Kiroyian, Kipaa ole Kitisia, Daniel ole Koisinget, Jackson and Kilowa ole Korduni, Richard ole Supeet, Lois ene Lanet, Shadrack ole Metui, Isaac and Solomon ole Mutunkei, Josephat Mutuko, Malano ole Nkampus, Peter ole Nkoyo, Daudi Oloomoni, Johnson ole Pilenaanka, Musa and Joseph ole Shenaai, and Solitei ole Supeet. *Ashe naleng!* I also thank Hewlett-Packard (Fort Collins, USA) and many individual contributors for donating a wealth of electronic and photographic equipment that remained with the photographers. The Maasai Mara study was part of the *Pastoral Transformations to Resilient Futures: Understanding Climate from the Ground Up* project, funded by the Feed the Future Innovation Lab for Collaborative Research on Adapting Livestock Systems to Climate Change at Colorado State University and the CGIAR Research Program on Climate Change, Agriculture and Food Security (CCAFS). This chapter has benefitted from the comments of two anonymous reviewers, as well as from insights had while participating in COST Action *Drylands Facing Change: Interdisciplinary Research on Climate Change, Food Insecurity, Political Instability* (CA 16233, European Cooperation in Science and Technology). In both Narok and Kajiado counties, I am grateful to all my informants and collaborators, the elders, the group ranch committees; and to Richard ole Supeet, Adil Bashir and their respective families for their always open-arms hospitality.

References

African Union (2010) Policy framework for pastoralism in Africa: securing, protecting and improving the lives, livelihoods and rights of pastoralist communities. Department of Rural Economy and Agriculture, African Union, Addis Ababa, Ethiopia

Altmann J, Alberts SC, Altmann SA, Roy SB (2002) Dramatic change in local climate patterns in the Amboseli basin, Kenya. African Journal of Ecology, 40(3):248

BurnSilver SB, Worden J (2008) Processes of fragmentation in the Amboseli ecosystem, Southern Kajiado District, Kenya. In: Galvin KA, Reid RS, Behnke RH, Hobbs NT (eds) Fragmentation in semi-arid and arid landscapes: Consequences for human and natural systems. Netherlands: Springer Netherlands, pp 225–253

Button GV, Peterson K (2009) Participatory action research: community partnership with social and physical scientists. In: Crate SA, Nuttall M (eds) Anthropology and climate change: From encounters to actions. Left Coast Press, Inc., pp 327–340

Carabine EA (2014) Resilience in the rangelands? Responses to change in the Amboseli social-ecological system of Kenya. Ph.D., Geography. University of Sheffield, Sheffield, UK.

Crate SA, Nuttall M (2009) Introduction: anthropology and climate change. In: Crate SA, Nuttall M (eds) Anthropology and climate change: from encounters to actions. Left Coast Press, Inc., pp 9–36

Ellis JE, Swift DM (1988) Stability of African pastoral ecosystems: alternate paradigms and implications for development. *Journal of Range Management, 41*(6):450–459

Freire P (1970) Pedagogy of the oppressed. Seabury Press, New York

Galaty JG (1998) The Maasai ornithorium: tropic flights of avian imagination in Africa. Ethnology *37*(3):227–238

Galvin KA (2009) Transitions: pastoralists living with change. Ann Rev Anthropol *38*:185–198

Galvin KA, Reid R, Nkedianye D, Njoka J, Roque de Pinho J, Kaelo D, Thornton PK (2013). Pastoral transformations to resilient futures: understanding climate from the ground up. Research brief RB-10-2013. Feed the future innovation laboratory for collaborative research on adapting livestock systems to climate change. Available: http://lcccrsp.org

Gichangi EM, Gatheru M, Njiru EN, Mungube EO, Wambua JM, Wamuongo JW (2015) Assessment of climate variability and change in semi-arid eastern Kenya. Climatic Change *130*:287–297

Goldman MJ (2006) Sharing pastures, building dialogues: Maasai and wildlife conservation in Northern Tanzania. Ph.D., Geography. University of Wisconsin-Madison, Madison, USA.

Goldman MJ (2007) Tracking wildebeest, locating knowledge: Maasai and conservation biology understandings of wildebeest behavior in Northern Tanzania. Environ Plann D Soc Space *25*:307–331

Goldman MJ, Riosmena F (2013) Adaptive capacity in Tanzanian Maasailand: changing strategies to cope with drought in fragmented landscapes. Glob Environ Chang *23*:588–597

Goldman MJ, Daly M, Lovell EJ (2015) Exploring multiple ontologies of drought in agro-pastoral regions of Northern Tanzania: a topological approach. Area *48*(1): 27–33

Gubrium A, Harper K (2013) Participatory visual and digital methods. Left Coast Press, Inc.

Hazzah L, Dolrenry S, Kaplan D, Frank L (2013) The influence of park access during drought on attitudes toward wildlife and lion killing behaviour in Maasailand, Kenya. Environ Conserv *40*(3):266–276

Herrero M, Ringler C, van de Steeg J, Thornton P, Zhu T, Bryan E, Omolo A, Koo J, Notenbaert A (2010) Climate variability and climate change and their impacts on Kenya's agricultural sector. International Livestock Research Institute, Nairobi, Kenya

Hodgson DL (2005) The church of women: gendered encounters between Maasai and Missionaries. Indiana University Press

Hollis AC (1905) The Maasai: Their Language and Folklore. Oxford: Oxford University Press.

IPCC (2012) Managing the Risks of Extreme Events and Disasters to Advance Climate Change Adaptation. Special Report of the Intergovernmental Panel on Climate Change. (C. B. Field, V. Barros, T. F. Stocker and D. Qin, eds.). Cambridge University Press, Cambridge.

Kenya Wildlife Service, African Conservation Center, & Amboseli Conservation Program (2009) The impact of the 2009 drought on wildlife, livestock and tourism in the Amboseli ecosystem: recommendations for prompt action and ecosystem restoration. In Proceedings of the Amboseli ecosystem workshop (p. 34). African Conservation Centre & the Dutch Government Institutional Support Program

Krätli S, Schareika N (2010) Living off uncertainty: the intelligent animal production of dryland pastoralists. Eur J Dev Res *22*(5):605–622

Krätli S, Huelsebusch C, Brooks S, Kaufmann B (2012) Pastoralism: a critical asset for food security under global climate change. Anim Front *3*(1):42–50

Lesorogol CK (2008) Land privatization and pastoralist well-being in Kenya. Dev Change *39*(2):309–331

Luseno WK, McPeak JG, Barrett CB, Little PD, Gebru G (2002) Assessing the value of climate forecast information for pastoralists: evidence from Southern Ethiopia and Northern Kenya. IRI Technical report NO. 02-03. New York City

Mariki, S. B., Svarstad, H., & Benjaminsen, or A. (2015). Elephants over the cliff: explaining wildlife killings in Tanzania. Land Use Policy *44*: 19–30

Marin A (2009) Riders under storms: contributions of nomadic herders' observation to analysing climate change in Mongolia. Glob Environ Chang *20*(1):162–176

Marion JS, Crowder JW (2013) Visual Research: A Concise Introduction to Thinking Visually. Bloomsbury Academic

Mol F (1996) Maasai language and culture dictionary. Diocese of Meru – Kolbe Press, Limuru

Mwangi M (2016) Diverse drought spatiotemporal trends, diverse etic-emic perceptions and knowledge: implications for adaptive capacity and resource management for indigenous Maasai-Pastoralism in the Rangelands of Kenya. Climate *4*(2):22.

Mworia JK, Kinyamario JI (2008) Traditional strategies used by pastoralists to cope with La Niña induced drought in Kajiado. Kenya African J Env Sci Technol *2*(1):10–14

Nkedianye D, De Leeuw J, Ogutu JO, Said MY, Saidimu TL, Kifuo SC et al (2011) Mobility and livestock mortality in communally used pastoral areas: the impact of the 2005–2006 drought on livestock mortality in Maasailand. Pastoralism Res Policy Pract *1*(1):17

Orlove B, Roncoli C, Kabugo M, Majugu A (2010) Indigenous climate knowledge in southern Uganda: the multiple components of a dynamic regional system. Clim Change *100*(2):243–265

Pas A (2018) Governing grazing and mobility in the Samburu Lowlands, Kenya. Sustainability *7*(2):41.

Raygorodetsky G, Green D (2010) Indigenous knowledge of a changing climate. Clim Change *100*(2):239–242

Reid RS, Fernandez-Gimenez ME, Galvin KA (2014) Dynamics and resilience of rangelands and pastoral peoples around the globe. Annu Rev Env Resour *39*:217–242

Roque de Pinho J (2009) "Staying together": people-wildlife relationships in a pastoral society in transition, Amboseli Ecosystem, Southern Kenya. Ph.D., Graduate Degree Program in Ecology. Colorado State University, Fort Collins, USA.

Roque de Pinho J (2013) Shooting climate change in the Maasai Mara: aesthetics and expectations in participatory filmmaking with Kenyan pastoralists. Anthropol Now *5*(2):74–86.

Roque de Pinho J, Galvin KA (2015) "Maasai voices on climate change (and other changes too)": participatory video and communication about environmental changes in the East African rangelands. In: Brites MJ, Jorge A, Correia Santos S (eds) Metodologias participativas: Os media e a educação. LabCom Books, pp 235–248. http://www.livroslabcom.ubi.pt/book/139

Roque de Pinho J, Galvin KA, Boone RB, Zahran S, Bucini G (2009) "God (rain) is not coming": climate change and challenges to pastoralist weather forecasting knowledge in Maasailand, Kenya. In: Brown DG, Höll BS, Mauser W, Robinson DT (eds) Though choices – land use under a changing climate: opportunities for joint German-US research activities in the field of land use and global change. The German National Committee on Global Change Research (NKGCF), Kiel Institute for the World Economy, Kiel, Germany, Berlin, p 25.

Roque De Pinho J, Grilo C, Boone RB, Galvin KA, Snodgrass JG (2014) Influence of aesthetic appreciation of wildlife species on attitudes towards their conservation in Kenyan agropastoralist communities. PLoS ONE *9*(2): e88842. https://doi.org/10.1371/journal.pone.0088842

Rutten M (2016) Dying Cows Due to Climate Change? Drought Can Never Finish the Maasai Cattle, Only the Human Mouth Can (Maasai Saying). In T. O. T. Tvedt (Ed.), A History of Water - Water and Food: From Hunter-Gatherers to Global Production in Africa (Series 3., pp. 299–336). I.B. Tauris.

Savo V, Lepofsky D, Benner JP, Kohfeld KE, Bailey J, Lertzman K (2016) Observation of climate change among subsistence-oriented communities around the world. Nat Clim Change *6*:462–473

Scoones I (1995) New Directions in Pastoral Development in Africa. In: Scoones I (ed.). Living with uncertainty: new directions in pastoral development in Africa. ITDG, London, pp 1–35

Sousa J, Luz AL (2018) "The tides rhyme with the moon": the impacts of knowledge transmission and strong spring tides on rice farming in Guinea-Bissau. Hum Ecol *46*(2):147–157

Speranza CI, Kiteme B, Ambenje P, Wiesmann U, Makali S (2010) Indigenous knowledge related to climate variability and change: insights from droughts in semi-arid areas of former Makueni District, Kenya. Clim Change *100*(2):295–315

Thornton PK, van de Steeg J, Notenbaert A, Herrero M (2009) The impacts of climate change on livestock and livestock systems in developing countries: a review of what we know and what we need to know. Agr Syst *101*:113–127

Wang CC (1999) Photovoice: a participatory action research strategy applied to women's health. J Womens Health *8*(2):185–192

Wang C, Burris MA (1997) Photovoice: concept, methodology, and use for participatory needs assessment. Health Educ Behav *24*(3):369–387

Wang C, Burris MA, Ping XY (1996) Chinese village women as visual anthropologists: a participatory approach to reaching policymakers. Soc Sci Med *42*(10):1391–1400

Wangai P, Muriithi JK, Koenig A (2013) Drought related impacts on local people's socioeconomic life and biodiversity conservation at Kuku Group Ranch, Southern Kenya. Int J Ecosyst *3*(1):1–6

Watson EE, Kochore HH (2012) Religion and climate change in Northern Kenya: new moral frameworks for new environmental challenges? J Study Relig Nat Cult *6*:319–343

Western D (1973) The structure, dynamics and change of the Amboseli ecosystem. Ph.D. University of Nairobi, Nairobi

Western D (2010) The worst drought: Tipping or turning point? Swara *3*:17–20

Western D, Manzolillo Nightingale DL (2005) Environmental changes and the vulnerability of pastoralists to drought: a case study of the Maasai in Amboseli, Kenya. In: UNEP (ed) Africa environment outlook case studies: human vulnerability to environmental change. Nairobi, UNEP, pp 35–50

Zwaagstra L, Sharif Z, Wambile A, de Leeuw J, Said MY, Johnson N, et al (2010) An assessment of the response to the 2008–2009 drought in Kenya – a report to the European Union Delegation to the Republic of Kenya. Nairobi, Kenya

Chapter 9
Operationalizing Local Ecological Knowledge in Climate Change Research: Challenges and Opportunities of Citizen Science

Victoria Reyes-García, Álvaro Fernández-Llamazares, David García-del-Amo, and Mar Cabeza

Abstract Current research on the local impacts of climate change is based on contrasting results from the simulation of historical trends in climatic variables produced with global models against climate data from independent observations. To date, these observations have mostly consisted of weather data from standardized meteorological stations. Given that the spatial distribution of weather stations is patchy, climate scientists have called for the exploration of new data sources. Knowledge developed by Indigenous Peoples and local communities with a long history of interaction with their environment has been proposed as a data source with untapped potential to contribute to our understanding of the local impacts of climate change. In this chapter, we discuss an approach that aims to bring insights from local knowledge systems to climate change research. First, we present a number of theoretical arguments that give support to the idea that local knowledge systems can contribute in original ways to the endeavors of climate change research. Then, we explore the potential of using information and communication technologies to gather and share local knowledge of climate change impacts. We do so

V. Reyes-García (✉)
Institució Catalana de Recerca i Estudis Avançats (ICREA), Cerdanyola del Vallès, Spain

Institut de Ciència i Tecnología Ambientals (ICTA), Universitat Autònoma de Barcelona, Barcelona, Spain
e-mail: Victoria.Reyes@uab.cat

Á. Fernández-Llamazares · M. Cabeza
Global Change and Conservation Lab, Organismal and Evolutionary Biology Research Programme, Faculty of Biological and Environmental Sciences, University of Helsinki, Helsinki, Finland

Helsinki Institute of Sustainability Science (HELSUS), University of Helsinki, Helsinki, Finland

D. García-del-Amo
Institut de Ciència i Tecnología Ambientals (ICTA), Universitat Autònoma de Barcelona, Barcelona, Spain

M. Welch-Devine et al. (eds.), *Changing Climate, Changing Worlds*, Ethnobiology, https://doi.org/10.1007/978-3-030-37312-2_9

through the examination of a citizen science initiative aiming to collect local indicators of climate change impacts: the LICCI project (www.licci.eu). Our findings illustrate that citizen science can inspire new approaches to articulate the inclusion of local knowledge systems in climate change research. However, this requires outlining careful approaches, with high ethical standards, toward knowledge validation and recognizing that there are aspects of local ecological knowledge that are incommensurable with scientific knowledge.

Keywords Climate change · Co-production of knowledge · Downscaling · Ethnoclimatology · Indigenous and local knowledge · Information and communication technologies

9.1 Introduction

While there is well-established evidence that climate is changing (IPCC 2013), we still have a poor understanding of how climate change differentially impacts physical and biological systems across the globe (Stott et al. 2010). Scientists acknowledge that downscaling global models to resolutions that are relevant for policy-making is challenging. This is so because of the uncertainties introduced by both the downscaling techniques and the coarse-scale data that drive global models (Zou et al. 2010). Social scientists argue that a better understanding of local climate change also requires a thorough analysis of its impacts on local social-ecological systems (Adger et al. 2013; Crate 2011). Local communities experience climate change impacts differently, not only because such impacts are place-specific but also because local economic systems and culture mediate climate change impacts on social-ecological systems (Barnes et al. 2013).

In their quest to better understand local climate change impacts, both natural and social scientists are challenged by the scarcity of data. This has resulted in climate scientists calling for the exploration of new data sources (Rosenzweig and Neofotis 2013). Within this context, several authors have argued that local ecological knowledge has an untapped potential to contribute to furthering our understanding of local impacts of climate change (Barnes et al. 2013; Savo et al. 2016). Researchers have documented many instances in which Indigenous Peoples and local communities (IPLC) with a long history of interaction with their environment have developed complex knowledge systems that allow them to detect changes in local weather and climatic variability as well as the impacts of such changes in the physical and the biological systems on which they depend (Fernández-Llamazares et al. 2015; Orlove et al. 2000; Reyes-García et al. 2016). Moreover, researchers also claim that – beyond their potential to detect changes on physical and biological systems – insights from local knowledge systems can improve our understanding of how climate change impacts local socioeconomic systems and livelihoods (Ford et al. 2016; IPCC 2014; Nakashima et al. 2012). Given that the way in which people perceive changes influences how they respond to them (Weber 2011), understanding the perceived impacts of climate change on local social-ecological systems could contribute to the development of appropriate adaptation and mitigation policies (Ford et al. 2016; Tengö et al. 2014).

Despite this potential, insights from local knowledge systems continue to be largely absent from both climate change research and global climate policy fora (Ford et al. 2016; IPCC 2014). Several reasons explain this absence. First, scholars have argued that it is difficult to discern whether locally reported impacts to physical and biological systems can be really attributed to climate change, as the complexity of such systems and confounding drivers of change (e.g., geological processes, land-use change, or human demography shifts) make attribution difficult (IPCC 2014; Rosenzweig and Neofotis 2013). While the problem of attribution is common to any research on climate change impacts, it seems to be aggravated when assessing localized impacts through local knowledge. Second, even scholars who recognize the potential value of local knowledge systems acknowledge that such knowledge systems are epistemologically different from scientific knowledge (Ford et al. 2016). Researchers working on local ecological knowledge have argued that, while this type of knowledge is based on factual and direct observations of biophysical phenomena, it is also largely perceptual, inherently tacit, and held in embodied experiential forms (i.e., not articulated in a form easily accessible to others; Garay-Barayazarra and Puri 2011; Orlove et al. 2010). Given these characteristics, it is difficult for local people to express their observations of climate change impacts in a way that can be compared with instrumental data (Reyes-García et al. 2018b). Such epistemological differences make it difficult to engage insights from these knowledge systems within the quantitative approach prevalent in climate change research (Adger et al. 2013). Finally, the spatial and temporal resolution of local knowledge tends to be less well defined than that of instrumental observations. For example, by nature of their mobile lifestyle, hunter-gatherers or nomadic pastoralists can gather ethnoclimatic information over areas larger than the ones covered by weather stations, which can result in a spatial mismatch between both measures (Fernández-Llamazares et al. 2017; Marin 2010).

In the face of these challenges, in this chapter, we discuss an approach that aims to bring insights from local knowledge systems to climate change research. First, we present the theoretical arguments that give support to the idea that local knowledge systems can contribute in original ways to the endeavors of climate change research. Then, we explore the potential use of information and communication technologies (ICTs) to gather and share local knowledge of climate change impacts. We do so through the examination of a citizen science initiative aimed at collecting local ecological knowledge on climate change impacts around the world: the Local Indicators of Climate Change Impacts (LICCI) project (www.licci.eu). Specifically, we discuss the opportunities and challenges of this approach to meaningfully gather local knowledge for understanding local climate change impacts.

9.2 The Theory: The Complementarity of Local and Scientific Knowledge Systems in Climate Research

Researchers have long debated the epistemological status of local knowledge systems in relation to mainstream scientific knowledge (e.g., Brokensha et al. 1980). This debate is also lively in climate change research, since some authors argue that there has been little effort to create synergies between scientific and local knowledge systems (Ford et al. 2016). On one side of the debate, many scientists consider scientific validation of local knowledge as a prerequisite for the inclusion of such knowledge in many research frameworks (see the discussion in Agrawal 1995). Along these lines, most previous studies aiming to bring local knowledge to climate research have often attempted to compare scientific data with local observations of climate change (e.g., Chaudhary and Bawa 2011; Marin 2010). For example, the authors of a recent meta-analysis of local indicators of climate change by resource-dependent societies around the world state that the local observations included in the study had been *verified (…), demonstrating that the observations from individual case studies are consistent with climate data* (Savo et al. 2016:462). While considering both knowledge systems as valid, these studies tend to evaluate the accuracy of local knowledge systems through scientific validation, implicitly confining local and scientific knowledge to discrete categories (Simpson 2004).

On the other side of the debate, other authors argue that knowledge generated by different knowledge systems should be viewed as valid in their own right (Pyhälä et al. 2016; Sutherland et al. 2013; Tengö et al. 2014). In that view, conflicting or contradictory evidence between – or within – different knowledge systems should not be neglected, dismissed, or concealed, but rather considered as a starting point for further knowledge generation. This recognition is precisely the cornerstone of the multiple evidence base (MEB) approach, which has been proposed as a mechanism to create synergies across knowledge systems in sustainability science (Sutherland et al. 2013; Tengö et al. 2014). The MEB approach attempts to bring together different knowledge systems on an equal platform where validation of knowledge occurs primarily within rather than across knowledge systems and where the knowledge generated by each system is viewed as useful in itself (Schweizer 2006; Tengö et al. 2014, 2017). This is important because, when local knowledge becomes *scientized* and assimilated within Western epistemic frameworks, it is often deprived of its local meaning (Mistry and Berardi 2016). As a result, a narrow understanding of what constitutes valid data can easily lead to the exclusion of essential insights from other knowledge systems (Klein et al. 2014; Mistry and Berardi 2016). Moreover, using the validation methods of a certain system to evaluate the reliability of knowledge from another system compromises the quality and/or integrity of the knowledge being "validated" (Berkes 2008; Nakashima and Roué 2002), which can be disempowering for local knowledge holders (Nadasdy 1999; Pyhälä et al. 2016). Alternatively, the identification of complementarities and contradictions between different knowledge systems can generate synergies that contribute to an enriched understanding of a process (Tengö et al. 2014, 2017). This approach has provided an overall framework for integrating different knowledge systems in science-policy arenas and agreements such as the

Intergovernmental Science-Policy Platform on Biodiversity and Ecosystem Services (IPBES) or the Convention on Biological Diversity (CBD) but has yet to be integrated into climate research (Tengö et al. 2017).

Following this second line of thought, here we argue that climate research would benefit by focusing on bridging different knowledge systems and experimenting with the co-production of knowledge in its quest to understand climate change impacts. Previous research suggests that there are two specific aspects for which local knowledge systems could complement scientific knowledge on the local impacts of climate change. First, local knowledge systems have the potential to fill spatial and temporal gaps in instrumental climatic data. And second, local knowledge systems could provide rich and fine-grained accounts of climate change impacts, beyond standard climatic measures (Reyes-García et al. 2019b). We briefly discuss these two potential contributions.

9.2.1 Local Knowledge Systems Can Fill Spatial and Temporal Gaps in Instrumental Observations

To date, the main tools available to project changes in climate are global circulation models (GCMs), i.e., mathematical representations of atmospheric, oceanic, and biotic processes that govern the global climate system. In combination with emission scenarios, GCMs are commonly used to project future climates and thus have become standard in all IPCC assessments (e.g., IPCC 2013). However, these climate models are best used to represent global and continental processes, operating at scales that are coarser than those at which impacts are observed and managed (e.g., Giorgi et al. 2009; Rummukainen 2016). Although resolution is increasing with computer power, GCMs have limited capacity to simulate climate variability and extreme events (e.g., Huntingford et al. 2003; Schaeffer et al. 2005) or key regional or local climate aspects, a situation that led to calls for higher resolution in climate projections (Mearns et al. 2001). Consequently, the last decade has seen a burst of efforts to increase the model's resolution through diverse downscaling approaches (IPCC 2007; Maraun et al. 2010; Rummukainen 2010). For example, the IPCC report (2013) largely took into account regional climate models (RCMs), i.e., models of biophysical processes nested within GCMs but allowing for further resolution by taking into account regionally relevant processes (e.g., orography, land-sea contrasts).

The performance of climate models (i.e., GCMs, RCMs) is often evaluated by simulating historical trends in climatic variables and then comparing them against independent observations mostly obtained from standardized weather stations. A major caveat of this approach is that instrumental data from standardized weather stations is patchy, both in space and time. Thus, there are important geographical gaps on weather station coverage, particularly in Central Africa, the Amazon, Southern and Southeast Asia, and the Arctic (Fig. 9.1). Similarly, we lack historical climate records for a large part of the globe (New et al. 2002). Gaps in instrumental data can be addressed in a variety of ways. For example, climatologists have used data interpolation techniques (e.g., Hijmans et al. 2005). Such techniques, however,

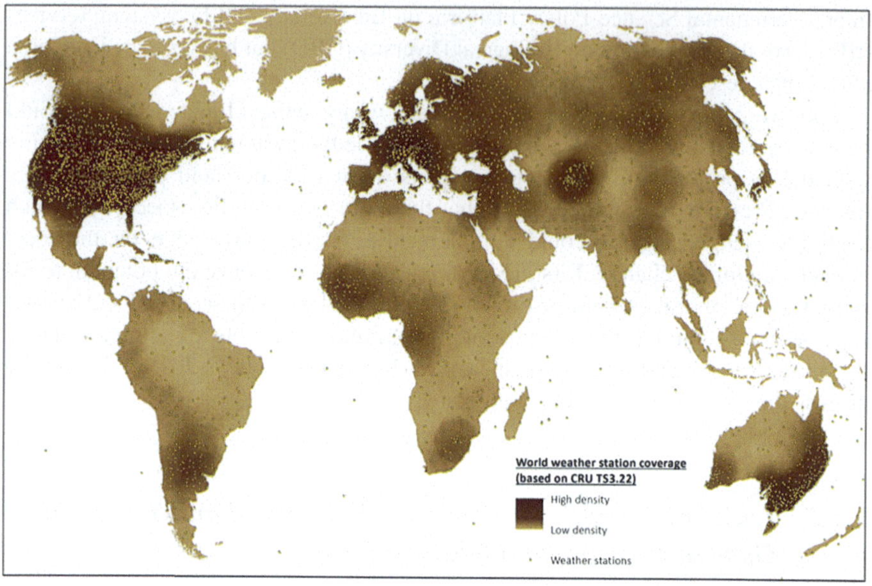

Fig. 9.1 World distribution of weather stations. (Source: Fernández-Llamazares 2015)

bring added uncertainties to modeling results (Brohan et al. 2006). An alternative approach has been to use data sources that are prior to instrumental records such as newspapers, ships' logs, weather diaries, or ancient manuscripts providing information on phenomena such as flowering and harvest dates or grain prices (Grau-Satorras et al. 2019; Lawrence 2009; Whitfield 2001).

Thus, spatial gaps in instrumental observations could be complemented with information from local knowledge systems as (1) such gaps do overlap with areas inhabited by IPLC with a long history of interaction with their environment and (2) the proxies used from data prior to instrumental records resemble closely the information encoded in local knowledge systems (Green et al. 2010; Turner and Clifton 2009). Indeed, the first studies on this line have already contributed to climate research in several ways. For example, in the Arctic, Inuit reports of changes in weather predictability (based on linkages among winds, animal behavior, and ice conditions) led scientists to examine the particularities of the local climatic system finding evidence of a strong drop in temperature in the local spring afternoons, somewhat at odds with changes on a larger scale (Weatherhead et al. 2010). In the same vein, research among the Tsimane, an Amazonian hunter-horticultural society, shows that their observations of climatic variability, although at odds with predictions from global models, were congruent with predictions from local meteorological stations (Fernández-Llamazares et al. 2017), suggesting that local knowledge can be instrumental to enhance our understanding of local climate in regions for which climate station data are meager, at best. In sum, collective memory encoded in local knowledge systems could be an alternative data source when evaluating climate models, downscaling projections, and/or interpolating gridded climate data.

9.2.2 *Local Knowledge Systems Can Provide Information on Impacts on Biophysical Systems*

A recent review of local indicators of climate change impacts has found that most previous research has focused on reports of local indicators of changes in the climatic system (i.e., temperature and precipitation) (Savo et al. 2016). While there are instances in which IPLC present accurate perceptions and documentation of change in the climatic system, research also shows that, often, climate change is better perceived and described through its impacts on the physical and the biological systems (Reyes-García et al. 2016).

People with a long history of resource dependence seem to be in a unique position to provide first-hand descriptions of the complex responses to climate change of physical and biological systems (Gearheard et al. 2011; Krupnik 2009; Moller et al. 2004). Indeed, previous work has already documented the richness of local knowledge systems regarding changes in local physical and biological systems (see Reyes-García et al. (2016) for a review). For instance, IPLC can provide a rich set of indicators referring to changes in the hydrological system and the cryosphere, including changes in snow cover, sea ice, lake ice, river ice, glaciers, ice sheets, and permafrost (e.g., Alessa et al. 2008; Gearheard et al. 2006; Laidler 2006; Nichols et al. 2004; Pearce et al. 2010; Riseth et al. 2011). Local observations of climate change impacts have also reported changes on coastal systems and particularly in relation to sea-level rise (Crona et al. 2013). Reports of impacts on the soil systems include changes in soil moisture (Kassie et al. 2013) or soil erosion (Boillat and Berkes 2013). Similarly, IPLC have reported many changes in biological systems, with a large emphasis on changes in terrestrial systems – often referring to specific species – and in seasonal events (e.g., Eisner et al. 2013; Krupnik 2009; Lantz and Turner 2003; Mallory et al. 2003; Prober et al. 2011; Tam et al. 2013).

This potential contribution of local knowledge systems to climate research seems even more important after the IPCC Fifth Assessment Report stressed that climate research should go beyond the climatic system and also include the biophysical system (IPCC 2013). Considering a general lack of scientific documentation on how temperature changes and other climatic variability affect many of the biophysical systems of the world, local knowledge systems could become an alternative data source.

9.3 Challenges and Opportunities of Citizen Science

While theoretical arguments could give support to the idea that bringing insights from local knowledge systems into climate research could result in the generation of new knowledge, few empirical examples exist aiming to materialize this theoretical approach. In this section, we explore the potential of one of them, citizen science, to bring insights from local knowledge systems into climate research. We discuss both opportunities and challenges of the approach based on ongoing efforts to operationalize it.

Previous studies aiming to generate synergies between local and scientific knowledge on climate change have mostly relied on ethnographic fieldwork to identify local observations of impacts generated by climate change (Crona et al. 2013; Galloway-McLean 2010; Reyes-García et al. 2019b). The approach, while informative, is highly time-consuming and – ultimately – under the researcher's control. To achieve a more faithful representation of local knowledge, methodologies based on the sharing of information by knowledge holders themselves would be more appropriate (Berkes 2017; Ford 2012; Ludwig 2001). In that sense, the use of citizen science provides an alternative to the standard ethnographic fieldwork, as this tool holds the potential both to promote the exchange of knowledge among an extended community of users and to make available local knowledge to the society at large, including climate researchers.

Citizen science was born at the beginning of the twentieth century and has flourished since then (Shirk et al. 2012). Citizen science has evolved from the use of nonscientist participants to conduct specific scientific tasks, like reporting abundance and species distribution, to citizens' participation in the design of scientific projects (Devictor et al. 2010; Dickinson et al. 2012) and recently to the creation of *citizen observatories* (Grainger 2017). Some citizen science projects have aimed at collecting local knowledge on climate change. For example, the Australian Government Bureau of Meteorology built a Web-based project where aboriginal communities can enter their weather calendars (http://www.bom.gov.au/iwk/; Lefale 2010). Some of these initiatives have tried to create bridges between different knowledge systems contributing information on climate change. One of such initiatives is a project by the National Institute of Water and Atmospheric Research of New Zealand (NIWA) which aims to bring together Māori ecological knowledge and practices with Western scientific methodologies of climate observations, research, assessment, and response to human-induced climate change (https://www.niwa.co.nz/te-kūwaha; King et al. 2008). Similarly, the Alaska Native Tribal Health Consortium (ANTHC) has developed an online platform where scientists and Alaskan Indigenous Peoples submit and share observations about unusual environmental and weather events (Local Environmental Observer (LEO) Network: https://www.leonetwork.org). While citizen science initiatives might offer the potential to create synergies between local knowledge holders and climate change researchers, they are not free from challenges. In what comes, we explore some of the challenges we have encountered while designing a platform to collect local observations of climate change under the framework of the LICCI project (*Local Indicators of Climate Change Impacts*, www.licci.eu).

LICCI is an ERC-funded project that aims to bring insights from Indigenous and local knowledge (ILK) to climate research. The LICCI project got inspiration from CONECT-e (www.conecte.es), a citizen science initiative aiming to bring local knowledge systems into efforts to document, protect, and share Spanish traditional ecological knowledge and practices (Calvet-Mir et al. 2018; Reyes-García et al. 2018a, a). To document traditional knowledge, CONECT-e asks users to contribute their own knowledge and talk to other knowledge holders and enter their knowledge. To protect knowledge, the Creative Commons Attribution-ShareAlike 4.0 International License (CC BY-SA 4.0) protects all the content of CONECT-e platform, so no patents can be issued on knowledge published in it. To share knowledge, the CONECT-e team works with different groups of interest (e.g., technical and primary education students, elders in rural adult schools, farmers and home gardeners, members of seed network associations) to engage them in activities for knowledge transmission. The CONECT-e team has tried to bridge the gap between knowledge systems in very specific ways. For example, in the design of the section on landraces, the CONECT-e team has closely worked with the Spanish Seed Network, Red de Semillas "Resembrando e Intercambiando" (RdS; http://www.redsemillas.info/), a civil society organization that promotes the commons management of landraces and farmer's varieties in the agri-food system. This alliance brings interested groups of citizens to the design of the project, resulting in a data collection tool tailored to the needs of the participating groups. Following Tengö et al. (2014), the CONECT-e platform is designed to facilitate knowledge validation within the knowledge system. Thus, all data published can be validated through a peer-to-peer validation inspired by the Wikipedia peer governance system (Kostakis 2010), and also common in other citizen science websites like Encyclopedia of Life (https://eol.org) or Natusfera (https://natusfera.gbif.es/). Specifically, the platform is designed in a way that registered users can "like" or "agree" with information posted in the platform, thus generating a type of review among peers holding similar knowledge.

We are applying these same principles in the development of a global platform to collect local observations of climate change impacts. The platform, which mostly targets IPLC whose livelihoods show direct dependence on the environment, will collect people's observations of how climate change is impacting the physical, biological, and socioeconomic systems on which they depend. The proposed design would guide users to enter information in relation to one or several elements potentially impacted by climate change (e.g., water, ice and snow, weather, seasons, birds, plants, mammals, health, transportation) in a way that the information can be compared across sites (see Reyes-García et al. 2019b). All information entered will be place-specific, but with a resolution that does not allow geographical tracking. Information will include previous reports from IPLC on climate change impacts documented in the literature, selected case studies from the LICCI team, and information from any citizen in the world. The idea is that information collected in this platform and based on local knowledge systems could complement the information provided by instrumental data on climate change. This information would cover not only impacts on biophysical systems but also the social perception of such changes, i.e., how rural communities perceive climate change impacts in physical, biological, and sociocultural systems.

Despite the potential, we have encountered two important challenges in our efforts to develop this platform: finding relevant interest groups and bridging the technology gap with local knowledge holders. First, unlike with other domains of knowledge (e.g., landraces), it is difficult to identify citizen's interest groups on climate change impacts. We have established contacts with UNESCO's Local and Indigenous Knowledge Systems program (LINKS) as well as with the Local Communities and Indigenous Peoples Platform (LCIPP) Facilitative Working Group (FWG) of the United Nations Framework Convention on Climate Change (UNFCCC), initiatives aiming to promote ILK's inclusion in global climate science and policy processes. However, differently from the Spanish Seed Network, these initiatives do not directly engage with the wide diversity of IPLC around the world, but rather with some of their representatives, for which it migh be difficult to motivate local participation through them. Consequently, the design of the section has mainly been done by the LICCI team, who based the selection of groups of impacts on a literature review of previous impacts reported by IPLC around the world (Reyes-García et al. 2016, 2019b). The approach is challenging because research suggests that not including citizens in the design of the project has important implications for the potential later use of this information by citizens themselves (Turreira-García et al. 2018). Second, targeted local knowledge holders are typically IPLCs highly dependent on the environment for subsistence, often with few skills in the use of ICTs. To bridge this technology gap, we are working with a group of trained partners (i.e., PhD students, postdocs, and practitioners), who can provide technological support to local knowledge holders. This alliance would make possible not only the gathering of local knowledge in the LICCI platform but also the in situ transmission of knowledge. Such an approach is being increasingly applied to revitalize traditional oral narratives among Indigenous Peoples (Fernández-Llamazares and Cabeza 2017; Iseke and Moore 2011; Ryan 2015) but again poses the challenge that it requires intensive fieldwork. The issue is, then, whether bridging the technology gap might be prohibitive for such type of projects.

9.4 Conclusion

The speed and magnitude of current global changes make standard scientific approaches ineffective to properly grasp the local impacts of climate change. The time needed to collect ground data on climate change impacts on all physical, biological, and socioeconomic variables would be so long that, by the time data collection is finished, the system might have changed completely. In the quest to provide alternative data sources, in this chapter, we have presented the theoretical arguments that support the idea that local knowledge systems can contribute in original ways to the endeavors of climate change research and then explored the potential of using ICTs to gather and share local knowledge of climate change impacts. Specifically, we have examined the opportunities and challenges of citizen science initiatives to bridge scientific and local knowledge on the local impacts of climate change.

From our examination, we argue that citizen science platforms might facilitate the co-production of new knowledge articulating communication between users, reducing time in fieldwork, and allowing simultaneous data collection in different geographical locations. Moreover, these initiatives can help connect global discourses of climate change with local contexts, contributing to bringing to daylight the knowledge, perspectives, and voices of Indigenous Peoples and local communities on the frontlines of climate change. The approach, however, is not without challenges as citizen science projects are inherently empirical and applying them in relation to local knowledge systems can incur in several epistemic, normative, methodological, and ethical pitfalls that require careful attention. Such challenges need to be consciously addressed from project's initial stages to generate knowledge that is useful and legitimate for a broad range of actors across different knowledge systems.

Acknowledgments Research leading to this chapter has received funding from the Spanish government through a grant of the Ministry of Economy and Competitiveness (CSO2014-59704-P) and from the European Research Council under an ERC Consolidator Grant (FP7-771056-LICCI). García-del-Amo and Reyes-García acknowledge financial support from the Spanish Ministry of Economy and Competitiveness, through the "María de Maeztu" program for Units of Excellence in R&D (MdM-2015-0552). Fernández-Llamazares and Cabeza acknowledge financial support from the Academy of Finland (grant agreement nrs. 311176 and 257686).

References

Adger W, Barnett J, Brown K, Marshall N, O'Brien K (2013) Cultural dimensions of climate change impacts and adaptation. Nat Clim Chang 3:112–117

Agrawal A (1995) Dismantling the divide between indigenous and scientific knowledge. Dev Chang 26(3):413–439. https://doi.org/10.1111/j.1467-7660.1995.tb00560.x

Alessa LN, Kliskey AA, Williams P, Barton M (2008) Perception of change in freshwater in remote resource-dependent Arctic communities. Glob Environ Chang 18(1):153–164

Barnes J, Dove M, Lahsen M, Mathews A, McElwee P, McIntosh R et al (2013) Contribution of anthropology to the study of climate change. Nat Clim Chang 3:541–544

Berkes F (2008) Sacred ecology. Routledge, New York

Berkes F (2017) Environmental governance for the anthropocene? Social-ecological systems, resilience, and collaborative learning. Sustainability 9(7):1232. https://doi.org/10.3390/su9071232

Boillat S, Berkes F (2013) Perception and interpretation of climate change among Quechua farmers of Bolivia: indigenous knowledge as a resource for adaptive capacity. Ecol Soc 18(4):21. https://doi.org/10.5751/es-05894-180421

Brohan P, Kennedy JJ, Harris I, Tett SFB, Jones PD (2006) Uncertainty estimates in regional and global observed temperature changes: a new data set from 1850. J Geophys Res 111:D12106. https://doi.org/10.1029/2005JD00654

Brokensha D, Warren D, Werner O (1980) Indigenous knowledge systems and development. University Press of America, Lanham

Calvet-Mir L, Benyei P, Aceituno-Mata L, Pardo de Santayana M, López-García D, Carrascosa M et al (2018) Contributions of traditional agroecological knowledge as a digital commons to agroecological transitions. The case of the CONECT-e platform. *Sustainability* 10:3214. https://doi.org/10.3390/su10093214

Chaudhary P, Bawa KS (2011) Local perceptions of climate change validated by scientific evidence in the Himalayas. Biol Lett 11(8):767–770

Crate SA (2011) Climate and culture: anthropology in the era of contemporary climate change. Annu Rev Anthropol 40(40):175–194. https://doi.org/10.1146/annurev.anthro.012809.104925

Crona B, Wutich A, Brewis A, Gartin M (2013) Perceptions of climate change: linking local and global perceptions through a cultural knowledge approach. Clim Chang 119(2):519–531. https://doi.org/10.1007/s10584-013-0708-5

Devictor V, Whittaker RJ, Beltrame C (2010) Beyond scarcity: citizen science programmes as useful tools for conservation biogeography. Divers Distrib 16:354–362. https://doi.org/10.1111/j.1472-4642.2009.00615.x

Dickinson JL, Shirk J, Bonter D, Bonney R, Crain RL, Martin J et al (2012) The current state of citizen science as a tool for ecological research and public engagement. Front Ecol Environ 10(6):291–297. https://doi.org/10.1890/110236

Eisner WR, Hinkel KM, Cuomo CJ, Beck RA (2013) Environmental, cultural, and social change in Arctic Alaska as observed by Iñupiat elders over their lifetimes: a GIS synthesis. Polar Geogr 36:1–11

Fernández-Llamazares Á (2015) Indigenous knowledge of a changing environment: an ethnoecological perspective from Bolivian Amazonia. PhD thesis, Autonomous University of Barcelona

Fernández-Llamazares Á, Cabeza M (2017) Rediscovering the potential of indigenous storytelling for conservation practice. Conserv Lett 11:1–12. https://doi.org/10.1111/conl.12398

Fernández-Llamazares Á, Méndez-López E, Díaz-Reviriego I, McBride M, Pyhälä A, Rosell-Melé A, Reyes-García V (2015) Links between media communication and local perceptions of climate change in an indigenous society. Clim Chang 131(2):307–320. https://doi.org/10.1007/s10584-015-1381-7

Fernández-Llamazares A, García R, Díaz-Reviriego I, Cabeza M, Pyhälä A, Reyes-García V (2017) An empirically-tested overlap between local and scientific knowledge of a changing climate. Reg Environ Change 17(6):1673–1685. https://doi.org/10.1007/s10113-017-1125-5

Ford JD (2012) Indigenous health and climate change. Am J Public Health 102(7):1260–1266. https://doi.org/10.2105/AJPH.2012.300752

Ford JD, Cameron L, Rubis J, Maillet M, Nakashima D, Willox AC et al (2016) Including indigenous knowledge and experience in IPCC assessment reports. Nat Clim Chang 6(4):349–353. https://doi.org/10.1038/nclimate2954

Galloway-McLean K (2010) Advance guard: climate change impacts, adaptation, mitigation and indigenous peoples–a compendium of case studies. United Nations University, Institute for Advanced Studies (UNU-IAS), Traditional Knowledge Initiative, Darwin

Garay-Barayazarra G, Puri RK (2011) Smelling the monsoon: senses and traditional weather forecasting knowledge among the Kenyah Badeng farmers of Sarawak, Malaysia. Indian J Tradit Knowl 10:21–30

Gearheard S, Matumeak W, Angutikjuaq I, Maslanik J, Huntington HP, Leavitt J et al (2006) "It's not that simple": a collaborative comparison of sea ice environments, their uses, observed changes, and adaptations in Barrow, Alaska, USA, and Clyde River, Nunavut, Canada. Ambio 35(4):204–212

Gearheard S, Aporta C, Aipellee G, O'Keefe K (2011) The Igliniit project: Inuit hunters document life on the trail to map and monitor arctic change. Can Geogr 55(1):42–55. https://doi.org/10.1111/j.1541-0064.2010.00344.x

Giorgi F, Jones C, Asrar GR (2009) Addressing climate information needs at the regional level: the CORDEX framework. World Meteorol Organ Bull 58(3):175–183

Grainger A (2017) Citizen observatories and the new earth observation science. Remote Sens 9(2):153. https://doi.org/10.3390/rs9020153

Grau-Satorras M, Otero I, Gómez-Baggethun E, Reyes-García V (2019) Prudent peasantries: multilevel adaptation to drought in early modern Spain (1600–1715). Environment and History. https://doi.org/10.3197/096734019X15463432086964

Green D, Billy J, Tapim A (2010) Indigenous Australians' knowledge of weather and climate. Climate Change 100:337–354. https://doi.org/10.1007/s10584-010-9803-z

Hijmans RJ, Cameron SE, Parra JL, Jones PG, Jarvis A (2005) Very high resolution interpolated climate surfaces for global land areas. Int J Climatol 25:1965–1978

Huntingford C, Jones RG, Prudhomme C, Lamb R, Gash JH, Jones DA (2003) Regional climate-model predictions of extreme rainfall for a changing climate. Q J R Meteorol Soc 129:1607–1621

IPCC (2007) Climate change 2007: the physical science basis. Contribution of Working Group I to the Fourth Assessment Report of the Intergovernmental Panel on Climate Change [Solomon, S., D. Qin, M. Manning, Z. Chen, M. Marquis, K.B. Averyt, M. Tignor and H.L. Miller (eds.)]. Cambridge University Press, Cambridge

IPCC (2013) Climate change 2013: the physical science basis. Contribution of Working Group I to the Fifth Assessment Report of the Intergovernmental Panel on Climate Change [Stocker, T.F., D. Qin, G.-K. Plattner, M. Tignor, S.K. Allen, J. Boschung, A. Nauels, Y. Xia, V. Bex and P.M. Midgley (eds.)]. Cambridge University Press, Cambridge/New York, 1535 pp, https://doi.org/10.1017/CBO9781107415324

IPCC (2014) Climate change 2014: impacts, adaptation, and vulnerability. Part A: Global and sectoral aspects. Contribution of Working Group II to the Fifth Assessment Report of the Intergovernmental Panel on Climate Change [Field, C.B., V.R. Barros, D.J. Dokken, K.J. Mach, M.D. Mastrandrea, T.E. Bilir, M. Chatterjee, K.L. Ebi, Y.O. Estrada, R.C. Genova, B. Girma, E.S. Kissel, A.N. Levy, S. MacCracken, P.R. Mastrandrea, and L.L. White (eds.)]. Cambridge University Press, Cambridge/New York

Iseke J, Moore S (2011) Community-based indigenous digital storytelling with elders and youth. Am Indian Cult Res J 35:19–38

Kassie BT, Hengsdijk H, Rotter R, Kahiluoto H, Asseng S, Van Ittersum M (2013) Adapting to climate variability and change: experiences from cereal-based farming in the central rift and kobo valleys, Ethiopia. Environ Manag 52(5):1115–1131. https://doi.org/10.1007/s00267-013-0145-2

King DNT, Skipper A, Tawhai WB (2008) Māori environmental knowledge of local weather and climate change in Aotearoa–New Zealand. Clim Chang 90(4):385. https://doi.org/10.1007/s10584-007-9372-y

Klein J, Hopping KA, Yeh ET, Nyima Y, Boone RB, Galvin KA (2014) Unexpected climate impacts on the Tibetan Plateau: local and scientific knowledge in findings of delayed summer. Glob Environ Chang 28(1):141–152

Kostakis V (2010) Identifying and understanding the problems of Wikipedia's peer governance: the case of inclusionists versus deletionists. First Monday 15:1–11

Krupnik I (2009) "The way we see it coming": building the legacy of indigenous observations in IPY 2007-2008. In: Krupnik I, Lang MA, Miller SE (eds) Smithsonian at the poles: contributions to international polar year science. Smithsonian Institution Scholarly Press, Washington, DC, pp 129–142

Laidler GJ (2006) Inuit and scientific perspectives on the relationship between sea ice and climate change: the ideal complement? Clim Chang 78:407–444

Lantz TC, Turner NJ (2003) Traditional phenological knowledge (TPK) of aboriginal peoples in British Columbia. J Ethnobiol 23(2):263–286

Lawrence A (2009) The first cuckoo in winter: phenology, recoding, credibility and meaning in Britain. Glob Environ Chang 19:173–179. https://doi.org/10.1016/J.GLOENVCHA.2009.01.006

Lefale PF (2010) Ua 'afa le Aso Stormy weather today: traditional ecological knowledge of weather and climate. The Samoa experience. Clim Chang 100(2):317–335. https://doi.org/10.1007/s10584-009-9722-z

Ludwig D (2001) The era of management is over. Ecosystems 4(8):758–764. https://doi.org/10.1007/s10021-001-0044-x

Mallory ML, Gilchrist HG, Fontaine AJ, Akearok JA (2003) Local ecological knowledge of Ivory Gull decline in Arctic Canada. Arctic 56(3):293–298

Maraun D, Wetterhall F, Ireson AM, Chandler RE, Kendon EJ, Widmann M et al (2010) Precipitation downscaling under climate change: recent developments to bridge the gap between dynamical models and the end user. Rev Geophys 48:RG3003. https://doi.org/10.1029/2009RG000314

Marin A (2010) Riders under storms: contributions of nomadic herders' observations to analysing climate change in Mongolia. Glob Environ Chang 20(1):162–176

Mearns LO, Hulme M, Carter TR, Leemans R, Lal M, Whetton P (2001) Climate scenario development, Chapter 13. In: Houghton J et al (eds) IPCC Third Assessment Report. The science of climate change. Cambridge University Press, Cambridge, pp 739–768

Mistry J, Berardi A (2016) Bridging indigenous and scientific knowledge. Science 352:1274–1275

Moller H, Berkes F, Lyver PO, Kislalioglu M (2004) Combining science and traditional ecological knowledge: monitoring populations for co-management. Ecol Soc 9(3)

Nadasdy P (1999) The politics of TEK: power and the "integration" of knowledge. Arct Anthropol 36:1–18

Nakashima DJ, Roué M (2002) Indigenous knowledge, peoples and sustainable practice. In: Munn T (ed) Encyclopedia of global environmental change. Wiley, Chichester, pp 314–324

Nakashima DJ, Galloway-McLean K, Thrulstrup HD, Ramos Castillo A, Rubis J (2012) Weathering uncertainty: traditional knowledge for climate change assessment and adaptation. UNESCO, Paris

New M, Lister D, Hulme M, Makin I (2002) A high-resolution data set of surface climate over global land areas. Clim Res 21:1–25

Nichols T, Berkes F, Jolly D, Snow NB (2004) Climate change and sea ice: local observations from the Canadian Western Arctic. Arctic 57(1):68–79

Orlove B, Chiang J, Cane M (2000) Forecasting Andean rainfall and crop yield from the influence of El Niño on Pleiades visibility. Nature 403:68–71

Orlove B, Roncoli C, Kabugo M, Majugu A (2010) Indigenous climate knowledge in southern Uganda: the multiple components of a dynamic regional system. Clim Chang 100(2):243–265

Pearce T, Smit B, Duerden F, Ford JD, Goose A, Kataoyak F (2010) Inuit vulnerability and adaptive capacity to climate change in Ulukhaktok, Northwest Territories, Canada. Polar Rec 46(237):157–177. https://doi.org/10.1017/s0032247409008602

Prober SM, O'Connor MH, Walsh FJ (2011) Australian aboriginal peoples' seasonal knowledge: a potential basis for shared understanding in environmental management. Ecol Soc 16(2):12

Pyhälä A, Fernández-Llamazares A, Lehvävirta H, Byg A, Ruiz-Mallén I, Salpeteur M, Thornton TF (2016) Global environmental change: local perceptions, understandings, and explanations. Ecol Soc 21(3):25. https://doi.org/10.5751/ES-08482-210325

Reyes-García V, Fernández-Llamazares A, Gueze M, Garces A, Mallo M, Vila-Gomez M et al (2016) Local indicators of climate change: the potential contribution of local knowledge to climate research. Wiley Interdiscip Rev Clim Change 7(1):109–124. https://doi.org/10.1002/wcc.374

Reyes-García V, Aceituno-Mata L, Benyei P, Calvet-Mir L, Carrascosa M, Pardo-de-Santayana M, J. Tardio for CONECT-e, IECTBA and Red de Semillas (2018a) Governing landraces and associated knowledge as commons. From theory to practice, Chapter 12. In: Girad F, Frison C (eds) The commons, plant breeding and agricultural research. Challenges for food security, agrobiodiversity and law. "Earthscan Food and Agriculture" series. Routledge, New York, pp 197–209

Reyes-García V, Fernández-Llamazares Á, Guèze M, Gallois S (2018b) Does weather forecasting relate to foraging efforts? An empirical test among three hunter-gatherer societies. Weather Clim Soc 10:163–177. https://doi.org/10.1175/WCAS-D-17-0064.1

Reyes-García V, Benyei P, Calvet-Mir L (2019a) Traditional agricultural knowledge as commons, Chapter 11. In: Vivero-Pol JL, Ferrando T, de Schutter O, Mattei U (eds) Routledge handbook of food as a commons. Routledge, New York, pp 173–185

Reyes-García V, García-del-Amo D, Benyei P, Fernández-Llamazares Á, Gravani K, Junqueira AB, Labeyrie V, Li X, Matias DM, McAlvay A, Mortyn PG, Porcuna-Ferrer A, Schlingmann A, Soleymani-Fard R (2019b) A collaborative approach to bring insights from local observations of climate change impacts into global climate change research. Curr Opin Environ Sustain 39:1–8

Riseth JA, Tommervik H, Helander-Renvall E, Labba N, Johansson C, Malnes E et al (2011) Sami traditional ecological knowledge as a guide to science: snow, ice and reindeer pasture facing climate change. Polar Rec 47(242):202–217. https://doi.org/10.1017/s0032247410000434

Rosenzweig C, Neofotis P (2013) Detection and attribution of anthropogenic climate change impacts. Wiley Interdiscip Rev Clim Change 4(2):121–150. https://doi.org/10.1002/wcc.209

Rummukainen M (2010) State-of-the-art with regional climate models. Wiley Interdiscip Rev Clim Change 1:82–96. https://doi.org/10.1002/wcc.8

Rummukainen M (2016) Added value in regional climate modeling. Wiley Interdiscip Rev Clim Change 7(1):145–159

Ryan JC (2015) The virtual and the vegetal: creating a "living" biocultural heritage archive through digital storytelling approaches. Glob Media J 9(1):49–58

Savo V, Lepofsky D, Benner JP, Kohfeld KE, Bailey J, Lertzman K (2016) Observations of climate change among subsistence-oriented communities around the world. Nat Clim Chang 6(5):462–473. https://doi.org/10.1038/nclimate2958

Schaeffer M, Selten FM, Opsteegh JD (2005) Shifts in means are not a proxy for changes in extreme winter temperatures in climate projections. Clim Dyn 25:51–63

Schweizer T (2006) Epistemology: the nature and validation of anthropological knowledge. In: Bernard HR (ed) Research methods in anthropology: qualitative and quantitative approaches. Rowman, Altamira, pp 39–88

Shirk J, Ballard H, Wilderman C, Phillips T, Wiggins A, Jordan R et al (2012) Public participation in scientific research: a framework for deliberate design. Ecol Soc 17(2):29. https://doi.org/10.5751/ES-04705-170229

Simpson L (2004) The colonial context for the indigenous experience of climate change. In: Helander E, Mustonen T (eds) *Snowscapes, dreamscapes: snowchange book on community voices of change*. Tampere Polytechnic Publications, Tampere, pp 25–29

Stott PA, Gillett NP, Hegerl GC, Karoly DJ, Stone DA, Zhang X et al (2010) Detection and attribution of climate change: a regional perspective. Wiley Interdiscip Rev Clim Change 1(2):192–211. https://doi.org/10.1002/wcc.34

Sutherland WJ, Gardner TB, Haider LJ, Dicks LV (2013) How can local and traditional knowledge be effectively incorporated into international assessments? Oryx 48(1):1–2

Tam BY, Gough WA, Edwards V, Tsuji LJS (2013) The impact of climate change on the well-being and lifestyle of a First Nation community in the western James Bay region. Can Geogr 57(4):441–456. https://doi.org/10.1111/j.1541-0064.2013.12033.x

Tengö M, Brondizio ES, Elmqvist T, Malmer P, Spierenburg M (2014) Connecting diverse knowledge systems for enhanced ecosystem governance: the multiple evidence base approach. Ambio 43(5):579–591. https://doi.org/10.1007/s13280-014-0501-3

Tengö M, Hill R, Malmer P, Raymond CM, Spierenburg M, Danielsen F, Elmqvist T, Folke C (2017) Weaving knowledge systems in IPBES, CBD and beyond – lessons learned for sustainability. Curr Opin Environ Sustain 26:17–25

Turner NJ, Clifton H (2009) “It’s so different today”: climate change and indigenous lifeways in British Columbia, Canada. Glob Environ Chang 19(2):180–190. https://doi.org/10.1016/j.gloenvcha.2009.01.005

Turreira-García N, Lund JF, Dominguez P, Carrillo-Anglés E, Brummer M, Dunn P, Reyes-García V (2018) What’s in a name? Unpacking “participatory” environmental monitoring. Ecol Soc 23(2):24

Weatherhead E, Gearheard S, Barry R (2010) Changes in weather persistence: insights from Inuit knowledge. Glob Environ Chang 20(3):523–528

Weber EU (2011) Climate change hits home. Nat Clim Chang 1(1):25–26

Whitfield J (2001) The budding amateurs. Nature 414(6864):578–579. https://doi.org/10.1038/414578a

Zou L, Zhou T, Li L, Zhang J (2010) East China summer rainfall variability of 1958–2000: dynamical downscaling with a variable-resolution AGCM. J Clim 23(23):6394–6408. https://doi.org/10.1175/2010jcli3689.1

Chapter 10
Using Local Observations of Climate Change to Identify Opportunities for Community Conversations in Southern Appalachia

Brian J. Burke, Meredith Welch-Devine, Stephanie Rzonca, and Chad Steacy

Abstract Climate change and exurban development present significant social, economic, and environmental challenges in Southern Appalachia. Addressing those challenges—whether to prevent them, mitigate them, or prepare for them—will require individual action and collective action at community and regional scales. However, the coordination necessary for such action will be difficult to achieve in a region long opposed to regulation, suspicious of newcomers, and characterized by increasing social diversity. One particularly salient difference that is likely to shape collective responses to climate change is the distinction between "newcomers" and "multigenerationals," descendants of people who have lived in the region for generations. In this chapter, we draw on nearly 80 interviews to address three questions: What indicators and consequences of climate change do people observe in their everyday lives and view as relevant? How does one's connection to this landscape shape the indicators and consequences they observe and care about? And what differences exist in how people theorize the causes of climate change? By examining this diversity of climate knowledges and climate cultures in Southern Appalachia, we hope to identify complementarities, bridges, and provocations that might help natural resource managers and community members identify effective and inclusive responses to climate change.

Keywords Local ecological knowledge · Climate cultures · Weather · Biodiversity · Gardening · Farming · Adaptation · Collective action · Collaboration · Conflict · Appalachia

B. J. Burke (✉)
Department of Sustainable Development, Appalachian State University, Boone, NC, USA
e-mail: burkebj@appstate.edu

M. Welch-Devine
The Graduate School, University of Georgia, Athens, GA, USA

S. Rzonca
Rutherford County Economic Development, Rutherfordton, NC, USA

C. Steacy
Department of Geography, University of Georgia, Athens, GA, USA

M. Welch-Devine et al. (eds.), *Changing Climate, Changing Worlds*, Ethnobiology, https://doi.org/10.1007/978-3-030-37312-2_10

10.1 Introduction

Nonscientific environmental knowledge—including local ecological knowledge (LEK), traditional ecological knowledge (TEK), and indigenous knowledge (IK)—is increasingly seen as a linchpin for effective resource management and climate action. Numerous scholars, for example, argue that most climate change adaptation will be driven not by science-based international policies but rather by small-scale action grounded in local environmental knowledge, cultural meanings, and values (Batterbury 1996; Brace and Geoghegan 2011; Richards 1993; Roncoli et al. 2002; Roncoli 2006). In the absence of meaningful national policy within the United States, the same may also be true of climate change mitigation actions (Kousky and Schneider 2003; Krause 2011; Lutsey and Sperling 2008). Local environmental knowledge may be an especially important basis for action in rural, conservative regions of the United States where many people have limited trust in science and government. To understand and support climate action in these contexts, we advocate increased attention to lay theories of environmental change and governance, which situate climate change "in the context of social life" (Fiske et al. 2014: 9) and in relation to culturally specific value systems.

This chapter illustrates our efforts to develop a detailed understanding of nonscientific environmental knowledge in Western North Carolina, the United States. Western North Carolina is part of Southern Appalachia, a mountainous landscape of dense temperate forests and heavy rainfall throughout the year. Topographic diversity, microclimatic diversity, and the presence of rare habitats have made this one of the most biodiverse regions of the United States, with many endemic species. Southern Appalachian forests also provide important ecosystem services, including high-elevation refuges for climate-vulnerable species and water provisioning for the Southeast. Climate change is provoking slight warming in this region, but more importantly, it is significantly increasing the *variability* of precipitation, leading to drier dry years, wetter wet years, and more episodes of extreme storms, flooding, and landslides (Laseter et al. 2012). Long-term ecological research suggests that these climatic changes may have a variety of environmental repercussions, including changes in seasonal patterns of soil moisture and streamflow; changes in forest composition in favor of more resilient tree species; changes in plant, insect, and bird populations as seasonal mutualisms grow increasingly out of sync; declines in trout and salamander populations due to warmer streams; salamander extinctions due to drying and warming of niche habitats; and increased likelihood of certain plant and ant invasions (Gragson et al. 2008b, 2014).

These climate change impacts interact with and often exacerbate the impacts related to exurbanization. In Southern Appalachia, exurbanization is driven largely by "amenity migration"—the movement of second homeowners and retirees to rural areas known for their natural beauty, recreational opportunities, and comfortable lifestyles. Amenity migration affects ecosystems through land subdivision, development, and changes in private land use, as well as through effects on governance and cultural values (Abrams et al. 2012). It also often generates social segregation or discord between long-term residents and newcomers (Abrams et al. 2012; Boucquey et al. 2012; Cadieux 2011; Cockerill and Groothuis 2014; Nesbitt and Weiner 2001). Some residents of Western North Carolina are concerned that more significant climatic change in other parts of the Southeast may lead to a new wave of in-migration, prompting further development, environmental degradation, and social division (Rice et al. 2015). In tandem, climate change and exurbanization are expected to increase forest patchiness and mountain slope instability, degrade stream quality, and generate human and nonhuman vulnerabilities with high likelihood of extinction for especially sensitive species.

Addressing the challenges of climate change and exurbanization—whether to prevent them, mitigate them, or prepare for them—will require household action and collective action at community and regional scales (Gragson et al. 2008b). However, the coordination necessary for such action will be difficult to achieve in a region long opposed to regulation and suspicious of newcomers (Gustafson et al. 2014; Vercoe et al. 2014). Responses to these challenges will depend on how people in Southern Appalachia understand the environment and coordinate across differences. To support community action, our previous work has examined the context for environmental decision-making (Burke et al. 2015) and attempted to build productive and mutually impactful dialogue between ecological scientists and nonscientists in the region (Rice and Burke 2017; Rice et al. 2015). Our present research examines lay theories of environmental change more systematically and across a broader swath of rural Western North Carolina.

We begin this chapter by describing the historical and contemporary dynamics that have shaped Southern Appalachia's landscape and people. These political-ecological dynamics are inextricably bound up with people's ways of understanding environmental change (as we argue more fully in Welch-Devine and Burke forthcoming). One important effect of this history has been the construction of local identities, especially those of "multigenerational residents" and "newcomers." We then turn to our empirical research on diverse theorizations of climate change, drawing on nearly 80 interviews to address three questions: What indicators and consequences of climate change do people observe in their everyday lives and view as relevant? How does one's connection to this landscape shape the indicators and consequences they observe and care about? And what differences exist in how people theorize the causes of climate change? By examining this diversity of climate knowledges and climate cultures in Southern Appalachia, we hope to identify complementarities, bridges, and provocations that might help natural resource managers and community members identify effective and inclusive responses to climate change.

10.2 Environmental Change, Knowledge Systems, and Vulnerabilities in Southern Appalachia

Many people in Southern Appalachia distinguish between "multigenerational residents" and "newcomers" and describe these two groups as inhabiting entirely separate social circles. Because these categories are so meaningful to local residents—and because these two groups are assumed to have different relationships to the land and the place, different priorities and values, different endowments of knowledge to draw on, and different vulnerabilities—they are one of the main ways we have stratified our research sample.[1] To understand contemporary dynamics of local environmental knowledge and climate responses, it is helpful to see how these knowledges have emerged from a changing socioecological landscape.

Our first group of research subjects—the multigenerational residents of Southern Appalachia—traces their ancestry to early European settlers, and their environmental knowledge reflects multigenerational experience of the place. These residents can detail ancestral home places and changes in the land over the centuries (authors' unpublished research; Rice et al. 2015; Veteto and Carlson 2014; Wagner 2002), and they voice cultural meanings and environmental values that were established through multigenerational engagement with the land (Evans 2013; Howell 2002; Hufford 2002; Wagner 2002). For some, family traditions and knowledge extend back to European settler economies, including an Appalachian leather industry that particularly altered regional political ecologies by culling several million deer in the early- and mid-eighteenth century and supporting migration to the mountains (Davis 2000; Newfont 2012). Government-supported Cherokee removal opened land for new farms and settlements (Otto 1989) and turned Appalachian forests into livestock grazing commons for European-descended settlers (Davis 2000; Newfont 2012).

Virtually all of these multigenerational residents have detailed family knowledge of post-Civil War Appalachia, where land speculation and timbering sparked new changes. By 1909, Southern Appalachia supplied nearly 40% of the timber for the nation's energy, communication, industrial, and transport sectors (Newfont 2012). Industrial logging generated unprecedented erosion, stream pollution, fire, and flooding (Newfont 2012) and exacerbated inequalities by incentivizing land concentration by absentee owners (Dunaway 1996; Salstrom 1994). At the start of the twentieth century, the ancestors of our multigenerational residents struggled under the pressures of ecological decline, population growth, reduced farm size, and enclosure of forest commons (Gragson et al. 2008a). Among many multigenerational families, the sense of being treated as an "internal periphery" has generated deep suspicion of outsiders and their environmental schemes, which likely affects

[1] We initially divided our sample across three residential groups ("multigenerational residents" whose families had lived in the region for more than two generations, "new natives" whose families had lived in the region for one to two generations, and "newcomers who had moved to the region during their lifetime) and further subdivided each of those residential categories into two different ways of engaging with nature (livelihood-centered engagements and recreation-centered engagements). We made two changes to this during preliminary analysis. First, after finding that the "new native" category was not sufficiently distinct from either of the other residential categories, we decided to conduct most analysis only on multigenerationals and newcomers. This of course suggests that there may be a spectrum of knowledges rather than two poles, but it is analytically useful to focus on the poles. Second, we decided to split the livelihood category into those whose livelihoods are oriented around the use of nature (e.g., farming or logging) and those whose livelihoods are oriented around environmental education, conservation, and nature-based tourism.

the politics of environmental knowledge and governance to this day (Merrifield 1989; Weaver 1996).

The environmental devastation caused by industrial logging prompted creation of the US Forest Service in 1911, the acquisition of nearly 400,000 acres by 1930, and significant reforestation. Post-World War II timber operations created jobs but offended many local people: they were typically managed by “scientific foresters” whose utilitarian language for timber contrasted with local perceptions of forests, they were often justified as part of civilizing missions for this “backward” region, and the goal of maximal extraction threatened livelihoods and violated local norms for multiuse forest commons (Newfont 2012). The 1940s and 1950s are also noted as the time when American chestnuts (*Castanea dentata*) were nearly extinguished due to an invasive disease. Chestnut shaped forest canopies, created large gaps for secondary growth, provided important habitat, offered food to people and livestock, and supplied the leather tanning industry. The loss of the chestnut was acutely felt by many local residents, both human and nonhuman (Welch-Devine et al. forthcoming).

The cultural legacies of the last three centuries are sometimes reproduced through livelihood practices such as farming, hunting, forestry, and gathering of non-timber forest product, but in many cases, they continue despite abandonment of these livelihoods. It is unclear, however, exactly how these waves of ecological disturbance and regeneration affect multigenerational residents’ perspectives on contemporary issues. For example, does their long-term connection to the land make them more concerned about change and more aware of subtle ecological shifts, or might their knowledge of historical regeneration dampen their concerns for contemporary climate-induced change? And how do their experiences with outsiders shape their concerns and governance possibilities today?

Our second demographic group—newcomers—arrived in the last 50 years with a new wave of social, economic, and environmental change. The Appalachian Regional Commission built highways to stimulate development, and certain sectors, including tourism and service, grew as much as 600% between 1970 and 1991 (Gragson et al. 2008a). In one county, out-of-staters own at least 43% of residences (Gustafson et al. 2014). These newcomers have a fundamentally different relationship with the land, lacking both the intergenerational knowledge of the region and (often) the livelihood connection to natural resources (Gragson and Bolstad 2006). Exurban migrants are generally wealthier and more educated than long-term residents (Falk and Lyson 1988), which can exacerbate social divisions (Riebsame et al. 1996), and newcomers tend to settle on upland ridges away from longer-term residents (Wear and Bolstad 1998). Their arrival has contributed to an economic shift among long-term residents away from resource-based livelihoods and toward the service sector. For example, Macon County, NC, saw farming, forestry, fishing, hunting, and mining employment drop from 36% in 1969 to 4% in 2007 (Smith et al. 2012). Similar trends have characterized Ashe, Avery, and Watauga counties. Despite the economic transition toward service-sector jobs, many residents maintain close livelihood links to nature through new farming opportunities (Christmas trees, vineyards, and agritourism), outdoor recreation, hunting, fishing, wild harvesting, and horticulture.

Researchers at the Coweeta Long Term Ecological Research (LTER) program have examined the ecological impacts of this wave of newcomers, concluding that home construction on mountain slopes, road building, and land-use decisions can significantly increase runoff, erosion, sedimentation, slope instability and landslides, and stream nitrate levels (Kirk et al. 2012; Webster et al. 2012), and land management practices leave discernible decades-long legacies on nutrient cycling, soil microbiota, and invasibility by non-native species (Gragson and Bolstad 2006; Kuhman et al. 2013). It is unclear how these impacts will interact with climate change, but exurban development will likely exacerbate vulnerabilities to extreme weather events, droughts, floods, and landslides (Wei et al. 2012) and may affect the vegetative diversity, ecological functional redundancy, deep soils, dense forests, and riparian cover necessary for ecological resilience (Kloeppel et al. 2003; Laliberté et al. 2010).

A second impact of these economic and demographic changes is a growing diversity of viewpoints on social and ecological issues (Gosnell and Abrams 2011; Gustafson et al. 2014; Jones et al. 2003). The environment, concepts of nature, and what constitutes suitable forms of development can be hotly contested in this region, with newcomers and multigenerational residents frequently finding themselves on opposing sides (Cockerill and Groothuis 2014; for other regions, see Abrams et al. 2012; Boucquey et al. 2012; Cadieux 2011; McLeod et al. 1999; Nesbitt and Weiner 2001; Ryan 1998). However, there is also precedent for newcomers bringing organizational and political skills that enable collaboration with longtime residents around shared concerns (Fortmann and Kusel 1990; Jobes 2000). New migrants often want to "close the gates behind them" in order to preserve the region as it was when they arrived (Cockerill and Groothuis 2014; Riebsame et al. 1996), but the changes they usher often create discord. Long-term residents complain that land prices are rising, that newcomers build their houses in unsuitable locations, and that newcomers and multigenerational residents don't mix. Environmental governance is a significant source of tension.

Bad blood aside, the policy context and land tenure make individual perceptions of environmental change, social norms, and collective action particularly important in this region. Land ownership is highly fragmented, with large areas of public land (as high as 80% in some counties, over 40% in most) interspersed with small private landholdings. Because public lands are fragmented by and interspersed with private lands—and because local regulations generally favor landowner autonomy—individuals' perceptions and decisions significantly affect environmental outcomes. All land-use regulations and environmental policies are contentious. There is little to no climate policy at the state and local levels, and few national climate regulations have relevance here. The precise environmental effects of climate change will therefore be shaped significantly by the nature of ongoing residential development and land management decisions.

Western North Carolinians' ability to develop effective collective responses to climate change will have significant and unequal effects. Although exposure to environmental hazards will likely be lower in Southern Appalachia than in warmer, drier, and more coastal areas of the Southeast, regional vulnerabilities are high.

Western North Carolina scores high in three of the most significant demographic predictors of vulnerability—poverty, age, and employment in the service industry (HVRI/SOVI n.d.)—and is among the worst in the country in terms of "inherent resilience" (Cutter et al. 2016). As Cutter et al. (2016) argue, while economic capital is the primary driver of resilience in urban areas, community cohesion is the most important driver of disaster resilience in rural areas. The multigenerational/newcomer divide seems troubling in this regard, and multigenerational residents are likely to suffer most from climate impacts because of their more limited economic resources.

10.3 Methods

This chapter is based on 68 semi-structured interviews and nine free-list interviews that were conducted with residents of Macon, Watauga, Ashe, and Avery counties, North Carolina, the United States. Interviewees were contacted via snowball sampling, starting with diverse initial references to maximize diversity. The sample was stratified to ensure representation of two residential groups (people whose families have been in the region for three or more generations and people who moved to the region in their lifetimes) and three livelihood groups (people who interact with nature primarily through resource-use livelihoods, primarily through conservation-oriented livelihoods, and primarily through recreation). The sample thus includes six residential-livelihood pairings or subgroups: multigenerational users, multigenerational conservationists, multigenerational recreationists, newcomer users, newcomer conservationists, and newcomer recreationists. The number of interviewees per pairing differed substantially—we interviewed many more resource users in the multigenerational group and many more recreationists in the newcomer group—but we believe that these disparities approximate actual differences in newcomer and multigenerational livelihoods (Table 10.1). The sample size is sufficient to analyze differences between the two residential groups, but not to reach robust conclusions comparing residential-livelihood pairings.

Semi-structured interviews used people's life course as an entry point for discussing how they interact with the environment and what changes they have observed. By contrast, the free-list interviews began by asking people about the changes they have seen in particular realms of biota, using this as an entry point to discuss personal interactions and changes over the life course. Except when

Table 10.1 Description of the research sample

Residential group	37 Multigenerational	38 Newcomers
Gender	12 Female/25 male	18 Female/20 male
Livelihood group	19 Resource users	10 Resource users
	8 Conservation livelihoods	7 Conservation livelihoods
	10 Recreationists	21 Recreationists

interviewees brought it up specifically, climate change was not raised until the end of the interview. Interviews were audio recorded, transcribed, and coded using Dedoose qualitative data analysis software. All references to climate, weather, and environmental changes were then extracted, and any changes explicitly linked to climate change were tabulated. Indicators and consequences of climate change were then grouped into six categories (defined post hoc by the researchers): weather, water, gardening/harvesting/farming, wild flora and fauna, recreation, and other. People's beliefs about the causes of climate change were categorized as mostly human, human and natural, mostly natural, or unsure; when informants provided additional details, these were listed and categorized as well.

There is one important limitation to these data. We did not list all possible changes and ask people if they had observed them. Rather, we allowed interviewees to raise the changes that came to mind first. Absence of a comment therefore does not mean lack of observation of any particular change. Rather, these data offer a sense of which observations are most salient. In the remainder of this chapter, we will therefore describe interviewees' observations as a reflection of their knowledge and concern combined, not as an exact index of "pure" knowledge and observation. This limits our ability to exactly map the distribution of lay knowledge and to compare it with formal scientific perspectives, but it is not necessarily a problem given our interest in sparking productive conversations about the issues that matter to stakeholders.

10.4 Results: How People See Climate

10.4.1 The Challenges of Discussing and Responding to Climate Change

Before detailing residents' observations of environmental change, it is worth sharing their reflections on the impediments to discussing and addressing climate change. Importantly, these impediments are not only a result of denialism or skepticism but also a response to people's keen sense that climate change truly is a "wicked problem" with multiple drivers, unforeseeable consequences, and politically volatile impacts. We found three clusters of responses, which can be represented by the following quotes.

Perspective 1: We're lucky here. Mac78, a newcomer who retired in the region, shared a common sentiment: "It's hard to say [what local climate impacts might be], but it's got to be gradual… because we have such a rich, moist environment. We're lucky. We don't see the fires. We don't see the melting icebergs. So, we don't have those immediate impacts." A newcomer farmer echoed this. Climate change has not yet had unambiguously negative impacts, and farmers need to be concerned with so many other, more immediate concerns:

> I would say right now it's an academic concern. It's not really a farming concern yet. And climate has certain natural variations, which are pretty great. And so, the small but steady rises in temperatures are mostly—it's going to take years to really see changes. But, I mean, I've read articles. I mean, they're growing grapes in England again, and they're—so changes are occurring, not all for the bad. Some for the good. A warmer climate's probably helpful in Canada, probably not so helpful in other places. So, I think most farmers are aware of that. But it's, like, not something uppermost on your mind. (Wat40)

Although these quotes could be seen as a form of denialism (Norgaard 2011), we think that misrepresents this perspective, which is actually rooted in an understanding of the region's unique ecological and topographic conditions and is consistent with scientific projections for regional climate impacts.

Perspective 2: There are so many factors. Many residents also underscored the complexity of global climate change. Mac27, a young newcomer involved in conservation work and small-scale farming, explained that it was difficult to anticipate the local effects of climate change because "global climate is immensely complicated; we're just understanding the factors." Even for people who firmly believe in anthropogenic climate change, the complexity of the science and the cacophony of public debate seemed to generate high levels of uncertainty.

Perspective 3: It's too political. Finally, many interviewees lamented that climate change—and environmental stewardship more generally—is difficult to discuss because it has become so politically polarizing. For Mac49, a multigenerational resident with diverse livelihood connections to nature, the politics made it difficult to trust the science: "I would like to see science used as science and not as political tokens that people like to use it as." Another multigenerational resident who works in conservation was more concerned about how partisanship makes collective action impossible:

> One of the things that's disappointing to be a professional in the field and to work with wildlands and the environment and to be a huge proponent of protecting resources was I feel like the debate on climate change has become… politicized. And that has not helped anybody…. In my opinion, everybody should want clean water. I mean, really, they do. If you talk to somebody, they want clean water, they want clean air, they want parks, they want protection of wild areas that are pristine. They have a lot in common. But we don't celebrate what's in common, and we don't work together. We form two factions, and they go further apart. So, you can never work together, and I think that's a shame. The politics in my opinion has really hindered environmental protection. And it has also altered the debate on climate change…. (EP_Wat04)

Finally, even if residents overcome these impediments to *discussing* climate impacts, the multigenerational/newcomer divide remains a major hindrance to collective responses. As one long-term newcomer said:

> There's a big conflict between the people who have lived here for generations [and newcomers]. Their idea is that if people would just go away and leave them alone, they don't have to do anything. In other words… there's no *policy* that makes any sense to them. They think the problem is just too many people who come up here. And if less people come up, then they can still hunt wherever they want to hunt, they can go wherever they want to go. I was at a hearing once where a local guy was griping about setbacks off of the river, and restricting straight-piping of sewage into the river. And he said, "we've been peeing in this river for generations, and a little bit of pee doesn't hurt the river." And so, that's a prevalent

> feeling is that… all of this stuff out here will take care of itself if people would just go away and leave only the people who were here for generations. They know how to manage it. And the distrust of government and any kind of policy [is tied to that]. And it's that distrust of government that prevents any kind of meaningful policy. Now it just blows up into resentment of the people who are coming here even though the people who are coming here, it's what provides a livelihood. (Mac63)

Thus, despite its social, economic, and ecological importance, many informants simply do not discuss climate change with friends and neighbors. This lack of dialogue impedes collective action and probably exacerbates the sense that climate change is a taboo subject.[2] We believe that work like ours may help overcome some of these impediments to dialogue and action by demonstrating the breadth of community concern and by transforming climate conversations from national political debates to local concerns based on people's own legitimate knowledge and experiences. We turn now to what those knowledge systems look like.

10.4.2 Weather and Climate

For both multigenerational residents and newcomers, changes in the weather are by far the most commonly cited indicators or consequences of climate change (Fig. 10.1). This is unsurprising, given that weather can be felt or experienced more directly than many other climate change indicators, but this may also suggest that weather is a particularly salient concern. What is interesting, however, is that different residential groups observe different types of weather changes (Fig. 10.2). While both groups are attuned to general trends (e.g., a general increase in temperature or a general decrease in snowfall), newcomers are much more focused on extreme events, while multigenerational residents are much more focused on changes in seasonal, diurnal, and geographical patterns of weather (e.g., the timing and direction of storms and the consistency of seasons).

However, residential status alone does not fully explain this difference in observations. Table 10.2 provides a finer-grained analysis of how frequently people mention general trends, extreme events, and changing weather patterns by dividing interviewees into subgroups that reflect both residential and livelihood status. With regard to general weather trends (Table 10.2a), we see little difference among the residential/livelihood pairings, though multigenerational conservationists were slightly less likely to comment on general trends in weather. When examining who observed differences in seasonal or daily weather patterns (Table 10.2b), we see that virtually the entire difference between multigenerationals and newcomers is found among recreationists. In other words, multigenerational residents and newcomers who interact with nature primarily via their livelihoods (whether for conservation or

[2] For example, many interviewees expressed the belief that their neighbors are climate skeptics, despite having not discussed the issue with them. We think they would be surprised to find such high levels of acknowledgment of climate change across our interviews.

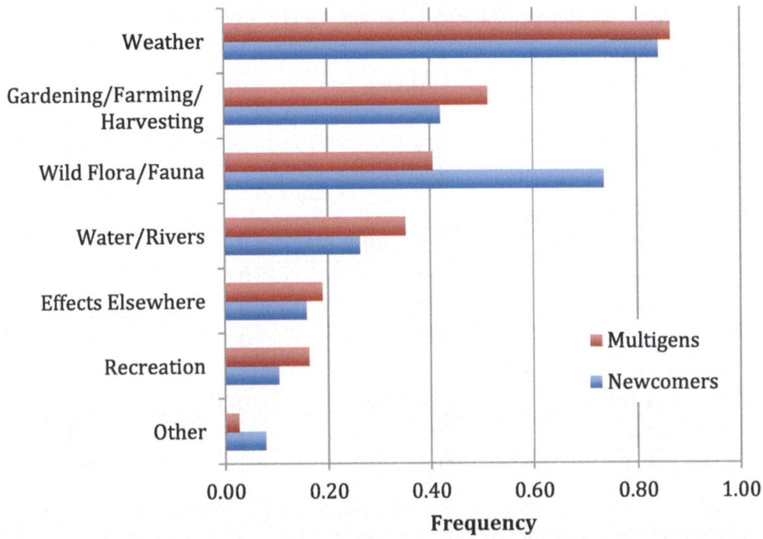

Fig. 10.1 What do people describe as indicators or consequences of climate change?

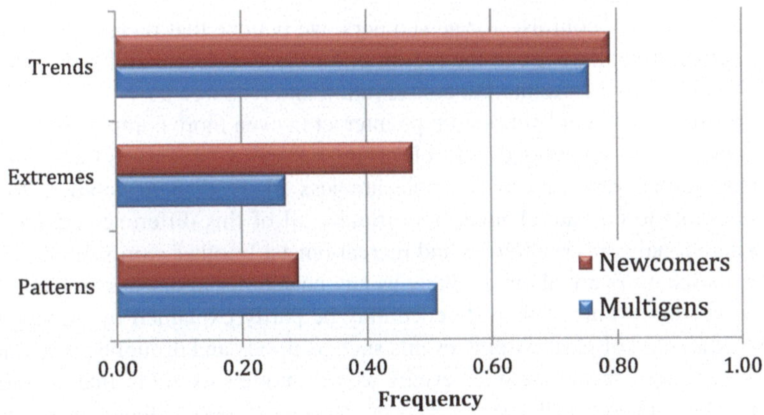

Fig. 10.2 Weather-related indicators of climate change, by type

use) are about equally likely to notice changes in weather patterns, but their recreational peers differ. Interestingly, they differ in opposite directions. Multigenerational recreationists were more likely than their livelihood peers to observe pattern changes while newcomer recreationists were significantly less likely to see these same pattern changes. This may result partly from differences in the types of recreation that multigenerationals and newcomers are drawn to, but it also suggests a possible difference in how these two populations integrate recreation-based knowledge. We suspect that for multigenerationals, recreation is additive; they are already part of social circles full of conservationists and resource users, so they have that knowledge, and recre-

Table 10.2 Different residential and livelihood groups observe different types of weather changes

	LH-CONS	LH-USE	REC	TOT
(a) *Likelihood of noticing trends*				
	LH-CONS	LH-USE	REC	TOT
MG	0.6	0.7	0.9	0.8
NEW	0.9	0.8	0.8	0.8
TOT	0.7	0.8	0.8	
(b) *Likelihood of noticing patterns*				
	LH-CONS	LH-USE	REC	TOT
MG	0.4	0.5	0.6	0.5
NEW	0.4	0.5	0.1	0.3
TOT	0.4	0.5	0.3	
(c) *Likelihood of noticing extremes*				
	LH-CONS	LH-USE	REC	TOT
MG	0.1	0.4	0.1	0.3
NEW	0.4	0.5	0.5	0.5
TOT	0.3	0.5	0.4	

MG = multigenerational residents, NEW = newcomers, LH-CONS = conservation-related livelihoods; LH-USE = resource-use livelihoods, REC = interacting with nature primarily through recreation

ation adds more. By contrast, for newcomers, we believe that recreational knowledge is substitutive; they are in the region primarily to recreate and lack historical ties and social networks related to conservation and resource use.

Residential status and livelihood type interact in even more complex ways when we examine the likelihood of observing extreme events (Table 10.2c). Multigenerational residents as a whole are less likely than newcomers to link extreme events to climate change, but virtually all of this difference comes from multigenerational conservationists and recreationists. In other words, multigenerational resource users are about as likely as any newcomer to discuss an increase in extreme weather events. This difference may be partly explained by multigenerationals' awareness of past extreme events such as floods and droughts, which leads them to see more recent weather events as yet another example that weather is unpredictable and they will survive. Several interviewees noted that they will adapt as their ancestors always have. However, the difference with regard to weather extremes may also be due to newcomers' experiences elsewhere. A few interviewees explicitly attributed their concern about severe weather to their experiences in Florida, where there has been more discussion of climate change and hurricanes. As one newcomer recreationist said: "being a Floridian, somebody who worked in hurricane recovery, I am intensely aware of how dramatic those changes are and how vulnerable we are" (Mac63).

Surprisingly, people who work in conservation and resource management—who we might assume would be most exposed to research and conversation on climate—were the least likely to link climate change to overall weather trends and extreme events, and they were also less likely than resource users to notice changing weather patterns. This likely stems from their intensive focus on the specific local resources they manage and a professional reluctance to draw "hasty" conclusions about the controversial topic of climate change. As one newcomer conservationist told us:

> Well, I'm not going to deny climate change. In my job, usually we don't directly jump to that [conclusion], because what we're really trying to get folks to do and because of where we're at. Climate change is going to be, I believe, more pronounced in other areas [than] this region, and it also depends upon what species you're talking about. Are you talking about salamanders? Some species may be more sensitive and you may see more of a direct shift. Fish... so what if smallmouth [bass] are moving a little bit warmer, if they're not displacing other fish? (Mac29)

10.4.3 Other Indicators of Climate Change

Returning to Fig. 10.1, we see that the next most-cited indicators of climate change are related to gardening, farming, and harvesting (a category that includes observations such as changes in soil temperature, planting seasons, pasture growth, livestock disease, the economics of farming, and the abundance and timing of wild-harvested plants) and wild flora and fauna. By contrast, fewer than half of people in any demographic subgroup observed links between climate and water, recreation, or effects elsewhere.

Looking at the high-interest categories in more detail helps clarify who within the community is most observant of and/or concerned with these issues, and it also provides insights into how one's relationship to place affects environmental knowledge. A detailed breakdown of who linked climate change to changes in gardening, farming, and harvesting is provided in Table 10.3a. Multigenerational residents across the board were moderately observant of gardening/farming/harvesting activities, but only multigenerational resource users passed the 50% mark. This is surprising given the importance of subsistence gardening to nearly all multigenerational

Table 10.3 How residential and livelihood groups differ in their observations of climate-related changes to gardening/farming/harvesting and wild flora/fauna

(a) *Likelihood of commenting on climate-related changes in gardening, farming, and harvesting*				
	LH-CONS	LH-USE	REC	TOT
MG	0.4	0.6	0.4	0.5
NEW	0.7	0.8	0.1	0.4
TOT	0.5	0.7	0.2	
(b) *Likelihood of commenting on climate-related changes in wild flora and fauna*				
	LH-CONS	LH-USE	REC	TOT
MG	0.6	0.3	0.5	0.4
NEW	0.7	0.6	0.8	0.7
TOT	0.7	0.4	0.7	

MG = multigenerational residents, NEW = newcomers, LH-CONS = conservation-related livelihoods; LH-USE = resource-use livelihoods, REC = interacting with nature primarily through recreation

residents. By contrast, newcomer conservationists and newcomer resource users were significantly more observant of gardening/farming/harvesting changes. We suspect this is because newcomers actively seek out information about their new homes, in part because of their amenity focus and in part to compensate for their lack of experiential knowledge. Among newcomer recreationists, however, only 10% mentioned changes in farming/gardening/harvesting. These data reveal that, while newcomer and multigenerational populations, on average, find gardening, farming, and harvesting changes to be quite salient, the salience varies significantly within each group. Furthermore, the fact that newcomer status increases observations of these changes among resource users and conservationists but decreases observations among recreationists reinforces the conclusion that residential status and livelihood status interact in complex ways.

Overall, multigenerationals observed a larger range of climate-related impacts on gardening, farming, and harvesting, including higher ground temperatures, less reliable ramp harvests, more difficulty raising certain crops and animals, increased farm work and expense to compensate for weather variability, changes in maple syrup production, and lower quality and price for furs. Newcomers' observations were narrower, and there was more consensus around a few impacts, especially shifts in the gardening season, more crop pests or diseases, and fewer wild mushrooms or changes in the mushroom season. While we might generally expect long-term residents to be more homogenous and therefore to have higher levels of cultural consensus, in this case, newcomers' knowledge and concerns may be narrower because they are acquired largely through the same formal activities, media sources, and environmental programs, whereas multigenerationals' knowledge is differentiated more according to their specific jobs and family histories.

With regard to the category of wild flora and fauna, newcomers were almost twice as likely as multigenerational residents to report climate-related changes. Nonetheless, climate-related changes to wild flora and fauna received over 50% interest from five of the six residential-livelihood pairings. Across the board, resource users were the least likely to cite changes in this area, which is not surprising given that they engage with forests and rivers less than do conservationists and recreationists. Unlike the other categories mentioned above, this is the only major category in which newcomer recreationists were the most likely to comment. One key implication of these data is that we cannot lump all newcomers together as one group of amenity migrants and certainly not as ignorant ones. Newcomer recreationists seem most amenity-focused in that they are unconcerned with long-term water supplies and not attentive to gardening and harvesting, but this does not make them unobservant across the board. They are more concerned with wild flora and fauna than any other group (a characteristic that likely results from their amenity interests in the wild beauty of this region) and as concerned with extreme weather as any other group (a characteristic that likely results from their previous exposure to extreme events and discussions of climate change impacts on extreme events elsewhere).

Within wild flora/fauna, multigenerationals also notice far more changes and/or are more likely to discuss changes in detailed terms. For example, whereas many newcomers mentioned "changes in the forest" or "diseases affecting the forest," multigenerationals often discussed specific tree species. Nonetheless, both groups shared a concern for forest health, especially related to increased vulnerability to disease, increased pest habitat and non-native invasion (the most commonly cited observation of newcomers, at over 25%), changes in ground cover, blooming time for native wildflowers, and populations of specific tree species. Fairly large numbers of both groups (11% of multigenerationals and 21% of newcomers) also cited changes in bird populations and migrations, often as an indirect effect of forest health. Discussion of special-interest species (bear), hunted species (deer, turkey, grouse), and nuisance species (bear, coyote, rattlesnake, poison ivy) was less common than we expected.

Two categories of indicators received relatively little attention from any group, but they nonetheless round out our understanding of who is concerned with climate-related changes and why. Recreation-related indicators (such as the number of skiing or rafting days or the health of the industry) were, not surprisingly, mentioned most often by recreationists. What is perhaps more interesting is that more multigenerationals (including at least one from each livelihood category) cited concerns for recreation, whereas only newcomer recreationists did so. In many cases, the multigenerationals expressed concern primarily for the recreation *industry*, which has broad-based and long-term economic impacts on the region, especially for people working in the service industry. Effects on other locations (such as in polar regions or tropical rainforests) were mentioned far less than one might expect given popular media portrayals of climate change, and they were not mentioned at all by people working in conservation. This is consistent with our interpretation (above) that conservationists are especially focused on their realm of expertise, local environmental issues.

10.4.4 Causes of Climate Change

Because our larger project focuses on environmental changes generally, we did not want to risk biasing or alienating informants by asking them to commit to an answer about the specific causes of climate change. Thus, in the following analysis, we draw from a smaller sample of 57 informants, 20 of whom discussed causes generally and 37 of whom gave detailed explanations of the causes of climate change.[3] Because the 37 individuals who offered detailed explanations for climate change include only 19 multigenerational residents and 18 newcomers, we do not have sufficient representation across livelihood and recreational groups to conduct finer-grained analysis or reach firm conclusions.

[3] Note that there is some degree of selection bias in this sample; 49 of these 57 interviewees (86%) reported being certain that climate change exists versus 70% among those excluded from this subsample.

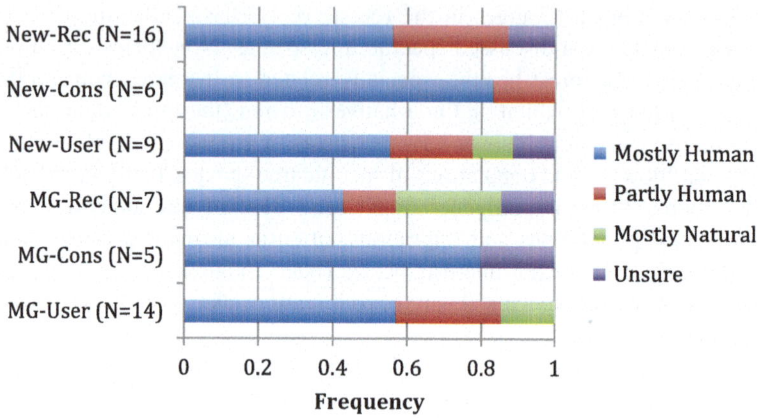

Fig. 10.3 Belief in causes of climate change

In the sample of 57, 90% said they believed that climate change is happening, and 81% believe humans bear some responsibility for those changes. As Fig. 10.3 shows, many informants (18 of the sample of 57, or 32%) expressed a belief that natural causes were significantly responsible for climate change. This conversation with Mac11 illustrates beliefs in the natural causes of climate change:

Interviewer: You seem really comfortable discussing climate change. What's your take on it?

Mac11: Well, I like to… No, it's changing. The climate is being affected. I mean we're here such a short period of time, but I do believe that the human race has pumped a lot of carbon into the atmosphere. A lot of scientists will say that that has something to do with climate change. I mean I'm not going to say these scientists are making it up. I think there's so many people on the earth that we are affecting the earth and we may be the tipping point that the earth can't self-correct if you will. And you know 100 years from now we will know whether we fouled up or not.

Interviewer: We certainly will.

Mac11: I'm smart enough to know too that there are cycles. You know I mean we had glaciers and ice age and all that, so yeah, we could be in one of those, somewhere in one of those cycles, but notwithstanding that, I believe that the humans are having an impact.

There is widespread belief, particularly among multigenerationals, that Southern Appalachia is prone to 100- or 200-year weather cycles. This focus on "natural cycles" is common in denialist communities and is an intentional misinformation strategy of denialist institutions (Oreskes and Conway 2010). As a result of the oppositional nature of national discourse and the fixation on belief versus denial, these types of claims would often be dismissed by mainstream environmentalists. Such a view may alienate people who, in a less polarized context, might be allies in climate action.

Residential status does not appear to affect acknowledgment of climate change or anthropogenic climate change. The multigenerational recreationist category was marked by a particularly low degree of belief (57%, compared to 78–100% among other groups), but our subsamples are too small to draw conclusions based on this.

Among the 37 interviewees who discussed causes in greater depth, 78% described changes to the atmosphere, energy, and pollution as the main causes of climate change.[4] Multigenerationals were more likely than newcomers to add secondary causes related to local land use (mentioned by 47% of multigenerationals versus 28% of newcomers) and other national and global issues (mentioned by 37% of multigenerationals and 28% of newcomers).[5] This focus on local land use by multigenerationals is not surprising given widespread concern with exurban development, and it is consistent with ecological research from the Coweeta Long Term Ecological Research program showing that regional development patterns can exacerbate climate change in two ways: independently of global climate patterns, through dynamics similar to urban heat islands, and by exacerbating the local consequences of global climate change.

Like belief in natural causes of climate change, attribution of climate change to regional development dynamics might seem naïve to many mainstream environmentalists and environmental scientists. A conversation with Mac29, a newcomer who works in conservation, illustrates that attending to local sources of environmental change does not necessarily entail ignoring global climate change. Looking at fish as an example, he said the data show some declines and some movement as streams warm, but it is hard to read the trends. There is a clear link between particularly wet or dry years and fish populations, and we know that climate change will lead to wetter wets and dryer dries, he said, so we can expect bigger problems in the future. However, there are so many other interactions with land use, especially with regard to streamside vegetation, that it's hard to focus exclusively on climate. When you remove even a handful of trees, you get radically warmer water, so it is not simply that vegetation loss exacerbates climate change but that "you can get your climate change effect [just] by removing trees, and these cold streams are warming up because you removed the shade."

We get a related, and much more common, expression of how local land use causes environmental change from Mac38, a multigenerational recreationist:

Interviewer: So you don't feel like farming or the outdoors are particularly under threat because of climate changes here. Would you say it's more management and preservation issues?

Mac38: If you could manage population.

Interviewer: Gotcha, gotcha. And population locally using these resources?

Mac38: Yeah. Obviously, the more... You take two counties over, Haywood County, every place you can build a house, a house is built. And because of that there's stream runoff and all that stuff. It affects erosion control, the quality of water, the French Broad... all that's affected.

---

[4] This category includes mentions of CO_2, greenhouse gases, fossil fuels, cars, electricity, pollution, industry, chemtrails (per one informant), and lifestyles (which was grouped here because they were most often described as fossil-fuel intensive).

[5] "Local land use" includes farming practices, forest management, development, an increase in asphalt or pavement, and "cutting the mountains" to build roadways. "Other national and global issues" included overpopulation, urbanization, policy, globalization of trade, broad cultural change, and Amazonian deforestation.

In some ways, this is similar to Wat40's argument that climate change is not "uppermost on your mind." For this perspective, it is not necessarily that exurbanization exacerbates climate change or that exurbanization can cause the same types of climate impacts as global change, but rather that exurbanization is having visible and significant effects *right now*, so it deserves more urgent attention.

10.5 Conclusion

As Coweeta LTER research shows, there is a great deal at stake in the present and future intersection of exurbanization and climate change (Gragson et al. 2008b, 2014). Several residents have argued that climate change raises concerns regarding future demographic shifts and increased development pressure. Furthermore, there are serious concerns about vulnerability and environmental justice, especially lower-income and multigenerational people's disproportionately high exposure to landslides and flooding. As one long-resident newcomer noted:

> So many people up here are very high altitude, and they're the ones less likely to be able to get out of the way faster. So, I worry for people on the mountains close to river areas. There's an awful lot of older people who bought retirement places in vulnerable areas with slides and flooding. And then also just being marooned, stranded when roads wash out. (Mac63)

There are many reasons to doubt that local communities have the adaptive capacity to respond effectively and justly to future environmental shocks, especially given social divisions, the paucity of public services and environmental policy, and the lack of sustained conversations to build collective action and collective norms regarding environmental governance and land management. Thus, while Mac38 and others may reasonably argue that exurbanization and regional land development pose greater threats to social and ecological well-being in the near term, we think there is great value in Hulme's (2009) insight that climate conversations offer significant opportunities for deep reflection about who we are, how we want to interact with the environment and one another, and how we want our ecological homes to look. Many of our interviewees expressed skepticism that climate mitigation was still feasible and that anything beyond adaptation or coping were reasonable goals. There is, however, more willingness to discuss climate change than we expected and, in fact, more than many informants expected, so opening discussions on climate change may be a way to generate some of the coming together that is necessary to build toward collective reflection and action.

Much of the reluctance that we see to discuss climate change appears, more accurately, to be a reluctance to discuss climate change *on the terms in which it is conventionally framed*. Our research, however, provides some insights into how an alternative climate conversation may be framed in order to engage more people in more open ways. First, even people who acknowledge anthropogenic climate change are often reticent to discuss the issue because the black-and-white terms of the national climate conversation do not permit nuance and expressions of even partial uncertainty. One solution is to seek pathways for solidarity without demanding conformity (Rice and Burke 2017; see Chap. 5). Another approach—beginning community conversations by detailing strategies for climate action that span the political spectrum—might help frame environmental protection as a multi-partisan concern, but it could also cement partisan divides.

Second, our research indicates that, in Western North Carolina, mitigation and adaptation efforts organized around weather, wild flora and fauna, and farming/gardening/harvesting are likely to receive the broadest base of support. Water, recreation, and global effects seem unlikely to be particularly fertile ground for climate conversations, though the economic importance of both water and recreation could quickly spark greater concern. When we analyzed observations and concerns at the level of residential-livelihood subgroups, we found that wild flora and fauna attracted more widespread attention than livelihood activities did (five out of six subgroups had over 50% likelihood of commenting on wild flora and fauna, compared to only three out of six subgroups for farming, gardening, and harvesting). While it has always been difficult to reconcile diverse and conflicting interests regarding forest management, the broad consensus around forest health and vulnerability to diseases and invasives seems promising, as does a multi-stakeholder forest management partnership that has sought to influence the Nantahala-Pisgah National Forest Management Plan over the last 5 years. In some ways, however, gardening and farming may be an easier place to start because of widespread interest in the agrarian character of this region and growing interest in local food. Multigenerational and livelihood-oriented concerns with farm economies may dovetail nicely with newcomers' interest in the rural idyll and food politics.

Third, this research suggests the importance of seeking diversity not just according to conventional demographic measures but also according to locally salient differences, in this case family residential history and job-related or recreation-related ways of engaging with the environment. These different ways of connecting to place do appear to shape people's observations and concerns in meaningful ways.

Finally, the significant focus among our interviewees on local indicators, effects, and causes of climate change may be more of a resource than an impediment. Many scholars have expressed concern that the global scale of climate change leaves people feeling disempowered and that communication based on distant impacts (e.g., to exotic species and island nations) may not resonate with all groups. In that context, Western North Carolinians' strong attention to regional impacts and causes seems to provide an "actionable" starting point for bringing people with diverse viewpoints together around issues that they experience first hand and about which they care deeply.

Acknowledgements We are deeply appreciative of the people of Southern Appalachia who opened their homes and their lives to us. We hope that this work will be of use as they chart their path forward.

This material is based upon work supported by the National Science Foundation under grants BCS 1558929 and 1558930, as well as DEB 1637522. Funding support was also provided through the ANR Young Researcher Program no. ANR-13-JSH1-0005-01.

References

Abrams JB, Gosnell H, Gill NJ, Klepeis PJ (2012) Re-creating the rural, reconstructing nature: an international literature review of the environmental implications of amenity migration. Conserv Soc 10(3):270–284

Batterbury S (1996) Planners or performers? Reflections on indigenous dryland farming in northern Burkina Faso. Agric Hum Values 13:12–22

Boucquey N, Campbell LM, Cumming G, Melitis ZA, Norwood C, Stoll J (2012) Interpreting amenities, envisioning the future: common ground and conflict in North Carolina's rural coastal communities. GeoJournal 77:83–101

Brace C, Geoghegan H (2011) Human geographies of climate change: landscape, temporality, and lay knowledges. Prog Hum Geogr 35(3):284–302

Burke BJ, Welch-Devine M, Gustafson S (2015) Nature talk in an Appalachian newspaper: what environmental discourse analysis reveals about efforts to address exurbanization and climate change. Hum Organ 74(2):185–196. https://doi.org/10.17730/0018-7259-74.2.185

Cadieux KV (2011) Competing discourses of nature in exurbia. GeoJournal 76(4):341–363

Cockerill K, Groothuis P (2014) Last settler's syndrome and resource use in Southern Appalachia. J Rural Community Dev 9(3):319–336

Cutter SL, Ash KD, Emrich CT (2016) Urban-rural differences in disaster resilience. Ann Assoc Am Geogr 106(6):1236–1252

Davis DE (2000) Where there are mountains: an environmental history of the Southern Appalachians. University of Georgia Press, Athens

Dunaway WA (1996) The first American frontier: transition to capitalism in Southern Appalachia, 1700–1860. University of North Carolina Press, Chapel Hill

Evans SR (2013) Exurbanizing water: stream management decision-making among Newcomer and Generational landowners in southern Appalachia. Ph.D. Dissertation, University of Georgia Department of Anthropology. Athens

Falk WW, Lyson TA (1988) High tech, low tech, no tech: recent industrial and occupational change in the South. State University of New York Press, Albany

Fiske SJ, Crate SA, Crumley CL, Galvin K, Lazrus H, Lucero L, OliverSmith A, Orlove B, Strauss S, Wilk R (2014) Changing the atmosphere. Anthropology and climate change. Final report of the AAA Global Climate Change Task Force. American Anthropological Association, Arlington

Fortmann L, Kusel J (1990) New voices, old beliefs: Forest environmentalism among new and longstanding rural residents. Rural Sociol 55(2):214–232

Gosnell H, Abrams J (2011) Amenity migration: diverse conceptualizations of drivers, socioeconomic dimensions, and emerging challenges. GeoJournal 76(4):303–322

Gragson TL, Bolstad PV (2006) Land use legacies and the future of southern Appalachia. Soc Nat Resour 19(2):175–190

Gragson TL, Bolstad PV, Welch-Devine M (2008a) Agricultural transformation of Southern Appalachia. In: Redman C, Foster D (eds) Agrarian landscapes in transition. Oxford University Press, New York, pp 89–121

Gragson TL and the co-Principal Investigators of the Coweeta Long Term Ecological Research Program (2008b) Southern Appalachia on the edge—exurbanization and climate interaction in the Southeast. Renewal proposal submitted to Long-Term Ecological Research Program, Division of Environmental Biology, National Science Foundation. Athens: University of Georgia Research Foundation. http://coweeta.uga.edu/publications/10462.pdf

Gragson TL and the co-Principal Investigators of the Coweeta Long Term Ecological Research Program (2014) LTER: the interacting effects of hydroclimate variability and human landscape modification in the Southern Appalachian Mountains. Renewal proposal submitted to Long-Term Ecological Research Program, Division of Environmental Biology, National Science Foundation. Athens: University of Georgia Research Foundation. Unpublished

Gustafson S, Heynen N, Rice JL, Gragson TL, Shepherd JM, Strother C (2014) Megapolitan political ecology and urban metabolism in southern Appalachia. Prof Geogr 66(4):664–675. https://doi.org/10.1080/00330124.2014.905158

Hazard and Vulnerability Research Institute/Social Vulnerability Index (n.d.) US count-level 2010–14 social vulnerability component summary. Columbia: University of South Carolina. http://artsandsciences.sc.edu/geog/hvri/sovi%C2%AE-0

Howell BJ (2002) Appalachian culture and environmental planning: expanding the role of the cultural sciences. In: Howell BJ (ed) Culture, environment, and conservation in the Appalachian South. University of Illinois Press, Urbana, pp 1–16

Hufford M (2002) Reclaiming the commons: narratives of progress, preservation, and ginseng. In: Howell BJ (ed) Culture, environment, and conservation in the Appalachian South. University of Illinois Press, Urbana, pp 100–120

Hulme M (2009) Why we disagree about climate change: understanding controversy, inaction, and opportunity. University of Cambridge Press, Cambridge

Jobes PC (2000) Moving nearer to heaven: the illusions and disillusions of migrants to scenic rural places. Praeger, Westport

Jones RE, Fly JM, Telley J, Cordell HK (2003) Green migration into rural America: the new frontier of environmentalism? Soc Nat Resour 16(3):221–238

Kirk RW, Bolstad PV, Manson SM (2012) Spatio-temporal trend analysis of long-term development patterns (1900–2030) in a Southern Appalachian County. Landsc Urban Plan 104(1):47–58

Kloeppel BD, Clinton BD, Vose JM, Cooper AR (2003) Drought impacts on tree growth and mortality of Southern Appalachian forests. In: Greenland D, Goodin DG, Smith RC (eds) Climate variability and ecosystem response at long-term ecological research sites. Oxford University Press, New York, pp 43–55

Kousky C, Schneider SH (2003) Global climate policy: will cities lead the way? Clim Pol 4:359–372

Krause RM (2011) An assessment of the greenhouse gas reducing activities being implemented in US cities. Local Environ Int J Justice Sustain 2:193–211

Kuhman TR, Pearson SM, Turner MG (2013) Why does land-use history facilitate non-native plant invasion? A field experiment with Celastrus orbiculatus in the Southern Appalachians. Biol Invasions 15(3):613–626

Laliberté E, Wells JA, DeClerck F, Metcalfe DJ, Catterall CP, Queiroz C, Aubin I, Bonser SP, Ding Y, Fraterrigo JM, Morgan JW, Merlos DS, Vesk PA, Mayfield MM (2010) Land use intensification reduces response diversity in plant communities. Ecol Lett 13(1):76–86

Laseter SH, Ford CR, Vose JM, Swift LW Jr (2012) Long-term temperature and precipitation trends at the Coweeta Hydrologic Laboratory, Otto, North Carolina, USA. Hydrol Res 43:890–901

Lutsey N, Sperling D (2008) America's bottom-up climate change mitigation policy. Energy Policy 36(2):673–685

McLeod D, Woirhaye J, Menkhaus DJ (1999) Factors influencing support for rural land use control: a case study. Agric Resour Econ Rev 28(1):212–214

Merrifield J (1989) Putting the scientists in their place: participatory research in environmental and occupational health. In: Highlander Research and Education Center: an approach to education presented through a collection of writings. Highlander Research and Education Center, New Market, pp 265–296

Nesbitt JT, Weiner D (2001) Conflicting environmental imaginaries and the politics of nature in Central Appalachia. Geoforum 32:333–349

Newfont K (2012) Blue ridge commons: environmental activism and forest history in Western North Carolina. University of Georgia Press, Athens

Norgaard KM (2011) Living in denial: climate change, emotions, and everyday life. MIT Press, Cambridge, MA

Oreskes N, Conway EM (2010) Merchants of doubt: how a handful of scientists obscured the truth on issues from cigarette smoke to global warming. Bloomsbury Press, New York

Otto J (1989) The Southern frontiers, 1607–1860: the agricultural evolution of the colonial and antebellum South. Greenwood Press, New York

Rice JL, Burke BJ (2017) Building more inclusive solidarities for socio-environmental change: lessons in resistance from Southern Appalachia. Antipode 50(1):212–232

Rice JL, Burke BJ, Heynen N (2015) Knowing climate change, embodying climate praxis: experiential knowledge in southern Appalachia. Ann Assoc Am Geogr 105(2):253–262. https://doi.org/10.1080/00045608.2014.985628

Richards P (1993) Cultivation: knowledge or performance? In: Hobart M (ed) An Anthropological Critique of Development: The Growth of Ignorance. Routledge, 61–78

Riebsame WE, Gosnell H, Theobald DM (1996) Land use and landscape change in the Colorado Mountains 1: theory, scale, and pattern. Mt Res Dev 16(4):395–405

Roncoli C (2006) Ethnographic and participatory approaches to research on farmers' responses to climate predictions. Clim Res 33(1):81–99

Roncoli C, Ingram K, Kirshen P (2002) Reading the rains: local knowledge and rainfall forecasting among farmers of Burkina Faso. Soc Nat Resour 15:411–430

Ryan RL (1998) Local perceptions and values for a Midwestern river corridor. Landsc Urban Plan 42:225–237

Salstrom P (1994) Appalachia's path to dependency: rethinking a region's economic history 1730–1940. University of Kentucky Press, Lexington

Smith JW, Moore RL, Anderson DH, Siderelis C (2012) Community resilience in Southern Appalachia: a theoretical framework and three case studies. Hum Ecol 40:341–353

Vercoe RA, Welch-Devine M, Hardy D, Demoss JA, Bonney SN, Allen K, Brosius JP, Charles D, Crawford B, Heisel S, Heynen N, de Jesús-Crespo RG, Nibbelink N, Parker L, Pringle C, Shaw A, Van Sant L (2014) Acknowledging trade-offs and understanding complexity: exurbanization issues in Macon County, North Carolina. Ecol Soc 19:23. http://www.ecologyandsociety.org/vol19/iss1/art23/

Veteto JR, Carlson SB (2014) Climate change and apple diversity: local perceptions from Appalachian North Carolina. J Ethnobiol 34(3):359–382. https://doi.org/10.2993/0278-0771-34.3.359

Wagner MB (2002) Space and place, land and legacy. In: Howell BJ (ed) Culture, environment, and conservation in the Appalachian South. University of Illinois Press, Urbana, pp 121–132

Wear DN, Bolstad PV (1998) Land use changes in southern Appalachian landscapes: spatial analyses and forecast evaluation. Ecosystems 1:575–594

Weaver BJ (1996) "What to do with the mountain people?" The darker side of the successful campaign to establish the Great Smoky Mountains National Park. In: Cantrill JG, Oravec CL (eds) The symbolic earth: discourse and our creation of the environment. University of Kentucky Press, Lexington, pp 151–175

Webster JR, Benfield EF, Cecala KK, Chamblee JF, Dehring CA, Gragson TL, Cymerman JH, Jackson CR, Knoepp JD, Leigh DS, Maerz JC, Pringle C, Valett HM (2012) Water quality and exurbanization in Southern Appalachian streams. In: Boon PJ, Raven PJ (eds) River Conservation and management: catchment conservation, ecosystem integrity, and threats to river systems. Wiley, Hoboken, pp 91–107

Wei W, Clark JS, Vose JM (2012) Application of a full hierarchical Bayesian model in assessing streamflow response to a climate change scenario at the Coweeta Basin, NC, USA. J Resour Ecol 3(2):118–128

Welch-Devine M, Burke B (forthcoming) Understanding climate change: toward fruitful engagement between ethnoecology and political ecology

Welch-Devine M, Burke B, Steacy C, Rzonca S (forthcoming) Seeing the forest for the trees: exurbanization and the diversification of environmental observations, values, and theories

Chapter 11
Understanding Climate from the Ground Up: Knowledge of Environmental Changes in the East African Savannas

Kathleen A. Galvin, Trevor Even, Robin S. Reid, Jesse Njoka, Joana Roque de Pinho, Philip Thornton, and Kirk Saylor

Abstract Africa makes a relatively minor contribution to global greenhouse gas emissions compared with developed nations, yet the African continent will be increasingly vulnerable to climate change processes in the coming decades. Critical challenges include meeting basic needs for food, water, shelter, and other necessities without undermining biodiversity and ecosystem services. Coordination efforts to address multiple climate-related stressors have generally occurred at the national level and taken an external approach, with national governments favoring collaboration with foreign-based NGOs and other international institutions over working with lower levels of government. However, the involvement of actors at the local level correlates with decisions that are better adapted to local social-cultural and environmental contexts, reducing implementation costs and increasing trust, thereby increasing the equity and efficacy of decentralized approaches. This chapter examines indigenous and local knowledge of climate change. It addresses climate and environmental change from the perspectives of Kenyan pastoralists who identified a myriad of environmental issues that occur and interact at different scales. They also identified ways

K. A. Galvin (✉) · K. Saylor
Department of Anthropology and Geography, Colorado State University, Fort Collins, CO, USA
e-mail: Kathleen.Galvin@colostate.edu

T. Even
Graduate Degree Program in Ecology, Colorado State University, Fort Collins, CO, USA

R. S. Reid
Department of Ecosystem Science and Sustainability, Colorado State University, Fort Collins, CO, USA

J. Njoka
African Drylands Institute for Sustainability, University of Nairobi, Nairobi, Kenya

J. R. de Pinho
Center for International Studies, University Institute of Lisbon, Lisbon, Portugal

P. Thornton
CGIAR Research Program on Climate Change, Agriculture and Food Security (CCAFS), International Livestock Research Institute, Nairobi, Kenya

M. Welch-Devine et al. (eds.), *Changing Climate, Changing Worlds*, Ethnobiology, https://doi.org/10.1007/978-3-030-37312-2_11

forward at several scales from the local to the global. The continued functioning of ecosystems by and for local populations will depend critically upon sound policy, planning, and practice.

Keywords Indigenous and local knowledge · Kenya · Climate change · Environmental change · Pastoralists · Global change solutions

11.1 Introduction

A UNESCO conference on indigenous people and climate change in 2015 was held as a lead-up to COP21 (the Twenty-first Conference of the Parties), the major annual conference of the United Nations Framework Convention on Climate Change. The UNESCO conference, entitled "Resilience in a Time of Uncertainty: Indigenous peoples and climate change," brought together over 60 speakers and an audience of over 650 people in plenary and parallel sessions and side events (UNESCO 2015). The Paris Climate Agreement followed the conference. The agreement document considered indigenous peoples and knowledge, but the language was vague and did not lead necessarily to action (Ford et al. 2016). Further, it was unclear how the understanding of the environment at the local scale was supposed to be integrated into policies. The Intergovernmental Science-Policy Platform on Biodiversity and Ecosystem Services (IPBES) framework (Diaz et al. 2015) attempts to integrate more fully different knowledges for biodiversity conservation, human well-being, and sustainable development. To that end, this chapter examines indigenous and local knowledge (ILK)[1] of climate change. It addresses climate change and environmental change from the perspective of Kenyan pastoralists who identified a myriad of environmental issues that occur and interact at different scales. They also identified ways forward at several scales from the local to the global.

11.1.1 Background

Africa's populations will face critical challenges in the coming decades in meeting basic needs for food, water, shelter, and other necessities without undermining biodiversity, ecosystems, and ecosystem services. Population increases caused by declining infant mortality and increasing life expectancy – despite declining fertility rates – raise the prospect of a net doubling in the size of many populations by mid-

[1] We use the definition of indigenous and local knowledge (ILK) provided by Diaz et al. (2015) and used by the Intergovernmental Science-Policy Platform on Biodiversity and Ecosystem Services: a cumulative body of knowledge, practice, and belief, evolving by adaptive processes and handed down through generations by cultural transmission, about the relationship of living beings (including humans) with one another and with their environment. It is also referred to by other terms such as indigenous, local, or traditional knowledge, traditional ecological/environmental knowledge (TEK), farmers' or fishers' knowledge, ethnoscience, indigenous science, and folk science (Diaz et al. 2015:13). We use indigenous knowledge and local knowledge interchangeably in the text.

century and a tripling by 2100 (HYDE v 3.2 2017). This scenario underscores a need for improved stewardship of environments – agrarian, pastoral, and otherwise – found across the continent and natural resources critical to sustaining plant and animal life. The future of both Africa's people and ecosystems is inextricably bound with the management of tightly coupled socialo-ecological systems involving food, water, and energy – which are becoming increasingly sensitive to anthropogenic disturbances from direct and indirect actions, and compounded by climatic perturbations, contributing to habitat decline and fragmentation (Archer et al. 2018).

Despite Africa's relatively minor contribution to global greenhouse gas emissions compared with the developed nations, social and ecological systems on the African continent will be increasingly vulnerable to climate change processes in the coming decades (Amegah et al. 2016). Overall, average temperatures are projected to increase, with a possible doubling in the frequency of days with temperatures measuring over 35 degrees centigrade (Alagidede et al. 2016; Coffel et al. 2018). Greater fluctuations in the amount and timing of rainfall will lead to both increasing aridification of the landscape and elevated risk of flooding events, with rain-fed agricultural systems being particularly vulnerable to perturbations such as ENSO, intensifying declines in plant growth and agricultural crop yields (Serdeczny et al. 2017; Fuller et al. 2018). Sea-level rise, possibly up to 1 meter by 2100 under a 4 degree centigrade warming scenario, would contribute to salt-water intrusion into surface water and groundwater supplies among other adverse effects (Serdeczny et al. 2017).

Coordination efforts to address multiple climate-related stressors have generally occurred at the national level and taken an externally oriented approach in the past, with national governments favoring collaboration with foreign-based NGOs and other international institutions over working with lower levels of government (Ford et al. 2015). This may be reflective of views held by the scientific community in the twentieth century that regarded indigenous knowledge as "inferior and deficient" and that are only recently beginning to recognize these knowledges as being "valuable and empirically sound" (Lauer 2017: 336). Unfortunately, actually incorporating ILK into governance and development is often marred by reinforcing existing power structures (Nadasdy 2005). That is not to diminish the importance of certain organizations with critical funding resources and technical capabilities, such as the International Livestock Research Institute (ILRI) and the International Water Management Institute (IWMI) to guide natural resource management efforts. However, the inclusion and involvement of actors at the local level may correlate with decisions that are better adapted to local social-cultural and environmental contexts, reducing implementation costs and increasing support and trust, thereby increasing the equity and efficacy of decentralized approaches (Schönhuth 2003; Sterling et al. 2017). A bottom-up approach abandons the idea that climate policy requires a universal framework and climate then separates into smaller more tractable problems that can be more easily solved by local actors (Rayner 2010). This approach is not opposed to grand targets for climate change, but achieving them is less problematic at local levels. Additionally, the building of social capital at the local level is critical for achieving meaningful stakeholder participation in community-based resource management schemes, with key variables being fairness of rules, dispute resolution procedures, inclusivity, sociability, and open dialogue (Musavengane and Simatele 2017). Conversely, the capture of resources and power by elites tends to undermine transparency, trust, and equity (Sterling et al. 2017; Galvin et al. 2018).

Climate change adaptation policies tend to be focused at the national level (the United Nations Framework Convention on Climate Change (UNFCC): https://unfccc.int/process-and-meetings/the-paris-agreement/the-paris-agreement) whereby each nation implements their efforts through "nationally determined contributions" through National Adaptation Plans of Actions (NAPA). There are few examples of inclusion of indigenous and local communities in creation and/or implementation of adaptation policies/programs (TEBTEBBA 2008). One example is the UN Convention on Biological Diversity (CBD) whose objective is the conservation of biological diversity and the fair and equitable sharing of its benefits. They state that national legislation should "respect, preserve and maintain knowledge, innovations and practices of indigenous and local communities embodying traditional lifestyles relevant for the conservation and sustainable use of biological diversity" (https://cbd.int/traditional/). The CBD supported the indigenous biodiversity outlook (Forest Peoples Programme 2016) where indigenous peoples and local communities (IPLCs) shared examples of engagement with policy reforms in building, for example, sustainable economies and resource management (the United States and Canada), in national development processes (Latin America), in valuing biological and cultural diversity (the Pacific), and in enhancing reindeer pastures (Russia), among other issues. For Africa, the Indigenous Peoples of Africa Coordinating Committee (IPACC) adopted a strategic plan on ways to strengthen the advocacy capacity of indigenous communities to influence decision-makers on the conservation of the environment and sustainable use of natural resources. IPACC's (https://www.ipacc.org.za/en/) objective is to make explicit the link between sustainable use of natural resources, protection of ecosystem services, and the survival of indigenous cultures and languages in Africa.

Though there is the call for indigenous people to work with decision-makers, in reality, many adaptation practices are supported by nongovernmental organizations (NGOs) both international and local, which interface with communities. Zampaligré et al. (2014), for example, examined the diversity of adaptation practices in Burkina Faso. They found that the Burkina Faso government supports activities that focus on cropping but not pastoralism. The national level one-size-fits-all approach demonstrates the gap between this approach and local-specific adaptations. This paper provides foundations to bridge the gap – asking what the local communities see as solutions.

It is not surprising that local peoples' observed climatic changes on plants and animals are widespread and the changes are already threatening food security (Savo et al. 2016). Many local peoples understand, experience, and see climate change, that is, through weather and environmental changes, what Rudiak-Gould (2013) calls "visibilism," and in the past have been masters of adaptation to these changes (e.g., Reid et al. 2014; Galvin 2009). This is because they are directly dependent on their environment for their livelihoods and changes in seasonality are not new to them (De Wit 2014; Weisser et al. 2014). In this chapter, we examine how livestock herders in Kenya perceive environmental changes, changes in both seasonality and ecology, and the impacts for livestock production. We also present locally derived solutions that span all scales of governance for coping with these changes.

11.2 Methods

This chapter draws upon findings from a series of focus groups, two community workshops, and interviews conducted as part of the Pastoral Transformations to Resilient Futures research program over two field trips in 2011. Sited in Narok and Kajiado counties, Kenya, this project sought to draw together pastoralists from across the country to examine how changes in climate, ecosystems, livelihoods, land tenure, and broader Kenyan society are interacting to shape pastoralist well-being and how local peoples are working to adapt to these changes. Participants were from Turkana, Garissa, Narok, and Kajiado counties (Fig. 11.1). As part of this process, we also engaged in numerous discussions on what local pastoralists envision for the future of the pastoralist way of life, including needed changes in top-down governance of pastoralist resources and locally driven pathways for building pastoralist resilience. Workshop participants were chosen and hosted by Reto-o-Reto, a Maasai-led NGO. All logistics including venues, focus group participants, and travel for locals

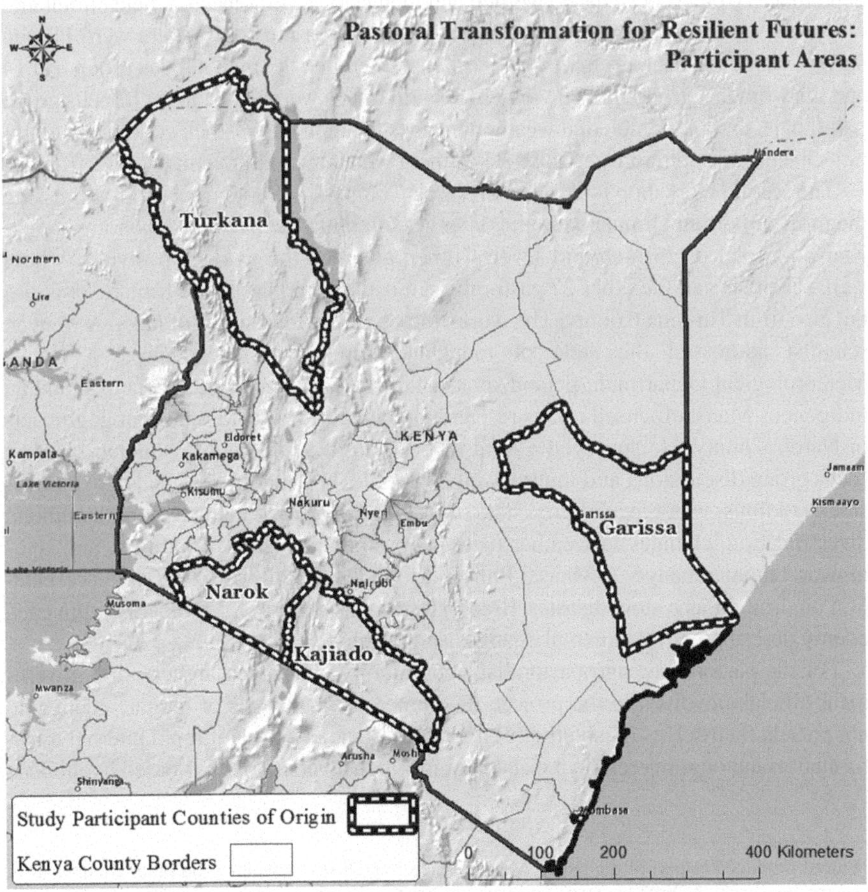

Map Credit: Trevor Even, Colorado State University Service Layer Credits: Tom Patternson - US Park Service Natural Earth I; World Resources Institute Kenya GIS Data; Source: US National Park Service

Fig. 11.1 Map showing Kenyan counties where project participants resided

were organized by Reto-o-Reto. Similar formats and questions for discussion occurred in both workshops and focus groups in each county. Three videos (one of them participatory) of pastoralists addressing environmental changes were products of the process[2] (Roque de Pinho 2013; Roque de Pinho and Galvin 2015).

The first workshop was conducted from 31 January to 1 February 2011 in Kitengela, Kajiado County. There were 30 people at the workshop, 10 women and 20 men. Nineteen were pastoralists, and 11 were from universities (University of Nairobi and Colorado State University), the Kenya Meteorological Department, international NGOs, and USAID (the funder of the project). There were presentations on climate trends and wildlife/livestock trends in the Kenyan drylands. These presentations were followed by breakout discussions of changes taking place in different counties and solutions to those issues from the view of the pastoralists who live in them. Pastoralists led all discussion groups; others in the meetings were scribes. Each of the four breakout groups (by county) presented their information followed by a discussion about issues, challenges, and successes from each pastoral region.

Gender-based focus group interviews took place on 2 February and another one in September 2011. At both dates, women and men's conversations were held simultaneously, and the project team split, so they attended one or the other. There were 11 men and six women in each group in February. In September, the men's focus group had 14 and the women's group had 18 women. Several issues were posed to each focus group participant, and these included weather changes, water and livestock condition, migration change, and solutions. We also asked them what they might want to know from us.

The second workshop took place in Narok County 5–6 September 2011 addressing the most important climate and other changes affecting the livestock economy across Kenya. A total of 25 men and seven women attended the workshop, five of whom were scientists and the other 27 pastoralists, mostly from Narok and Kajiado counties but also from Turkana County. The workshop opened with a series of talks. A Kenyan scientist addressed the state of rangelands under climate change, a Kenya Meteorological Department person spoke of climate change trends in Kenya, and an indigenous Maasai focused on his research on perceptions of climate change impacts in Narok County. He surveyed a total of 60 Maasai adults and conducted his own focus group discussions and interviewed 25 people. His survey included data on livestock holdings, water availability, state of pastures, livestock management, livelihood diversification, changes in weather patterns, pastures, livestock condition, and solutions to climate change. A Maasai PhD and founder of Reto-o-Reto gave a presentation on drought as a slow disaster. Breakout sessions occurred. Members within each county described environmental changes and developed solutions.

For the pastoralists, agropastoralists, community elders, researchers, and government officials involved in the project, the numerous challenges of climate change do not present themselves in isolation. Rather, they come at a time of rapid internal transformation and ongoing regional instability, in which a variety of land-based livelihoods

[2]Videos produced as part of the study: Maasai Voices on Climate Change (and other changes too): https://vimeo.com/73980798; Pastoralist Voices on Climate Change: https://www.youtube.com/watch?v=765l9XmMpDY; Of God, Rain and Motorbikes: https://vimeo.com/65117460

and time-tested pathways for subsistence are challenged. The results demonstrate multiple levels of this situation, paying special attention to participants' observations regarding changes in climate, local ecosystems, the livelihoods that depend upon them, and the broader social system as a whole (Fig. 11.2). Throughout this process, regional variations across the participant group are evident. Though Fig. 11.2 shows a summary of changes and the governance and spatial scales that occur, the variability among regions remains in the narrative. Finally, we present a variety of adaptive measures and other solutions suggested by program participants, ranging from top-down, policy-driven solutions to household-level adjustments in livelihood activity (Fig. 11.3). In this, we attempt to identify the various trajectories imagined therein and how these various paths interrelate. We combined the vast amount of information from the workshops, focus groups, and interviews in the results except when otherwise noted. This chapter tells the story of change and envisioned future of pastoralism, by pastoralists.

11.3 Results

The results of drivers of change are described in detail below. Figure 11.2 depicts the understanding that pastoralists bring to the complexity of changes across spatial and social/political organizational scales. The left-hand side hexagons show the approximate scales at which changes are occurring while showing the overlap in processes at different scales. The hexagons on the right hand generally show the institutional and social/political/ecological processes that interact across scales to affect changes.

11.3.1 Climate and Weather

Pastoralists around the world have long dealt with the vicissitudes of climate. Indeed, one of the defining traits of this lifeway, wherever it is found, is a variety of adaptations aimed at coping with high levels of rainfall and ecosystem variability; most important are high levels of mobility and the ability to respond to the shifting availability of grazing resources (Hobbs et al. 2008; Galvin 2009). As the climate changes, however, pastoralists in Kenya are finding themselves being forced to confront changes in both the character of their regional climate and the applicability of their traditional knowledge systems used to sense, understand, and predict its impacts.

The farthest-reaching of these changes comes in the form of rising temperatures. Although the region has been historically typified by warm weather year-round (between 25 and 30 °C, even in the coldest months), participants from across the various study regions noted that the last decade has been marked by increasing maximum and average temperatures, with several noting unusual increases in night-time temperatures. These observations were especially acute in the northwestern Turkana region, where participants observed maximum temperatures that at times

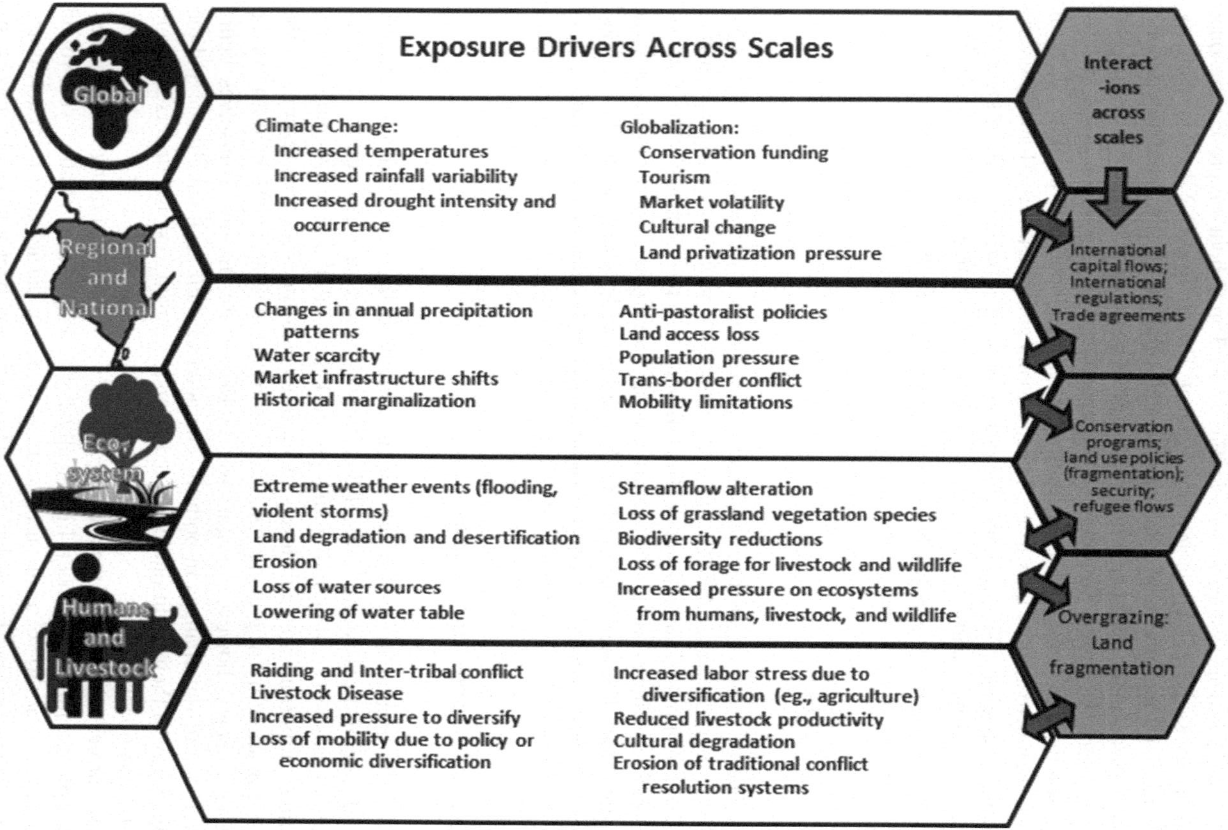

Fig. 11.2 Summary of changes that are occurring in Kenyan savannas and their interactions among different scales

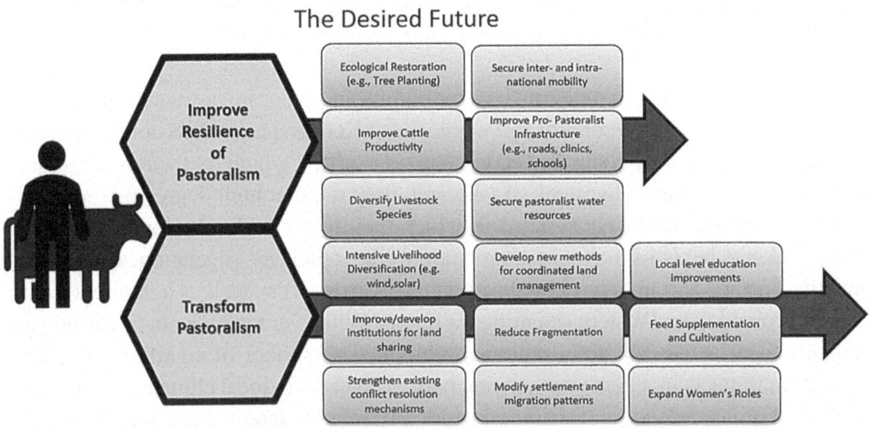

Fig. 11.3 Summary solutions for an improved future for pastoralism while also addressing processes that negatively affect a desired future

reached 40 °C. In all regions, however, increased temperatures linked to a variety of negative impacts on ecosystems, livelihoods, and social order, ranging from increased stress on cattle to changes in the reliability of local water systems.

Indeed, while rising temperatures were troublesome in their own right, for most participants, their most serious effects were felt indirectly, particularly through their impacts on the manner in which water cycled through local ecosystems, mainly via the rapid drying of landscapes that seems to follow even heavy rainfall. This effect was compounded by the fact that, in addition to tightening water balances, participants across the various study regions also observed significant changes in the predictability and character of local precipitation patterns. Normally characterized by two distinct rainy seasons (the "long rains," during the months of March through June, and the "short rains," which take place in the months of November and December), participants noted a feeling that neither rainy seasons could be relied upon, either in terms of its expected character or its probability of actually occurring. As one participant put it, the months of the local calendar – whose native names are primarily based upon their dominant weather and ecological impacts – could no longer be trusted. In each region, these changes manifested in different ways: in Turkana County, concerns focused mainly on long-term drought episodes, with rains failing entirely for years at a time. In Garissa County, observations of increasingly erratic rainfall during both rainy seasons were paired with concerns about both enduring drought conditions and, when rains did occur, widespread flooding. In Kajiado, increasing occurrence of drought simultaneously with increasingly worrisome evaporation rates following rain events was driven by both the loss of local vegetation cover and increasingly high temperatures. In Narok County, the wettest county in this study, broad observations regarding the irregularity of rains were coupled with concerns about intensifying rainfall patterns during the end-of-year short rains and the occurrence of increasingly violent and difficult to predict rainstorms throughout the year.

11.3.2 Ecosystems

Although recent efforts to modernize water pumps and other infrastructure have made some headway across the country as a whole, most Kenyan pastoralists continue to rely directly upon minimally controlled ecosystem service flows, both for their own resource needs and those of their livestock. As a result, they retain a high degree of sensitivity regarding changes in the ecosystems in which they live, enabled by long-established traditional mechanisms for ecosystem monitoring (such as vegetation species and quantity, surface and underground water, and soil erosion) as well as a variety of concepts borrowed from Western scientific traditions. Whatever the lexicon, however, participants from across all study regions highlighted a number of significant negative ecosystem impacts in recent years, owing both to changes in local climate and the ways in which human activity is reshaping historically reliable landscapes. Simply put, the region's rangeland ecosystems are under increasing strain from a variety of directions and are already undergoing significant transformation in many areas.

The majority of these impacts and transformations are related to water and the way it moves through the varied landscapes that pastoralists utilize. At the most basic level, there is simply less water available at any given time of the year, driven both by increased evaporation following rains and increased utilization by the nation's growing population. At the same time, recurring drought episodes, combined with more violent storms, have led many to observe ever-increasing rates of the development of bare, easily eroded soils. This, in turn, has led to increasingly degraded streambank conditions in many areas, both as a direct result of erosion and due to increased pressure from wildlife and livestock as they gather with increasing density around remaining surface water resources. The use of wells – usually dug by hand – also plays an increasingly ambivalent role, as its adoption intensifies (with more wells being dug, in general) at the same time as its effectiveness appears to be on the decline due to the lowering of local water tables, siltation following erosion events, and recurring drought episodes. These concerns were especially acute in Kajiado County, where siltation had begun to effect the functioning of dams.

These changes, working in tandem with other climate and land use factors, led many study participants to observe an array of negative impacts to local vegetation communities, in terms of both their overall abundance and health and the availability of species suitable for human and livestock use. For example, in the Garissa County, pastoralists face challenges relating to overstocking-driven erosion, deforestation due to overharvesting, and the replacement of useful grass species by more drought-tolerant shrubs. In Narok County, increasing soil compaction, losses of soil organic matter, and overuse of fertilizer by agriculturalists in the north of the county have led to a loss of grass cover near river banks, a reduction in palatable grass species in rangelands, and the drying of swamps normally critical for local wildlife. Loss of viable soil conditions and useful species is also evinced in Turkana and Kajiado counties, as well, with participants also noting shorter periods when the grass is green. Several Maasai elders also noted the disappearance of numerous useful forb species, as well as the reduction in the availability of tubers and flowering plants utilized by sheep and goat herds. Women, mainly from the Kajiado County, also noted the decreasing availability of various tall grass species useful for building materials and other household work. As one woman noted, "We have noticed that the grass no longer grows tall. There are too many cattle that are grazing just a few areas. In the past, we used the tall grasses to build small houses, but now we cannot find those grasses."

11.3.3 Livestock and Livelihoods

Pastoralist livelihoods are defined through their complex subsistence and cultural connections to livestock, and recent ecological, climatological, and land-use pressures have brought about an array of changes – some neutral, some plainly negative – for pastoralists across the various study areas. In all areas, livelihoods are in a state of transformation, with many households and communities engaging in higher than normal levels of migration, increasing economic diversification, and, as part of this latter shift, the growing economic role of women at a variety of scales (Fig. 11.2). Likewise, livestock herds, in many cases historically dominated by cattle, are also being diversified, with an increased emphasis on sheep, goats, and, in the arid north, camels. As part of this process, migration patterns are also changing in more subtle ways, as goats and sheep require more frequent access to water and are less able to engage in the long foraging treks formerly central to most male pastoralist activity. More broadly, market-driven activity is increasing in all but the least developed areas, with cattle being increasingly traded for currency. All of this comes amidst the backdrop of decades of efforts by colonial and national governments to privatize land and regularize herd movements, with many pastoralists relying increasingly upon the privately held lands of relatives as common lands become more fragmented and overutilized.

In sparsely developed and generally arid Turkana County, for example, stocking rates of cattle and other livestock are on the decline, with many pastoralists responding to degrading pasture conditions by simply migrating to other areas. Those livestock that remain generally have poor body condition and life expectancy and only fetch meager prices at the region's various isolated local markets. As a result, livelihood diversification is increasing, with many women diversifying into the production and sale of cosmetics and other small goods for sale at regional markets. Others in the area are shifting into more intensive agricultural production, mainly through the use of irrigation infrastructure near the Turkwel River. These efforts combine with increased fishing activity in Lake Turkana, and, for young people unwilling or unable to go to school, woodcutting and charcoal production. At the same time, many youths are being encouraged to attend school regularly, with many eventually making their way into national schools for secondary and post-secondary education in the hopes of leaving pastoralism behind but also eventually sending money home. A pastoralist elder noted, "So let us find a way of bringing education to every pastoralist child. When this youngster has twenty acres, and knows economics, he will know how to manage a loan, will know how to sell at market."

Similar changes are underway in Garissa, where declining cattle herd fitness due to inbreeding and harshening landscape conditions has led many to shift their emphasis away from pastoralist livelihoods. Here, women are playing an ever more prominent role through trade of handcrafts, camel milk, and other domestic products, as well as through their central role in the area's growing number of beekeeping operations. Others in the area are also increasingly engaging in small-scale agriculture to offset caloric losses due to herd decline, with food resources further supplemented through purchases at local markets (Fig. 11.2).

In the south, participants voiced a more intense set of concerns, as tensions between humans, wildlife, livestock, and their environment are especially acute. In Kajiado County, where land privatization and a tendency toward permanent settlement have been particularly intense, many pastoralists also reported severely challenged herds, including generally poor herd health and body condition, low reproductive rates, and increasing exposure to a variety of diseases due to the need for constant movement outside of traditional grazing areas. And while many in this region reported increasing diversification of livelihoods, herd composition, and the growing integration of cellular phone technology into traditional resource scouting practices, many also felt that these measures were not enough. In some areas, reports of dangerous food insecurity and the constant pressure to sell off what little private lands families possessed were intermixed with concerns voiced over growing rates of intertribal conflict, theft, and clashes with local conservation areas. While some in the area are already actively engaged with the region's growing ecotourism economy, many have yet to see benefits and instead experience only the exclusion from formerly grazed lands that are essential to tourism and biodiversity conservation schemes.

These concerns were echoed to a great degree in nearby Narok County, where rapid land fragmentation and urbanization (due to many forces such as agriculture, fences around community-based conservation areas and private lands, tourism, and infrastructure development; see Galvin et al. 2008) have left those who remain facing a number of severe challenges. Here too, cattle condition is poor, with calves growing slowly, infertility on the rise among females during drought periods, increasing disease occurrence, and growing conflicts between peoples as herders are forced to move constantly to find new grazing resources. One pastoralist from this region noted, "Now the animals have no place to hide, rest, or give birth. This means increased pressure and this is why there is conflict." This, in turn, has led to overall decreases in pastoralist herd sizes, leaving many with little to no disposable income left over for childhood education or supplementary food. As a result, increasingly impoverished conditions were noted for many groups in the region, with some speculating that many had been trapped in a downward spiral of land loss and deepening economic privation. This has also led to what many participants saw as an increase in illnesses among herders, themselves, which in turn limited their ability to engage in the long-range movement their livelihoods more frequently required.

11.3.4 Social Change

Participants from all study sites told a variety of stories of social and cultural change, both in their own communities and across Kenya as a whole. In some cases, these derived directly from climate shifts. In others, global, regional, and historical economic transformations played a more dominant role. In many cases, however, the impetus for change came about at the confluence of the two. Indeed, the most unanimously noted problem across sites derived not from any singular factor, but rather the complex interaction of increasingly stressful drought episodes, growing populations, and widespread land privatization. Any one of these factors, individually, might increase pressure on landscapes, complicate mobility and herd migration, and reduce the overall viability of traditional pastoralist practices. Working together, however, they have prompted a variety of significant shifts in the way pastoralists survive, ranging from longer herd movements that isolate families and women from herders to a growing incidence of violent cattle rustling, raids, and intertribal conflict (see Galvin et al. 2014).

As a result, participants from all study areas echoed a common concern, one that in some ways transcends any singular social issue: namely, that the social fabric of pastoralist communities was degrading and that long-standing strategies of social support were less able to provide a buffering effect in times of hardship. In some areas, this manifested as more frequent fights and other intrafamilial stress, especially during droughts. In others, these conflicts extended to larger kin networks, with old agreements between distant relatives becoming more and more strained as grazing and water resources dwindle. In some areas, such as Garissa County, these challenges have led to pastoralists reaching out to more and more distant networks of resources and alliances, which, while providing some relief from resource stress, has brought with it other problems such as increased exposure to HIV and the need to cross already contested national borders.

These effects are compounded when water resources, impacted by increased usage, intensifying runoff, pond siltation, and more rapid drying, become clear signals that the landscape is stressed, and higher numbers of cattle and people aggregate around those few reliable resources that remain. And while the proliferation of man-made wells and machine-drilled boreholes has provided some relief, many also worry that overuse of these water sources is already stressing area's water tables and, potentially, opening yet another arena of future conflict.

At the same time, several more subtle but perhaps equally as influential impacts were noted. For example, several women participants noted that milk production – often a cornerstone of many pastoralist diets – has fallen alongside grass species biodiversity and its attendant negative effects on cattle health and fertility. Although troubling enough on its own, this dynamic comes at a time where the traditional pastoralist diet is also losing its appeal to younger generations, who find themselves more attracted to foods made available through widening regional markets and as a result more prone to seeking short-term profits from land and cattle sales. These changes are wrapped in a broader complex of shifts toward a more commercially oriented cattle-raising culture, in which cattle are supplanted as the main source of social prestige and cultural capital by currency, the ability to pay for children's education, and the possession of modern, permanent housing. However, education presented a particularly challenging paradox to many, with most voicing the concern that while education was critical to improving overall family viability and improving options for young people, it also strains household and herding labor resources and may ultimately draw those with educations away from the pastoralist way of life. And while many voiced the notion that for pastoralism to survive in some fashion, the number of people dependent upon rangelands had to be reduced, many also noted that losing cattle and giving up traditional migratory/nomadic behavior could amount to the overall loss of pastoralist culture.

Similarly, the changing economic and decision-making roles of women – who normally remain relatively stationary as young men follow the herds – left many wondering about how these changes might alter pastoralist culture, with little clarity as to its ultimate effects. Changing conditions have led to women playing an ever more important role in ensuring family and community viability, be it through economic diversification into trade, the organization of local cooperatives, or the increasingly modern management of water resources. One woman noted, "We need a market for our beadwork. Our beads do not disappear in a drought, they will last in our house for more than a year. Women are smart enough to count the number of beads needed to make complicated designs. There are too many middlemen between us and the buyers of our beads, they take too much of the profits." For some, this signified a clear path for future adaptation, in which educated, empowered women lead the way to new strategies for subsistence and interaction with surrounding markets. For others, it was seen as a challenge to existing power structures and yet another strain on the traditional systems of knowledge and authority (such as strong kin and clan ties to access sharing of livestock and decreasing conflict) normally relied upon to negotiate survival during periods of climatic difficulty.

11.4 Solutions

Despite – or perhaps because of – the seeming enormity of the challenges discussed above, questions focused on eliciting local solutions and paths to desired change provoked a wide-ranging set of responses, focused on both short- and long-term goals for improving pastoralists' well-being. These solutions ranged both in scale and character, with some aimed at regional and national policy and others at more local adaptations to the changing biophysical and economic climate (Fig. 11.3). In the following subsections, we review the most common themes from these discussions, focusing specifically on security, infrastructure, mobility assurance, ecosystem management, and livelihood diversification. However, it should be noted that while we treat each of these separately here, for workshop participants, they were seen as components of a broader strategy for working toward a more resilient future. Moreover, these solutions, aspirational though they may be, were a means of heading off the ongoing process of cultural degradation that pastoralists faced already and instead engaging in a wide-ranging program of "bottom-up" cultural change driven by pastoralists themselves. As a result, many participants across the various participant regions noted that, whatever the specific strategy, "the dream must be shared by all pastoralists" and must be driven forward through cooperative efforts rather than "top-down" administration or prohibition. Figure 11.3 shows the categories of changes that both contribute to resilience and transformation of pastoralism.

11.4.1 *Security*

In the northern regions, pastoralists face a number of threats from increasing livestock raiding activity, the movement of small arms across the South Sudan and Somali borders, and conflicts arising from reduced ecosystem resilience. As a result, improving regional and local security was a high priority for participants from these regions, as was recognition that solutions to these problems would likely require a multifaceted approach. For example, in Turkana County, where government and law enforcement presence are relatively sparse, participants suggested the expansion of government offices in the region and the development of infrastructure and agencies to support safe and legal cross-border migration. Some also suggested the development of a "cross-border coalition" of herders from neighboring countries to enable peaceable migration. In Garissa, participants suggested a range of strategies, including support for international treaties to stem the flow of small arms and policies to support stability in neighboring South Sudan and Somalia. More broadly, however, it was recognized that conflict and violence were generally driven by the increasing difficulty of making a decent livelihood in some pastoralist regions (see Galvin et al. 2014). As a result, it would seem, true security will rely on both broad-scale policy changes and more focused efforts to improve the resilience of pastoralists and aid in their long-term adaptation to changing conditions (Fig. 11.3).

11.4.2 *Infrastructure*

Participants also noted that many of their difficulties stemmed from their inability to connect with markets, to develop sustainable resources, or to gain access to services. As a result, infrastructure development also ranked high on the various lists of participants' hopes for their respective regions, albeit in different ways depending upon the local situation. In Turkana County, for instance, roads – be it for access to markets, access to health and other services, or for the provisioning of more effective governance – were noted as a critical need. In Garissa, on the other hand, the emphasis was less on transportation infrastructure than the development of local markets and other service provision sites such as banks and clinics. In both of the northern regions, the development of renewable energy resources was also emphasized, as both areas have ample potential for solar and wind development, and even small-scale installations near relatively permanent settlements were seen as potentially leading to the facilitation of other infrastructure and cultural development (Fig. 11.3). This was also coupled, in both northern regions, with a growing awareness of the need for more robust educational infrastructure at the local level so as to provide youth with the ability to contribute to local labor pools.

In Kajiado and Narok counties, participants saw a similar need as those to the north for the development of more robust cattle and other livestock markets, particularly with regard to processing facilities for leather and meat. However, participants from the south also noted the need for a number of "soft" infrastructural developments that relied less upon physical capital and more upon social networks and developing connections to various types of scientific expertise. Chief among these were "disease-free zones" and other forms of livestock disease mitigation and prevention approaches, which were seen as requiring both expert scientific input and significant coordination between herders, themselves. Along these same lines, participants in Kajiado suggested a livestock early warning system that would allow pastoralists to become more aware when climate conditions and other factors conducive to disease spread were likely to occur. These suggestions were paired with broader ambitions toward pastoralist advocacy networks and other cooperative systems to better communicate local pastoralist needs to government officials and policy makers. Some participants noted that the index-based livestock insurance program (https://ccafs.cgiar.org/publications/index-based-livestock-insurance-ibli#.W0UgJf6Wx-g) supported by the Consultative Group on International Agricultural Research (CGIAR) Research Program, Climate Change, Agriculture and Food Security (CCAFS) could insure livestock keepers to protect against climate-related risks. No one had actually used it.

Participants also put forth a number of suggestions for improving access to reliable water resources, particularly in areas where existing wells were primitive and existing irrigation potential had yet to be realized. This included both the development, outright, of new well sites and irrigation infrastructure and efforts to improve existing water resource management in areas where primitive wells and ephemeral ponds were showing signs of overuse.

11.4.3 Mobility

Because of its centrality to the pastoralist way of life, participants from all regions voiced a definite need for efforts to ensure the free mobility of herds and herders both within Kenya and across nearby international borders. That said, without the ability to reduce ongoing processes of land fragmentation, this effort would unlikely be successful, as one of the main drivers of reduced mobility over recent decades has been the introduction of private property and other forms of exclusionary land use. As such, efforts to reduce fragmentation and ensure mobility were often seen as one and the same. Here, solutions to the issue proved diverse, with some suggesting the development of community banking systems to reduce the poor financial planning that often led to piecemeal land sales and others looking to national and international policy solutions designed to ensure free movement. This included, at the very least, work toward getting the Kenyan government to recognize the value and potential of pastoralist lifeways and the needs thereof[3].

---

[3]Since 2010–2011 when we conducted this study, the government of Kenya has passed the Community Land Act, No. 27 of 2016 (https://kwcakenya.com/download/community-land-act-2016/?wpdmdl=10313), that recognizes and secures community land rights in former group ranches and trust lands including grazing rights.

11.4.4 Sustainable Ecosystem Management

Participants also noted the broader need for the development of skills, policies, and practices that would allow for more sustainable management of the region's diverse ecosystems. This included, as mentioned above, the development of more robust energy production and water management infrastructure, the improvement of existing resource management networks among different pastoralist groups, and the development of mechanisms for more effective livestock disease management systems. Along these same lines, calls for better systems for invasive grass species control (including harvesting for biofuel) and rainwater management were also suggested. Part of this, as participants from Garissa suggested, would involve thorough assessments of local ecosystem service flows and mechanisms for helping area pastoralists to maximize their potential for long-term well-being through information sharing, education, and access to scientific expertise.

11.4.5 Livelihood Diversification

In and alongside the various other calls for change in the region, there flowed an increasing understanding of the need for the diversification of livelihoods in ways that reflected the broader goals and desires of pastoralists hoping to manage the transformation of their culture in sustainable and equitable ways. Foremost among these were calls for developing more robust cultural and ecological tourism systems and improving upon these mechanisms where they existed already (such as in Narok County). A pastoralist elder stated, "So if you want to sustain this new cow called tourism, you must manage and sustain the environment." Rather than cleaving to older, conservation-style models, however, here, participants emphasized approaches that allowed for the coexistence of herders and wildlife through careful habitat management and the simultaneous emphasis of livelihoods, for instance, through community-based conservancies. Other forms of diversification with less sweeping herd management implications were also noted, ranging from increasing the marketability of local beadwork to the shifting of purely symbiotic livestock management approaches toward more commercial and production-oriented schemes. In the latter case, this involved both the development of improved cattle breeds, the building of silage and cattle processing infrastructure, and the assurance of stable market access both locally and across the greater global markets. In some areas – particularly in the north – this also included the development of various dryland product industries, such as the cultivation of gum arabic, aloe, and other products.

Whatever the specific strategy, however, what was most broadly noted by participants in workshops and focus groups regardless of gender was the need for the recognition and empowerment of women as economic actors, both within families and villages as well as across the region. This included ensuring representation for women not only within decision-making spaces but also systems, such as banks and local endowment funds, that allowed women to engage in entrepreneurial efforts and to expand their ongoing efforts to supplement traditional food and resource management systems. Likewise, education for women was seen as an important step forward in this regard, particularly within localized schools that allowed women and girls to maintain their traditional roles while also building skills and gaining access to new knowledge networks. A pastoralist woman noted, “Today, we consider the educated to be rich because they are immune to drought and diseases; they can get out. An investment now in education pays back in the future. You could have 200 cows, but could lose them all in a drought. Twenty years ago, we defined wealth by the number of livestock, the number of children, and the number of wives.”

11.5 Discussion and Conclusion

The pastoralists who participated in this project understood clearly the myriad changes affecting their ecosystems and livelihoods and envisioned a suite of clear avenues for a desired future at several different spatial and governance scales (Figs. 11.2 and 11.3). What does this mean in terms of developing strategies of national-to-local coordination efforts? There is a pressing need for devolving authority and resources to the subnational government level (county/district and sub-district) to promote engagement and involvement of local populations who better understand ongoing environmental change while also addressing climate change at the national level (Friis-Hansen 2017). Environmental change has many drivers including climate change, history, and power differences. These research results show the direct effects of environmental changes through documentation of change through the narratives presented here and the videos accompanying the project. They also show pathways to policy and cultural changes, and that trust in the solutions and the process is essential for success. Do pastoral perspectives empower local experiences and knowledge? The information presented here certainly shows that locals have insights that speak to climate change and actionable solutions to issues of environmental change in the Kenyan drylands (see Sillitoe 2003).

There are concerns about existing adaptation strategies based on the adequacy of current climate models for anticipating impacts, policies, and programs that fail to include relevant local knowledge and the sufficiency of financial resources to fund necessary adaptation policies (Adenle et al. 2017). The management of large rangeland districts, which support about one-quarter of the national population and the vast majority of livestock herds and wildlife, will be critical to the future of the nation, as economic development further concentrates in the core agricultural zone spanning Lake Victoria to Mount Kenya. Water resources management practices, including the increasing diversion of surface waters from rivers for irrigation and power generation, will further compound climate-driven water stress in the outlying areas by decreasing availability needed by humans, livestock, and wildlife. This may be particularly important in the Tana River watershed, which provides the only reliable source of surface water to downstream inhabitants of the rangelands of

Garissa County. Projected increases in human settlements across Kenya will affect the ability of pastoral peoples to find adequate forage and water for their herds, which already consume a greater share of water compared to wildlife populations, and with the dairy yields of such herds declining in many areas.

The viability of a pastoral way of life depends upon the maintenance of upstream rivers and forests. Under pressure from climate, continued population growth (at rates above global averages), and economic development, the diverse assemblage of plants, animals, and birds, many unique to particular African locales, faces threats from habitat degradation and fragmentation (Biggs et al. 2004). This underscores a need for sustainable development-oriented measures to maintain environmental quality and species viability while alleviating human challenges in terms of health (including malnutrition, vector-borne diseases, and other contributors to health) and household income associated with climate-amplified environmental change processes (Amegah et al. 2016). The role of legally recognized indigenous and community lands within these counties (recognized in Kenya's new constitution) is important to the future of these pastoral populations (Dubertret and Wily 2015). The empowerment of rural women would also contribute significantly to efforts to transform the system, with the Danish Institute for International Studies program's Climate Change and Rural Institutions studies (2012–2016) in Tanzania and Uganda providing a possible precedent for the region (Friis-Hansen 2017).

The multi-scalar nature of climate change and environmental change processes more broadly will require concerted efforts at multiple levels of governance to maintain economic and political stability among communities who are dependent upon natural resources to sustain crops and herds, based on observed trends (Fjelde and von Uexkull 2012; Hendrix and Salehyan 2012). Similarly, governance practices including development of an umbrella group to champion pastoralist issues in each region and including pastoralists issues in Kenya's National Climate Change Response Strategy[4] to promote adaptation to hydroclimatic extremes, especially severe drought events, will be critical to sustaining economic growth. Severe drought effects on crops and livestock will compel adaptation to water supplies to meet short-term food needs while also considering longer-term ecosystem sustainability (Connolly-Boutin and Smit 2016). It is clear from the information gathered from the pastoralists in this study that they are part of the solution. Several people in this work were leaders in their communities, so there is the likelihood that their voices are heard at county and national efforts to address climate and other changes affecting their communities (e.g., one person became the governor of his county after this project was completed). As one Maasai elder stated, "There is always a saying, where there's a will there's a way. And if people can go to the moon and come back, then I think land for this pastoralist system can be managed. This is the only planet we have; we have to be very careful with it!" The continued functioning of ecosystems and underlying biodiversity by and for local populations will depend critically upon sound policy, planning, and practice just as local/indigenous participants in this project know.

[4] Government of Kenya. National Climate Change Response Strategy 2010 http://www.environment.go.ke/wp-content/documents/complete%20nccrs%20executive%20brief.pdf

Acknowledgments Sincere thanks to all the participants who provided knowledge in the workshops, focus groups, and interviews. Thank you to the team that made this project possible including David Nkedianye and Dickson ole Kaelo, Mohammed Khalif, and Gregory Akall. This project was funded by the USAID Climate Change Livestock Collaborative Research Support Program that changed its name in 2010 to Feed the Future Innovation Lab for Collaborative Research on Adapting Livestock Systems to Climate Change.

References

Adenle AA, Ford JD, Morton J, Twomlow S, Alverson K, Cattaneo A, Cervigni R, Kurukulasuriya P, Huq S, Helfgott A, Ebinger JO (2017) Managing climate change risks in Africa-a global perspective. Ecol Econ 141:190–201

Alagidede P, Adu G, Frimpong PB (2016) The effect of climate change on economic growth: evidence from Sub-Saharan Africa. Environ Econ Policy Stud 18(3):417–436

Amegah AK, Rezza G, Jaakkola JJK (2016) Temperature-related morbidity and mortality in Sub-Saharan Africa: a systematic review of the empirical evidence. Environ Int 91:133–149

Archer E, Dziba L, Mulongoy KJ, Maoela MA, Walters M, Biggs RO, Cormier-Salem MC, DeClerck F, Diaw MC, Dunham AE, Failler P (2018) Summary for policymakers of the regional assessment report on biodiversity and ecosystem services for Africa of the intergovernmental science-policy platform on biodiversity and ecosystem services. IPBES secretariat, Bonn

Biggs R, Bohensky E, Desanker PV, Fabricius C, Lynam T, Misselhorn AA, Musvoto C, Mutale M, Reyers B, Scholes RJ, Shikongo S, van Jaarsveld AS (2004) Nature supporting people: Southern African millennium ecosystem assessment. Integrated report. Council for Scientific and Industrial Research, Pretoria

Coffel ED, Horton RM, de Sherbinin A (2018) Temperature and humidity based projections of a rapid rise in global heat stress exposure during the 21st century. Environ Res Lett 13(1):014001

Connolly-Boutin L, Smit B (2016) Climate change, food security, and livelihoods in sub-Saharan Africa. Reg Environ Chang 16(2):385–399

De Wit S (2014) Changing patterns of rain or power? How an idea of adaptation to climate change travels up and down to Maasailand, Northern Tanzania. Working Papers of the Priority Programme 1448 of the German Research Foundation WP 11, pp 1–22. http://www.spp1448.de/fileadmin/media/galleries/SPP_Administration/Working_Paper_Series/SPP1448_WP11_de_Wit.pdf

Diaz S, Demissew S, Carabias J, Joly C, Lonsdale M, Ash N, Larigauderie A, Adhikari JR, Arico S, Báldi A, Bartuska A (2015) The IPBES conceptual framework—connecting nature and people. Curr Opin Environ Sustain 14:1–16

Dubertret F, Wily LA (2015) Percent of indigenous and community lands. Data file from LandMark: the global platform of indigenous and community lands. http://www.landmarkmap.org

Fjelde H, von Uexkull N (2012) Climate triggers: rainfall anomalies, vulnerability and communal conflict in Sub-Saharan Africa. Polit Geogr 31(7):444–453

Ford JD, Berrang-Ford L, Bunce A, McKay C, Irwin M, Pearce T (2015) The status of climate change adaptation in Africa and Asia. Reg Environ Chang 15(5):801–814

Ford JD, Cameron L, Rubis J, Maillet M, Nakashima D, Willox AC, Pearce T (2016) Including indigenous knowledge and experience in IPCC assessment reports. Nat Clim Chang 6(4):349

Forest Peoples Programme, the International Indigenous Forum on Biodiversity and the Secretariat of the Convention on Biological Diversity (2016) Local biodiversity outlooks. Indigenous peoples'and local communities' contributions to the implementation of the strategic plan for biodiversity 2011-2020. A complement to the fourth edition of the Global Biodiversity Outlook. Moreton-in-Marsh, England

Friis-Hansen E (2017) Decentralized governance of adaptation to climate change in Africa. CABI Publishing, Boston

Fuller TL, Clee PRS, Njabo KY, Tróchez A, Morgan K, Meñe DB, Anthony NM, Gonder MK, Allen WR, Hanna R, Smith TB (2018) Climate warming causes declines in crop yields and lowers school attendance rates in Central Africa. Sci Total Environ 610:503–510

Galvin KA (2009) Transitions: pastoralists living with change. Annu Rev Anthropol 38:185–198

Galvin KA, Reid RS, Behnke RH, Hobbs NT (eds) (2008) Fragmentation of semi-arid and arid landscapes. Consequences for human and natural systems. Springer, Dorthecht

Galvin KA, Reid RS, Beeton TA (2014) East African pastoralism and the governance of grazing land: case studies from Kenya. In: Barnes G, Child B (eds) Adaptive cross-scalar governance of natural resources. Earthscan by Routledge, New York, pp 264–282

Galvin KA, Beeton TA, Luizza MW (2018) African community-based conservation: a systematic review of social and ecological outcomes. Ecol Soc 23(3):39. https://doi.org/10.5751/ES-10217-230339

Hendrix CS, Salehyan I (2012) Climate change, rainfall, and social conflict in Africa. J Peace Res 49(1):35–50

Hobbs NT, Galvin KA, Stokes CJ, Lackett JM, Ash AJ, Boone RB, Reid RS, Thornton PK (2008) Fragmentation of rangelands: implications for humans, animals and landscapes. Glob Environ Chang 18:776–778

HYDE (2017) PBL Netherlands Environmental Assessment Agency. History database of the global environment. Basic driving factors. Population. https://themasites.pbl.nl/tridion/en/themasites/hyde/basicdrivingfactors/population/index-2.html

Lauer M (2017) Changing understandings of local knowledge in island environments. Environ Conserv 44(4):336–347

Musavengane R, Simatele D (2017) Significance of social capital in collaborative management of natural resources in Sub-Saharan African rural communities: a qualitative meta-analysis. S Afr Geogr J 99(3):267–282

Nadasdy P (2005) The anti-politics of TEK: the institutionalization of co-management discourse and practice. Anthropologica 47(2):215–232

Rayner S (2010) How to eat an elephant: a bottom-up approach to climate policy. Clim Pol 10(6):615–621

Reid RS, Fernández-Giménez ME, Galvin KA (2014) Dynamics and resilience of rangelands and pastoral peoples around the globe. Annu Rev Environ Resour 39:217–242

Roque de Pinho J (2013) Shooting climate change in the Maasai Mara: aesthetics and expectations in participatory filmmaking with Kenyan pastoralists. Anthropol Now 5(2):74–86

Roque de Pinho J, Galvin K (2015) "Maasai voices on climate change (and other changes too)": participatory video and communication about environmental changes in the East African rangelands. In: Brites MJ, Jorge A, Correia Santos S (eds) Metodologias participativas: Os media e a educação. LabCom Books, Lisbon, pp 235–248

Rudiak-Gould P (2013) We have seen it with our own eyes: why we disagree about climate change visibility. Weather Clim Soc 5:120–132

Savo V, Lepofsky D, Benner JP, Kohfeld KE, Bailey J, Lertzman K (2016) Observations of climate change among subsistence-oriented communities around the world. Nat Clim Chang 6(5):462

Schönhuth M (2003) Negotiating with knowledge for development interfaces: anthropology and the quest for participation. In: Sillitoe P, Bicker A, Pottier J (eds) Participating in development. Approaches to indigenous knowledge. Routledge, New York, pp 139–161

Serdeczny O, Adams S, Baarsch F, Coumou D, Robinson A, Hare W, Schaeffer M, Perrette M, Reinhardt J (2017) Climate change impacts in Sub-Saharan Africa: from physical changes to their social repercussions. Reg Environ Chang 17(6):1585–1600

Sillitoe P (2003) Globalizing indigenous knowledge. In: Sillitoe P, Bicker A, Pottier J (eds) Participating in development. Approaches to indigenous knowledge. Routledge, New York, pp 108–138

Sterling EJ, Betley E, Sigouin A, Gomez A, Toomey A, Cullman G, Malone C, Pekor A, Arengo F, Blair M, Filardi C (2017) Assessing the evidence for stakeholder engagement in biodiversity conservation. Biol Conserv 209:159–171

TEBTEBBA (2008) Indicators relevant for indigenous peoples: a resource book. Tebtebba Foundation, Baguio City. http://www.tebtebba.org/index.php/content/123-indicators-relevant-for-indigenous-peoples-a-resource-book

UNESCO (2015) Resilience in a time of uncertainty: indigenous peoples and climate change. Conference held 26 Nov 2015. (https://en.unesco.org/events/resilience-time-uncertainty-indigenous-peoples-and-climate-change)

Weisser F, Bollig M, Doevenspeck M, Müller-Mahn D (2014) Translating the 'adaptation to climate change' paradigm: the politics of a travelling idea in Africa. Geogr J 180(2):111–119

Zampaligré N, Dossa LH, Eva Schlecht E (2014) Climate change and variability: perception and adaptation strategies of pastoralists and agro-pastoralists across different zones of Burkina Faso. Reg Environ Chang 14(2):769–783

Chapter 12
Understanding Global Change: From Documentation and Collaboration to Social Transformation

Karen Pennesi

Abstract The conclusion to this book situates the chapters within four programs of anthropological research on climate change: (1) documentation of local impacts of and adaptations to climate change, (2) connections to socioeconomic and political contexts, (3) collaborations with nonanthropologists, and (4) activism and social transformation. The final section notes the persistent challenges to creating positive change and meaningful research outcomes. It highlights some examples of success and outlines future directions for politically engaged anthropological work around climate change.

Keywords Climate change · Collaborative research · Environmental anthropology · Democratization · Decolonizing research

12.1 Decolonizing Environmental Knowledge and Climate Change Research

Anyone who has spent time among people who make their living off the land—farmers, fishers, animal producers—can attest to their resourcefulness, their resilience in the face of hazards and changes, and their detailed knowledge of the natural environment in which they live and work. Some would argue that it is a matter of survival to be so attentive, creative, and cooperative. Others might point to the collective wisdom passed down over generations of trial, error, and innovation. In my own work with smallholder farmers in Northeast Brazil, I have listened to stories proudly told about persistence despite economic hardship, endurance through suffering in times of drought, and hard-won success after trying some new technique or crop. I was a witness to their labor, and I saw clearly the connections between moral worth and willingness to undertake difficult work (Pennesi 2015). What I also observed repeatedly was a deep sense of frustration among peasant farmers, some-

K. Pennesi (✉)
Department of Anthropology, University of Western Ontario, London, ON, Canada
e-mail: pennesi@uwo.ca

M. Welch-Devine et al. (eds.), *Changing Climate, Changing Worlds*, Ethnobiology, https://doi.org/10.1007/978-3-030-37312-2_12

times bordering on resignation, with the political and economic systems that constrained their pursuits and disadvantaged them in relation to large-scale agribusiness, city dwellers, or others with more money and influence. Over 10 years of doing research in the state of Ceará, I heard from many people that agropastoralism could be more successful there, even at the subsistence level, if there were better management of water resources and government policies to support infrastructure and social development appropriate to the semiarid climate. *Conviver com o semi-árido* ("living with the semiarid") is a phrase often used to describe policies and programs that are progressive in building capacity for sustainable production in rural areas of the Brazilian Northeast. In reviewing related literature—including the preceding chapters of this book—I have learned that the problem I see in Ceará is one familiar to those who work with peasant farmers or subsistence fishers in other parts of the world. Political, economic, and social structures limit the adaptive capacity of individuals and communities, increasing vulnerability to natural hazards such as droughts and floods. The obvious conclusion is that if natural disasters are a product of both environmental conditions and social factors (Blaikie et al. 1994), then mitigation or prevention of disasters must include social change (Ribot et al. 1996).

The question I originally set out to investigate in Ceará was why farmers were not using meteorological forecasts disseminated by the state-funded agency in their agricultural decision-making. I soon discovered that there was a competing source of predictions for the rainy season: people known as *profetas da chuva*, "rain prophets," who based their forecasts on empirical observations of plants, animals, birds, celestial bodies, and other indicators in their environment (Pennesi 2011). This local environmental knowledge was valued by the farmers who lived alongside the rain prophets but was more often dismissed by scientists and urban consumers of news reports as being merely folklore or superstition. At the same time, the government meteorological agency offering the official climate forecast was not trusted by farmers, who criticized the meteorologists' lack of local place-based knowledge as well as the agency's connection to the state, which they saw as both interfering and willfully negligent (Pennesi 2013). Each year, a deeply rooted and complex conflict between rural and urban, tradition and science, and citizen and state played out in discussions of the rainy season. The rain prophets represented a symbolic resistance to the dominance and arrogance of decontextualized scientific knowledge (Taddei 2012) and to the federal and state governments whose claims to be helping disguised structures that marginalized farmers. Over the last two decades, there has been a shift in public discourse toward acknowledging the wisdom and experience of local and traditional knowledge embodied by rain prophets (Taddei 2006). Efforts are being made at the regional level to valorize and document the work of rain prophets in annual public meetings, media reports, academic publications, documentary films, and theatrical productions and to encourage young people to carry on the traditional practices (Pennesi and Souza 2012). As Taddei (2012) argues, however, the positive attention toward rain prophets as cultural performers has not resulted in any kind of organized political movement that could make significant differences in the farmers' lives. The authority and legitimacy of scientific knowledge remain unquestioned at the level of state policy, while the rural population remains largely poor and dependent on governmental and nongovernmental programs for any measurable improvement in their situation.

This background brings me to the frameworks of decolonization and democratization that I use to organize this concluding chapter. I start with a quote by Chandra Talpade Mohanty (2003: 254) which links the two: "Decolonization involves both engagement with the everyday issues in our own lives so that we can make sense of the world in relation to hegemonic power, and engagement with collectivities that are premised on ideas of autonomy and self-determination, in other words, democratic practice." Decolonization and democratization emerge in response to colonial and authoritarian governance structures whose purpose is to amass land, resources, and power and which necessarily create subordinate populations. The goal of these movements is for indigenous and subjugated peoples to have ownership and control over management of natural resources, over their livelihoods, over their way of life, and over their knowledge and thoughts. In short, they are a response to the fast and slow violence that Burke, Welch-Devine, and Sourdril (Chap. 1) discuss in the introduction to this book. With democratization, there is a focus on equality and participation in decision-making in all domains of social life. In this book, these issues apply to adaptations and responses to environmental and climatic conditions. Decolonization involves movement toward independence in all spheres of activity, including research (Smith 2012). In this chapter, I am interested in how giving attention and credibility to indigenous/traditional/empirical knowledge decenters colonial and hegemonic epistemologies.

In their article entitled "Decolonization is not a metaphor," Eve Tuck and K. Wayne Yang (2012) insist that "decolonization specifically requires the repatriation of Indigenous land and life" (p. 21) and caution against using "decolonizing" to describe struggles against all forms of systemic social injustice (p. 17). The communities discussed in this book are not all indigenous peoples living in the settler colonial states that Tuck and Yang write about, but what is relevant here is their focus on the material aspects of decolonization and the vital importance of land. The colonial/capitalist transformation of land into property and resource, and the concentration of land as wealth in the hands of powerful minorities, underlies the vulnerability of many communities described here. Furthermore, the exploitation of land-based resources, such as fossil fuels, in the pursuit of wealth and power is widely understood to be a primary cause of global climate change (IPCC 2013). As the research in this volume demonstrates, those who derive their subsistence directly from the land and water tend to be among the internally marginalized groups who suffer social discrimination, leaving them particularly vulnerable to natural hazards and climate variations. Therefore, analyses of perceptions of and adaptations to global changes, including climate change, must be understood within the context of struggles for self-determination and fuller participation in political and economic life.

An anthropological approach is especially well suited to illuminating the interconnections among multiple dynamic systems and forces and tracing how different human groups act within these systems over time. With people at the center, anthropologists investigate ecosystems, atmospheric conditions, social structures, and cultural systems of meaning-making, in addition to physiological processes and characteristics. The international group of scholars contributing to this book elucidates both human and environmental perspectives on the topic of climate change, drawing on cultural and biological anthropology as well as botany (Salick et al.) and geography (Dervieux and Belgherbi). A wide range of qualitative and quantitative methods are employed in these ethnographic studies, including interviews (Katz et al.; Dervieux and Belgherbi; Roque de Pinho; Sourdril et al.; Galvin et al.; Salick et al.; Burke et al.), surveys and questionnaires (Katz et al.; Seara et al.), participant observation (Dervieux and Belgherbi; Sourdril et al.), group discussions (Katz et al.; Roque de Pinho; Galvin et al.), community workshops (Galvin et al.), creation of an interactive web site (Reyes-García et al.), free listing observations of local flora and fauna (Dervieux and Belgherbi; Burke et al.), pile sorts, calendar construction and mapping (Salick et al.), participatory photography and video making (Galvin et al.; Roque de Pinho; Salick et al.), accompanied field visits (Salick et al.), statistical analysis (Salick et al.; Seara et al.), and path analysis (Seara et al.). Such a variety of methods enable the rich and nuanced analyses that are characteristic of anthropology and are evidence of an evolving understanding of what it means to study people.

This chapter is organized around four programmatic categories of anthropological research on human responses to climatic and environmental changes. The first category stems from the foundational interest of anthropology in the relationship between people and their environment. It centers on documentation of various adaptive strategies in particular climates (e.g., Moran 2008) and more recently includes descriptions of "local observations in climate-sensitive places" (Crate and Nuttall 2016: 13; Reuter 2015). The second kind of anthropological research on climate change moves beyond documentation of local observations and perceptions to examine the broader socioeconomic and political contexts that shape experiences of changing environments (Casagrande et al. 2007; Pokrant and Stocker 2011). The third category of research is marked by collaboration with interdisciplinary teams and with local people as research partners (Ayers and Forsyth 2009; Crate and Nuttall 2009). Collaboration and participatory methodologies demonstrate that local and indigenous knowledge is valued and this kind of research supports the capacity of local people to contribute to their own understanding. The fourth category of anthropological research related to climate change aims to foster social transformation by working alongside local community members in service of their own goals and projects. In these research programs, scholars take a more activist stance in their work to influence policy and facilitate social change (e.g., Baer and Singer 2018; Crate and Nuttall 2016). In the remainder of this concluding chapter, I show how the previous chapters relate to each of these four categories of research. Recognizing that it is impossible to make meaningful generalizations across such diverse geographic, cultural, and social contexts as are represented in the preceding chapters, I nonetheless attempt to identify common threads while highlighting particular cases when relevant. This chapter concludes with a brief discussion of remaining challenges and argues for future anthropological work to further the goal of social transformation through decolonization and democratization. I argue for a politically engaged anthropology of climate change befitting the seriousness, urgency, and global reach of the topic.

12.2 Anthropology and Climate Change

12.2.1 *Documentation*

Anthropology has long held that adaptation to climate and environment is a fundamental human capacity. Numerous theories have been proposed over the last century to explain variation in human physiology, livelihoods, cultures, and political systems in relation to environmental conditions (Dove and Carpenter 2008). This earlier work teaches us that the effects of climate change must be understood from a human ecological perspective which considers how land use for settlement, agriculture, and other purposes influence the ecosystem, and at the same time, human activities are shaped by shifting environmental conditions (e.g., see Chaps. 1, 5, 7, 10, and 11). Anthropology offers unique insights to our collective knowledge of ongoing global changes and adaptations to them. Ethnographic fieldwork, attention to the historical context of contemporary climate debates, and an integrated view of human and natural systems are important aspects of anthropological research that can inform understandings of climate change (Barnes et al. 2013). Each of the preceding chapters has demonstrated these characteristics in its presentation of research examining local perceptions of how changing conditions affect relationships between plants, animals, humans, water, and land, in Africa (Zimbabwe, Kenya, Cameroon), Asia (Eastern Himalayan region), Europe (France, Spain), North America (Puerto Rico, Southern Appalachia), and South America (Brazil).

Providing portraits of what changing weather patterns and environments look like in different parts of the world, the authors of the preceding chapters contribute to a well-established anthropological literature (Barnes and Dove 2015; Crate and Nuttall 2009; Dove 2014; Jankovic and Barboza 2009; Lazrus 2012; Strauss and Orlove 2003). We learn that at the local level, climate change is felt and understood in ways very different from what is described in reports focusing on global-scale meteorological and geographic trends, such as rising temperatures, melting ice caps, rising sea levels, and increased carbon dioxide in the air. In documenting these experiences, this book deepens our understanding of the kinds of changes that are occurring and adds a human dimension to geographic and atmospheric descriptions. Importantly, the publication of these findings gives a voice to people who are often marginalized within their own societies. Below is a list of the impacts of climate change research participants have observed:

- Increase/decrease in rainfall/snowfall and more extreme events
- Changes to onset and duration of seasons
- Increase in seasonal temperatures
- Retreat of glaciers
- Changes in flora and fauna that make prediction of climatic phenomena difficult
- Difficulties in growing particular crops; less reliable harvests
- Vegetation in high altitudes dominated by warmer temperature species
- Proliferation of weeds

- Shifting of tree lines further up the mountains
- Increased incidence of diseased vegetation and insect attacks
- Changes in water quality, affecting life cycles and distribution of fish important to humans
- Loss of habitat for fish
- Wild animals searching for food in human settlements or dying of starvation

These observations are made with both bodies and minds, and the effects are felt in profound emotional and psychological ways—not just physical ways that can be counted and measured. Throughout the chapters, it is clear that these changes *matter*. They are significant to individuals as well as to communities and cultural groups, who are struggling to make a living and to live well according to their own cultural traditions and personal convictions.

This book presents insightful studies of the consequences of different adaptive responses for particular populations. For example, Chaps. 2 and 4 describe how adaptations to negative impacts of climate change present new risks, such as those pursuing livelihood activities that are more dangerous, less sustainable, or less profitable. Salick, Staver, and Hart (Chap. 4) observe that planting traditional crops and taking animals to graze higher up in the Himalayan mountains involve riskier travel, while moving to new livelihood activities such as agroforestry, tourism, and growing cash crops brings other problems. Meanwhile, Seara, Pollnac, and Jakubowski (Chap. 2) highlight the intersectionality of risk types where environmental changes and climate change exacerbate the socioeconomic vulnerability of fishers in Puerto Rico. They take into account the sociocultural and psychological factors that influence decisions and feelings, noting that some fishers are reluctant to give up fishing altogether to pursue other livelihoods, despite the risks and problems associated with it, because fishing is part of their identity and their social relationships. Depending on their age, education, level of experience, and satisfaction with fishing, some fishers adapt to reduced numbers of the usual fish in the usual places by pursuing new livelihoods, while others take risks finding new fishing grounds in deeper waters, and still others accept the dangers of scuba diving. These examples help us understand how sociocultural factors beyond rational decision-making shape adaptation in important ways. Studies such as these are essential if we aim to develop adaptation practices, programs, and policies that are effective and culturally appropriate.

12.2.2 Connections

The chapters in the current volume showcase the breadth of anthropological research on the complexities of climate change problems and solutions, which intertwine environment, culture, sociopolitical processes, local and global economic systems, and individual human experience. They demonstrate that climate change is not just about the environment, but about how the environment is implicated in human relationships and how people see themselves in the world.

Dervieux and Belgherbi (Chap. 3) reveal interconnections between climate change, environmental practices, religious beliefs, and politics. They describe discourses circulating among residents of communal lands near a protected area in Zimbabwe, in which angry ancestors cause droughts, deforestation, soil erosion, reduction of plant and animal species, and negative interactions with elephants, lions, and other animals. Research participants explain that the ancestors are angry because people no longer do ritual rainmaking ceremonies and violate taboos on using some plant and animal species. These transgressions occur because the villagers no longer have access to ancestral sites within the protected lands, the population has grown, and the spread of Christianity has discouraged traditional spiritual practices. The authors argue that the discourse in which climate change and its environmental effects are attributed to angry ancestors is effectively a political commentary on the marginalization of these rural people from their traditional lands as well as their exclusion from a national community-based natural management program. Here, we see how climate change is linked explicitly with appeals for democratization. Similarly, Salick, Staver, and Hart (Chap. 4) reported that many Tibetans believe climate change is caused by bad deeds and that good deeds will appease angry gods and restore balance. Compounding the material disadvantages that follow from disconnecting people from the land, we see how "the disruption of Indigenous relationships to land represents a profound epistemic, ontological, cosmological violence" (Tuck and Yang 2012: 5). These studies indicate that relationships between people and the rest of the natural world are understood to be inherently moral and political.

Two additional examples highlight the political dimension of human interactions with their environments. Sourdril and colleagues explain that the proliferation of wild flora in the rural area of Bas-Comminges, France, results both from a warmer, wetter climate and from changes in public infrastructure. In the past, government workers used herbicides to remove weeds in public green spaces. Nowadays, the management of the green spaces is left to community groups, and herbicides are restricted. Conflict arises because newcomers favor manual weed removal, while most locally born residents prefer to use herbicides. Furthermore, there is disagreement about which plants should be removed and which should be left for aesthetic reasons or harvested for various uses. In this way, "the weeds issue reveals not only the impacts of seasonal variations and climate changes but also profound changes in the organization of the institutions, local societies, and the French rural world" (Chap. 5).

Finally, Burke, Welch-Devine, Steacy, and Rzonca (Chap. 10) suggest that collective response to climate change in the Southern Appalachian region of the United States may be impeded by a general and long-standing lack of trust in science and government. Their work urges us to attend not only to local environmental knowledge but also to local attitudes and beliefs about how government and society function. Thus, this book's examination of human responses to climate change leads us to insights about human relationships and social structures.

12.2.3 Collaborations

Several contributors to this book have found ways to "move beyond descriptions of social and cultural effects of climate change" (Crate and Nuttall 2016: 7), by collaborating with participants to produce outputs that can be used by other academic and lay researchers. Making new data and ways of knowing available to researchers worldwide enables comparisons and the recognition of trends and patterns. For instance, the interactive Web-based platform created by Reyes-García, Fernández-Llamazares, García-del-Amo, and Cabeza (Chap. 9) as part of the LICCI project, along with the citizen scientists who shared their local and indigenous knowledge of climate change impacts, demonstrates how anthropological research collaborations can bridge local and global aspects of climate change. While acknowledging challenges arising from the incommensurability of some kinds of environmental knowledge and the continual need for widespread participation, this project delineates a path forward for research that combines local and scientific knowledge and that potentially involves people of all ages. Engaging "citizen scientists" increases relevance and promotes education and awareness of climate change. The project also has the potential to provide data for comparative analyses of changes over time and across space. Expanding the scope of this idea, some scholars have already begun to envision a worldwide database of local/indigenous environmental knowledge (Pennesi 2009). Knowledge exchange among researchers, local populations, governments, and other organizations is a critical component of democratization and for developing effective policies and adaptation strategies related to climate change.

Salick, Staver, and Hart (Chap. 4) synthesize multiple studies which they undertook over many years with various interdisciplinary collaborators and using multiple ethnographic and participatory methods in China, Bhutan, and Nepal. They describe the close connections between climate, people, and vegetation in their work on ethnobotany of the Eastern Himalayan region. The input from numerous research partners enabled them to design a predictive model of how human and vegetation responses to climate change affect each other, asserting that the overexploitation of medicinal plants, along with increased grazing and tourist activity at higher elevations, may strain the people, the animals, and the pastures. Their prediction can inform policy in the Himalaya and provide a starting point for developing collaborative solutions. Additionally, their model can serve as an example for interdisciplinary research partnerships in other climate regions.

In Kenya, Roque de Pinho (Chap. 8) worked with Maasai pastoralists who took photos and videos documenting the impacts of severe drought on the land, the animals, and, subsequently, the people. The participants then explained the photos and registered their complaints to government leaders and representatives of NGOs regarding fencing of private lands that hinder mobility of livestock and wildlife populations and land-use restrictions for conservation and urbanization. The herders used the photos as evidence of these and other problems, and the research context provided an opportunity to present their concerns to people in power. These collaborations represent steps toward democratization as knowledge is jointly produced and shared for the benefit of those most acutely affected by climatic changes.

It is encouraging to see from the preceding chapters that scholars and their collaborators are exploring ways to share local/indigenous knowledge of climate change and related effects. Some are also creating practical applications for their research. These collaborations succeed in considering social, economic, and environmental variables that affect vulnerability, resilience, and adaptation of different populations "to understand and develop approaches that are locally generated, owned and perpetuated" (Fiske et al. 2014: 70). Looking ahead, Malsale and colleagues (2019) provide helpful guidance on best practices for collecting and documenting traditional climate knowledge, including developing legal protection for traditional knowledge and intellectual property, establishing local partnerships with institutions and communities, developing trust and involvement of the community in the project, and adhering to local protocols for sharing information. While these guidelines were developed by scholars working in the Pacific, their focus on respect and collaboration is relevant to many of the cases describing place-based knowledge in this book. The research presented here highlights the inadequacy and inappropriateness of simply extracting and decontextualizing information, as was done in the past. Creating relationships and partnerships with research participants is becoming the norm. Still, legal protection for intellectual property does not seem to figure in the projects discussed by the authors here. Working toward this would further the goal of decolonizing knowledge by legitimizing a plurality of epistemologies. Future collaborative work in these areas will have the highest potential to benefit communities if their participation and knowledge are truly supported in these ways.

12.2.4 Activism and Social Transformation

The final report of the American Anthropological Association's Global Climate Change Task Force (Fiske et al. 2014) highlights the value of community-centered approaches and interdisciplinary research in which anthropologists studying climate change are involved. The authors of the report argue that it is by working collaboratively with local peoples and with scholars who have different skills and perspectives that anthropologists can contribute most effectively to adaptation efforts at the local and regional level. Their starting assumption, and mine, is that anthropologists should be taking action.

An illustrative example of this kind of politically engaged research in the current volume is the community-based project undertaken by Galvin and colleagues with Maasai pastoralists in Kenya. Together, they documented how changes in climate and environmental conditions (e.g., prolonged drought) led to compromised health of livestock as well as differences in animal and human behavior, which ultimately resulted in cultural change. The evident shift away from a traditional pastoral culture based on migration and structured social relations is explained not only in terms of access to available environmental resources, such as water and forage, but also in terms of economic activities and education. In light of this, the local groups proposed solutions that took into account both environmental factors and sociopolitical factors, suggesting the expansion of government offices, support for cross-border migration of people and animals, improved infrastructure, and the empowerment of women in spheres of education and finance. Here, we see how vulnerable communities can amplify their political voice in the pursuit of material improvements to their adaptive capacity by engaging in research partnerships with anthropologists and using the results to serve their own ends.

From the perspective of collaboration, it is heartening to learn from this group of researchers that local environmental knowledges and traditional or indigenous knowledge are beginning to be taken seriously as complements to scientific research on climate and climate change. Chapters 2, 7, 8, and 9 illustrate how local knowledge can provide data at a scale and in regions where scientific data does not exist, adding further insight to other similar studies (Strauss 2003). Other chapters contribute to the growing literature documenting ethnometeorological and ecological knowledge to enhance basic understanding or inform predictions (Jiri et al. 2016; Magalhães 1963; Orlove et al. 2002; Speranza et al. 2010). Katz, Lammel, and Bonnet (Chap. 7) describe microscale environmental knowledge about dew, lightning, winds, and other observations that *Ribeirinhos* use to predict weather and river levels in a floodplain of the Amazon River. Roque de Pinho (Chap. 8) provides a "schedule of rain" and explains how Maasai associate it with astronomical indicators of rain, in conjunction with their observations of wind direction, sunset color, and the behavior of wild animals and birds. Dervieux and Belgherbi (Chap. 3) mention similar observational practices for generating predictions. The interactive web site being designed by Reyes-García et al. (Chap. 9) will contain a large collection of images and information about plants and the management of particular ecosystems. This kind of fine-grained data is valuable for others interested in these particular cultures and ecosystems, as well as for those making comparisons to other parts of the world. Acceptance of local knowledge as legitimate is a step toward decolonization, when "traditionally, science has been hostile to indigenous ways of knowing" (Smith 2012: 265).

On the other hand, it appears that just when the value of such knowledge is finally being recognized outside of local communities, the number of practitioners is decreasing due to sociopolitical, religious, and cultural shifts. In some cases, the practices and environmental indicators used to create indigenous knowledge are becoming less reliable (e.g., Katz et al.; Galvin et al.). In other cases, traditional knowledge is lost because elders are not passing it on to youth, who are occupied with other concerns or lack access to the necessary natural resources (e.g., Dervieux and Belgherbi; Roque de Pinho). Although not emphasized by this book's contributors, it is also important to recognize how ideological shifts stemming from colonial systems that work to devalue, delegitimize, and even eliminate indigenous languages and cultural traditions have led to breaks in the intergenerational transmission of languages and traditional knowledge (Hill 2004). If people's languages and cultures are ridiculed, disallowed, or ignored; if their work brings insufficient reward; if their bodies and personhood are abused; and if their knowledge is discounted, then there arises an unwillingness to subject one's children and oneself to the same treatment by engaging in the same practices. Therefore, overcoming the subordinated group's belief in its own inferiority, or "decolonizing the mind" (Fanon

1963; Ngũgĩ wa Thiong'o 1986), is necessary for the continued development of indigenous environmental knowledge. Only when they have both symbolic and material value can indigenous and other marginalized knowledges, cultures, and communities flourish. Economic and cultural adaptations to changing climates are thus entangled with decolonization, language revitalization, and cultural continuity. The societies described in this book have generally been adaptable and resilient when facing past environmental fluctuations and uncertainties; however, it is increasingly clear from these studies that changes are happening much more quickly and uncertainties are growing. Exacerbating the problem, political and economic factors create vulnerable communities with reduced adaptive capacity. Anthropological research shows how crucial it is to illuminate the connections between a group's material conditions, its social position, and its geopolitical location and how it is situated in specific ecologies.

If social transformation is the goal of anthropologists seeking to make a difference, it is at this point that it becomes important to engage in critical assessment of the situation. This book shows the effects of colonialism in places like Kenya, Zimbabwe, and Brazil and the socioeconomic disparities inherent in capitalism around the world, including in the global North. The contributors outline historical and political processes which have marginalized people who engage in land-based livelihoods and especially indigenous peoples and their ways of knowing. Success in fishing, farming, and herding is directly affected by fluctuations and changes in the climate and environment, yet so many of the populations who depend on these activities lack adequate access to resources that would enable effective adaptive responses, and they do not have the political power to improve their status. The objective, then, cannot be simply to publish bits of indigenous or traditional knowledge which fill in gaps in the natural sciences or to secure a place for local collaborators to work alongside academic researchers. Understanding global change should inform work toward social change. As Linda Tuhiwai Smith (2012: 266–267) reminds us, "Indigenous knowledge extends beyond the environment… it has values and principles about human behavior and ethics, about relationships, about wellness and leading a good life." In other words, Smith continues, our aim should not be to solve isolated problems, but to ensure the well-being of whole communities and to find ways to support them as they transform themselves.

To do this, we need to move beyond doing research *on* local (subordinated, marginalized) people and how they adapt to the impacts of climate change or research *with* those same subordinated groups as they create adaptive strategies appropriate to their contextual constraints. The next step requires us to turn around, step up, and step aside. By "turning around" I mean to turn our attention to investigating why and how *dominant* social groups continue to engage in environmentally destructive activities and to reproduce unjust policies that create the difficult conditions in which subordinated groups are living. To make a real difference, we must examine both how effective social change happens and what impedes it. Then, we can focus our energy on teaching our own privileged communities how to do things more equitably, more sustainably, and more respectfully to reduce negative impacts on more vulnerable populations.

At the same time, we must “step up” to help build the capacity of subordinated peoples to achieve their own goals. In other words, scholars, governments, and others in powerful positions need to value indigenous/local/traditional/rural *people*, not just their knowledge. Valuing the people can mean providing training, salaries, and mentorship, not just a voice in the research process. One example is “Rising Voices: Collaborative Science with Indigenous Knowledge for Climate Solutions,” a program funded by the National Center for Atmospheric Research (NCAR) and the University Corporation for Atmospheric Research (UCAR) in the United States. To date, the program has held six workshops that “facilitate intercultural, relation-based approaches for understanding and adapting to extreme weather and climate events, climate variability and climate change” (UCAR 2019). Workshop participants include indigenous leaders and professionals as well as scholars, educators, students, and community leaders from around the world. They collaborate on “joint research aimed at developing optimal plans for community action towards sustainability” (UCAR 2019). Integral to this program’s success are the ongoing mentorship in multiple domains and the government funding provided to arrange and host the annual workshops, in addition to maintaining the network through regular communications, the web site, and the preparation of reports. The Rising Voices example shows that commitments of time, money, and relationship building are part of the decolonization and democratization process. Anthropologists, and social scientists in general, can “step up” to contribute their insights and their time, as well as their skills in grant-writing, to increase competencies in the communities where they work.

Valuing the people can also mean “stepping aside” as principal investigators and assisting as funders or volunteers in indigenous or community-led projects. Indigenous Climate Action (https://www.indigenousclimateaction.com/) in Canada and the College of Menominee Nation Sustainable Development Institute (http://sustainabledevelopmentinstitute.org/) in the United States are excellent examples of this kind of initiative. There are surely others elsewhere in the world; we need to find them and support them. Valuing the people by stepping aside can also mean observing from a distance and learning in a way that does not burden *them* with teaching *us*. Our questions are not always the most important ones.

12.3 Implications and Future Directions

Challenges remain in all four categories of anthropological research. In describing local knowledge and observations for outsiders, the specificity of local environments and cultures, the incommensurability of some forms of knowing, and the scalar mismatch of data types can make it difficult to generalize across contexts or to create models for integrating different kinds of environmental knowledge. If the aim is to disseminate this knowledge in a way that is useful to others, there is more work to be done on how to facilitate this kind of translation or exchange. But knowledge is not only produced for some collective benefit; it is also a form of resistance

(Smith 2012: 266). While anthropologists are getting better at supporting local people and collaborating with them, there is room to increase the degree to which research outputs and outcomes are actually beneficial to participants. One way to achieve this is to continue to find connections between experiences of climates changing and the social, cultural, political, and economic factors that shape those experiences. A deeper understanding of context can open doors to more meaningful participation. The fourth category of decolonizing and democratizing research is still only emergent. Settler colonialism is ongoing. Social divisions within communities and political-economic inequalities within nations are firmly entrenched, resisting efforts for systemic change. The purpose of anthropological research is to create new knowledge, gain deeper insights, and develop novel ways of thinking and doing things. Within the academy, researchers are obligated to publish in recognized scholarly venues, to teach, and to serve the discipline in various ways, leaving limited time for engagement in community-based activism. Moreover, as Andrew Walsh (2007: 215) notes, "given the current state of the productivity regimes in which many of us work, the greatest deterrent to the development of new ways of doing anthropology is the possibility that some efforts simply will not count." It can be difficult to take a back-seat role in research design and implementation when we are expected to show leadership and demonstrate evidence of our individual productivity. And of course, not all anthropologists are convinced that we should be engaging politically through our research.

Nevertheless, this book provides reasons to be optimistic. Collaborative research and participatory methodologies such as those used here yield important results that can increase the relevance of findings to policy makers and improve the effectiveness of adaptation strategies. For example, the work of Katz, Lammel, and Bonnet (Chap. 7) paves the way for proposing better environmental management strategies that combine local microscale knowledge with scientific knowledge of rainfall and temperature fluctuations at a larger scale in the Amazon basin. The Maasai participants who worked alongside Roque de Pinho, Galvin, and their colleagues are already in conversation with political leaders to advocate for improved policies and infrastructure, using their videos and photos to disseminate their environmental knowledge and observations of climate change. These projects build capacity among Maasai pastoralists as researchers and as contributors to political debates. Seara, Pollnac, and Jakubowski (Chap. 2) show that policy and educational strategies around climate change adaptation are more likely to be accepted and implemented by Puerto Rican fishers if they incorporate local environmental knowledge and consider how factors like age, education, attitudes, and experience influence perceptions. Burke, Welch-Devine, Steacy, and Rzonca (Chap. 10) are facilitating "climate conversations" that engage diverse perspectives associated with residential characteristics and forms of land use to build solidarity among Appalachian community members and natural resource managers. These conversations can eventually lead participants to develop effective collective responses to climate change, despite deep political divisions. Sourdril et al. (Chap. 5) describe how community members can work together to care for public spaces despite their conflicting perspectives on environmental issues. The predictions of Salick, Staver, and Hart (Chap. 4) for the

Eastern Himalaya, while dire, have potential to compel decision-makers to move in new directions. Nearly every chapter offers suggestions or a path forward for how local environmental knowledge can inform management of resources and human adaptations to changing climates.

This book helps us see the importance of developing adaptive strategies for climate change that take into account sociocultural factors influencing relationships between people and their environment, as well as relationships among people that are mediated by the environment. After several decades of research around the globe, we are now entering a time when the "locals" are becoming the researchers conducting studies in their own regions and communities, to evaluate adaptive strategies, programs, and policies related to global changes. This research illustrates how people are already adapting to ongoing changes and how there is a history of adaptation and resilience in these communities. In future, the anthropologist's role may most appropriately be one of accompanying people as they face changes in progress, collaborating alongside them when asked, and using our power to contribute to social change where we live.

References

Ayers J, Forsyth T (2009) Community-based adaptation to climate change. Environ Sci Policy Sustain Dev 51(4):22–31

Baer H, Singer M (eds) (2018) The anthropology of climate change: an integrated critical perspective, 2nd edn. Routledge, New York

Barnes J, Dove M (eds) (2015) Climate cultures: anthropological perspectives on climate change. Yale Uni Press, New Haven

Barnes J, Dove M, Lahsen M, Mathews A, McElwee P, McIntosh R, Moore F, O'Reilly J, Orlove B, Puri R, Weiss H, Yager K (2013) Contribution of anthropology to the study of climate change. Nat Clim Chang 3:541–544. https://doi.org/10.1038/nclimate1775

Blaikie P, Cannon T, Davis I, Wisner B (1994) At risk: natural hazards, people's vulnerability and disasters. Routledge, New York

Casagrande D, Hope D, Farley-Metzger E, Cook W, Yabiku S, Redman C (2007) Problem and opportunity: integrating anthropology, ecology, and policy through adaptive experimentation in the urban U.S. Southwest. Hum Organ 66(2):125–139

Crate S, Nuttall M (eds) (2009) Anthropology and climate change: from encounters to actions. Routledge, New York

Crate S, Nuttall M (eds) (2016) Anthropology and climate change: from actions to transformations, 2nd edn. Routledge, New York

Dove M (ed) (2014) The anthropology of climate change: an historical reader. Wiley Blackwell, Malden

Dove M, Carpenter C (eds) (2008) Environmental anthropology: a historical reader. Blackwell Publishing, Malden

Fanon F (1963) The wretched of the earth. Grove Press, New York

Fiske SJ, Crate SA, Crumley CL, Galvin K, Lazrus H, Lucero L, Oliver-Smith A, Orlove B, Strauss S, Wilk R (2014) Changing the atmosphere: anthropology and climate change. Final report of the AAA Global Climate Change Task Force. American Anthropological Association, Arlington, p 137

Hill J (2004) What is lost when names are forgotten? In: Sanga G, Ortalli G (eds) Nature knowledge. Berghahn Press, New York

IPCC (Intergovernmental Panel on Climate Change) (2013) Summary for policymakers. In: Stocker TF, Qin D, Plattner G-K, Tignor M, Allen SK, Boschung J, Nauels A, Xia Y, Bex V, Midgley PM (eds) Climate change 2013: the physical science basis. Contribution of working group I to the fifth assessment report of the intergovernmental panel on climate change. Cambridge University Press, New York

Jankovic V, Barboza C (eds) (2009) Weather, local knowledge and everyday life: issues in integrated climate studies. MAST, Rio de Janeiro

Jiri O, Mafongoya P, Mubaya C, Mafongoya O (2016) Seasonal climate prediction and adaptation using indigenous knowledge systems in agriculture systems in Southern Africa: a review. J Agric Sci 8(5):156–172

Lazrus H (2012) Sea change: island communities and climate change. Annu Rev Anthropol 41:285–301

Magalhães J (1963) Previsões Folclóricas das Sêcas e dos Invernos no Nordeste Brasileiro. Imprensa Universitária do Ceará, Fortaleza

Malsale P, Sanau N, Tofaeono TI, Kavisi Z, Willy A, Mitiepo R, Lui S, Chambers LE, Plotz RD (2019) Protocols and partnerships for engaging Pacific Island communities in the collection and use of traditional climate knowledge. Bull Am Meteorol Soc 99:2471–2489. https://journals.ametsoc.org/doi/10.1175/BAMS-D-17-0163.1

Mohanty CT (2003) Feminism without borders: decolonizing theory, practicing solidarity. Duke University Press, Durham

Moran EF (2008) Human adaptability: an introduction to ecological anthropology, 3rd edn. Westview Press, Boulder

Ngũgĩ wa Thiong'o (1986) Decolonising the mind: the politics of language in African literature. Heinemann, Portsmouth

Orlove B, Chiang J, Cane M (2002) Ethnoclimatology in the Andes: a cross-disciplinary study uncovers a scientific basis for the scheme Andean potato farmers traditionally use to predict the coming rains. Am Sci 90(5):428–435

Pennesi K (2009) Creating an online archive of traditional weather prediction indicators: notes from a roundtable discussion. In: Jankovic V, Barboza C (eds) Weather, local knowledge and everyday life: issues in integrated climate studies. MAST, Rio de Janeiro, pp 297–304

Pennesi K (2011) Making forecasts meaningful: explanations of problematic predictions in Northeast Brazil. Weather Clim Soc 3(2):90–105

Pennesi K (2013) Predictions as lies in Ceará, Brazil: the intersection of two cultural models. Anthropol Q 86(3):759–790

Pennesi K (2015) Constructing 'farmer' and 'state' identities in moral discourses about semi-subsistence agriculture in North-East Brazil. J Lat Am Stud 47(4):781–809

Pennesi K, Souza CRBD (2012) O Encontro Anual dos Profetas da Chuva em Quixadá, Ceará: A Circulação de Discursos na Invenção de uma Tradição (The annual meeting of rain prophets in Quixadá, Ceará: the circulation of discourses in the invention of tradition). Horizontes Antropológicos 38:159–186. https://doi.org/10.1590/S0104-71832012000200007

Pokrant BOB, Stocker L (2011) Anthropology, climate change and coastal planning. In: Kopnina H, Shoreman-Ouimet E (eds) Environmental anthropology today. Routledge, New York, pp 179–194

Reuter T (2015) Averting a global environmental collapse: the role of anthropology and local knowledge. Cambridge Scholars Publishing, Newcastle upon Tyne

Ribot J, Najam A, Watson G (1996) Climate variation, vulnerability and sustainable development in the semi-arid tropics. In: Ribot J, Magalhães A, Panagides S (eds) Climate variability, climate change and social vulnerability in the semi-arid tropics. Cambridge University Press, Cambridge, pp 13–51

Smith LT (2012) Decolonizing methodologies: research and indigenous peoples, 2nd edn. Palgrave Macmillan, New York

Speranza CI, Kiteme B, Ambenje P, Wiesmann U, Makali S (2010) Indigenous knowledge related to climate variability and change: insights from droughts in semi-arid areas of former Makueni District, Kenya. Clim Chang 2010(100):295–315

Strauss S (2003) Weather wise: speaking folklore to science in Leukerbad. In: Strauss S, Orlove B (eds) Weather, climate, culture. Berg, New York, pp 39–60
Strauss S, Orlove B (eds) (2003) Weather, climate, culture. Berg, New York
Taddei R (2006) Oráculos da chuva em tempos modernos: Mídia, desenvolvimento econômico e as transformações na identidade social dos profetas do sertão (Modern day rain oracles: media, economic development and transformations of the social identity of rain prophets). In: Martins KPH (ed) Profetas da Chuva. Tempo d'Imagem, Fortaleza, pp 161–170
Taddei R (2012) Social participation and the politics of climate in Northeast Brazil. In: Latta A, Wittman H (eds) Environment and citizenship in Latin America: natures, subjects and struggles. Berghahn Books, New York, pp 77–93
Tuck E, Yang KW (2012) Decolonization is not a metaphor. Decolonization Indigeneity Educ Soc 1(1):1–40
UCAR (University Corporation for Atmospheric Research) (2019) Rising voices. https://risingvoices.ucar.edu/
Walsh A (2007) Ethnographic alchemy: perspectives on anthropological work from Northern Madagascar. In: Field L, Fox RG (eds) Anthropology put to work. Berg Publishers, Oxford, pp 201–216

Index

M. Welch-Devine et al. (eds.), *Changing Climate, Changing Worlds*, Ethnobiology, https://doi.org/10.1007/978-3-030-37312-2

The manufacturer's authorised representative in the EU is Springer Nature Customer Service Centre GmbH, Europaplatz 3, 69115 Heidelberg, Germany. If you have any concerns regarding our products, please contact ProductSafety@springernature.com

Printed and bound by CPI Group (UK) Ltd, Croydon, CR0 4YY
22/09/2026
02221057-0008